The Crayfishes of Kentucky

By

Christopher A. Taylor
Illinois Natural History Survey
607 E. Peabody Drive
Champaign, IL 61820

and

Guenter A. Schuster
Department of Biological Sciences
Eastern Kentucky University
Richmond, KY 40475

Illinois Natural History Survey, Dr. David L. Thomas, Chief
A Division of the Illinois Department of Natural Resources, Joel Brunsvold, Director

Illinois Natural History Survey
Natural Resources Building
607 East Peabody Drive
Champaign, Illinois 61820

Printed by authority of the state of Illinois.
0053794–1.5M–11-04
US ISSN 0888-9546

Editors: Thomas E. Rice, Charles Warwick

Cover design: Thomas E. Rice
Cover photos: *Cambarus graysoni*, photo by C. Lukhaup; Cumberland Falls, Whitley County, Kentucky, photo by B. Burr.

Illustrations by C.A. Taylor unless otherwise noted.

Photos by authors unless otherwise noted.

Citation: Taylor, C.A., and G.A. Schuster. 2004. The Crayfishes of Kentucky. Illinois Natural History Survey Special Publication No. 28. viii + 219 pp.

ISBN: 1-882932-09-9

Funded cooperatively by the United States Department of Agriculture, U.S. Forest Service and Eastern Kentucky Power Cooperative, a Touchstone Energy Cooperative.

ILLINOIS NATURAL HISTORY SURVEY

ILLINOIS DEPARTMENT OF NATURAL RESOURCES

Touchstone Energy Cooperatives

Common and lowly as most may think the crayfish, it is yet so full of wonders that the greatest naturalist may be puzzled to give a clear account of it.

— Roesel von Rosenhof in Huxley's *The Crayfish, an Introduction to the Study of Zoology* (Huxley 1880).

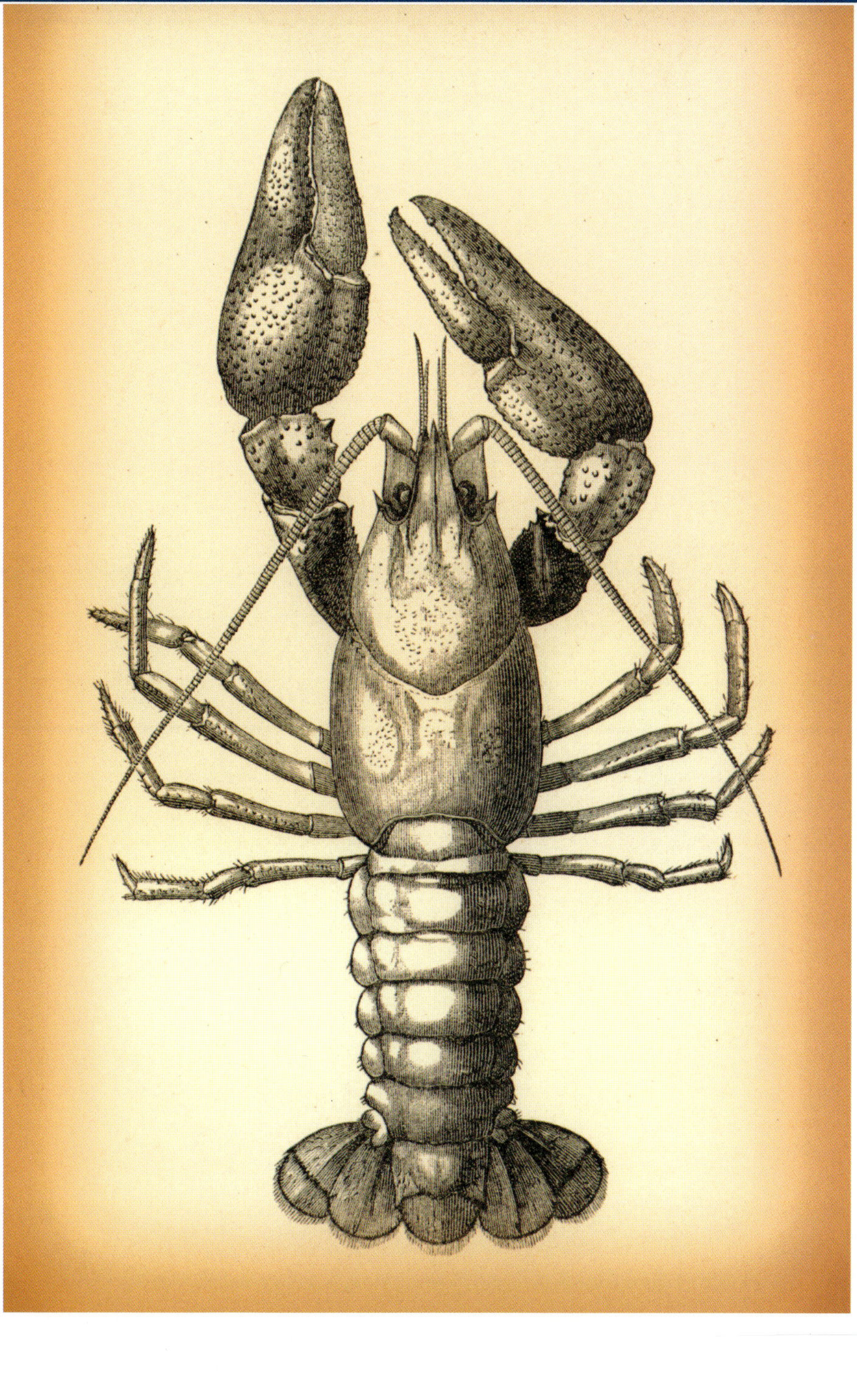

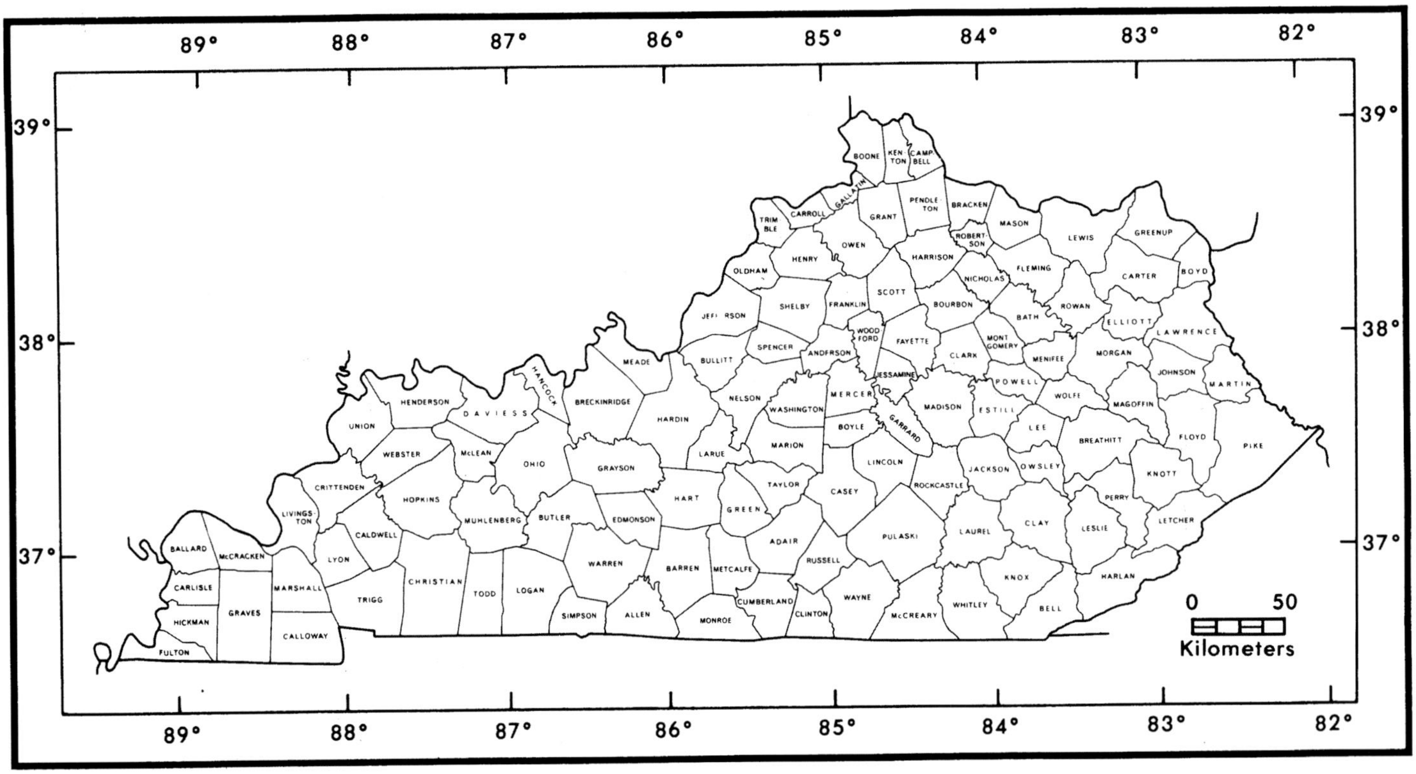
89°
88°
87°
86°
85°
84°
83°
82°
39°
38°
37°
BOONE
KEN TON
CAMP BELL
GALLATIN
CARROLL
TRIM BLE
GRANT
PENDLE TON
BRACKEN
MASON
ROBERT SON
LEWIS
GREENUP
OWEN
HENRY
OLDHAM
HARRISON
NICHOLAS
FLEMING
CARTER
BOYD
SHELBY
FRANKLIN
SCOTT
BOURBON
BATH
ROWAN
ELLIOTT
LAWRENCE
SPENCER
ANDERSON
WOOD FORD
FAYETTE
MONT GOMERY
MENIFEE
MORGAN
BULLITT
MEADE
HANCOCK
CLARK
JESSAMINE
POWELL
JOHNSON
MARTIN
HENDERSON
DAVIESS
BRECKINRIDGE
NELSON
MERCER
MADISON
ESTILL
WOLFE
MAGOFFIN
UNION
HARDIN
WASHINGTON
BOYLE
GARRARD
LEE
BREATHITT
FLOYD
PIKE
WEBSTER
McLEAN
OHIO
GRAYSON
LARUE
MARION
LINCOLN
JACKSON
OWSLEY
KNOTT
CRITTENDEN
HOPKINS
TAYLOR
CASEY
ROCKCASTLE
PERRY
LIVINGS TON
HART
GREEN
MUHLENBERG
BUTLER
EDMONSON
BALLARD
McCRACKEN
CALDWELL
ADAIR
PULASKI
LAUREL
CLAY
LESLIE
LETCHER
LYON
RUSSELL
MARSHALL
CARLISLE
CHRISTIAN
WARREN
BARREN
METCALFE
KNOX
HARLAN
TRIGG
TODD
LOGAN
GRAVES
HICKMAN
WAYNE
CUMBERLAND
WHITLEY
BELL
ALLEN
SIMPSON
MONROE
CLINTON
McCREARY
CALLOWAY
FULTON
0
50
Kilometers

Contents

Acknowledgments

The completion of this project could not have been achieved without the help of numerous individuals. Our only hope is that we have not forgotten anyone who has contributed over the many years. If we have, we offer our deepest regrets.

The collection of data used in this book could not have been accomplished without the field and museum assistance of a long list of persons. For field help we are grateful to E. Akers, S. Anderson, B. Bauer, R. Blanton, C. Bucher, M. Compton, B. Daniels, C. Distel, J. Evans, J. Flotemersch, J. Fry, K. Groves-Medlock, M. Gumbert, M. Hardman, T. Hagman, A. Hill, C. Isaac, V. Jenkins, J. Johansen, T. Jubenville, D.M. Jones, D.R. Jones, E. Kimsey, J. Kiser, J. Knouft, A. Kopp, L. Lawson, T. Lowe, C. Lukhaup, S. McMurray, P. McMurray, M. Moeykens, P. Moosman, T. Near, J. O'Keefe, L. Page, M. Paterson, D. Peake, L. Pintor, G. Pond, S. Powers, A. Rainwater-Bradshaw, J. Ray, M. Sabaj, D. Schuster, A. Shepard, N. Smith, W. Stamm, R. Steinberg, D. Swartley-Moosman, J. Tertuliani, M. Thomas, J. Thomas, Q. Tolliver, and M. Vukovich. Of those people, M. Compton, M. Hardman, D.R. Jones, J. Knouft, L. Lawson, J. O'Keefe, D. Peake, and M. Sabaj deserve special recognition.

Numerous specimens were deposited in the crustacean collections of the Illinois Natural History Survey and Eastern Kentucky University and used for this study. For those specimens we thank T. Cervone, Bernardin, Lochmueller and Assoc.; G. Weddle, Campbellsville University; J. O'Keefe, M. Gumbert, J. Settles, J. Young, East Kentucky Power Cooperative; P. Ceas, S. Harrel, Eastern Kentucky University; J. Crosby, L. Kornman, J. MacGregor, Kentucky Department of Fish and Wildlife Resources; B. McMurray, S. Rice, Kentucky Department of Transportation; J. Brumley, S. Call, M. Compton, R. Houp, S. McMurray, L. Metzmeier, D. Peake, R. Pierce, G. Pond, J. Schuster, C. Schneider, K. Smathers, M. Vogel, Kentucky Division of Water; R. Cicerello, E. Laudermilk, B. Palmer-Ball, Jr., Kentucky State Nature Preserves Commission; R. Dunlap, Mead-Westvaco Paper Corp.; R. Bouchard, Philadelphia Academy of Sciences; R. Cooper, B. Dickey, R. Ziemba, University of Kentucky; B. Burr, E. Wetzel, University of Southern Illinois at Carbondale; B. Harrel, R. Butler, U.S. Fish and Wildlife Service; V. Bishop, U.S. Forest Service; and R. Prins, Western Kentucky University.

For assistance with museum visits we are especially grateful to R. Lemaitre, K. Reed, and T. Chad Walter at the National Museum of Natural History, Smithsonian Institution. The efforts of A. Johnston at the Museum of Comparative Zoology and R. Thoma and B. Armitage at The Ohio State University and Ohio Biological Survey, respectively, are also recognized.

For assistance with graphics we are grateful for the efforts of F. Hutto, M. Jeffords, J. Mui, and R. Klein. R. Butler, L. Page and C. Skelton reviewed an earlier version of this manuscript and their efforts vastly improved its quality; for that we are grateful. We also appreciate the tireless efforts of Illinois Natural History Survey editors T. Rice and C. Warwick.

Partial funding for printing costs was graciously provided by the United States Department of Agriculture Forest Service and the East Kentucky Power Cooperative, the latter a member of the Touchstone Energy Alliance. Within the Unites States Forest Service, we would like to thank K. Leftwich and K. Russell, and at East Kentucky Power Cooperative we thank J. Hohman.

The senior author would like to thank the Illinois Natural History Survey's Center for Biodiversity for clerical and financial support during the course of this project. Special recognition is also due to L. Page for his enthusiastic support of the INHS Crustacean Collection and providing motivation to the senior author during his early years at INHS. The junior author would like to thank R. Clark, S. Jones, and B. Ramey, past and present chairs of the Department of Biological Sciences, Eastern Kentucky University (EKU) for their support. Special thanks from the junior author also go to D. Batch and B. Branson for many years of collecting crayfishes in Kentucky and establishing the crayfish collection at EKU.

Finally, we would like to acknowledge those institutions that continue to scrape together funding to support their biological collections. During tough budget times this can be a very trying endeavor. Collections are irreplaceable libraries of biological information and without their existence works such as this book could not be completed.

Cambarus dubius from Big Black Mountain, Letcher County, KY.

Introduction

Crayfishes, also known as crawfish, crawdads, or mudbugs, are a diverse and important component of freshwater aquatic and semi-aquatic ecosystems around the world. Their familiar form is recognizable by almost anyone who has spent time in and around lakes or creeks. Crayfishes are found natively on every continent except Africa and Antarctica and occur in almost every type of aquatic and semi-aquatic habitat.

This book is intended to provide an up-to-date list of known crayfish species from the state of Kentucky. Distribution maps, illustrations, photos, and a key are provided to assist those interested in identifying Kentucky crayfishes. Information on the habitat and biology of each species is also provided, as is a summary of the historical efforts of crayfish biologists, or astacologists, in the state. The material presented in the following pages should not, however, be taken as the final word on crayfishes in Kentucky. Future work will surely provide new information on the life histories and may yield new distributional records, especially for those species that are difficult to collect, such as the burrowers. We hope this book will serve as a catalyst for such scientific efforts.

Taxonomy, Diversity, and Distribution

Crayfishes are members of the Phylum Arthropoda, a large group of over 1 million named organisms that includes insects, spiders, and crustaceans. This group is characterized by a hard exoskeleton that is continuously shed or molted and regrown. In descending taxonomic order, crayfishes belong to the Subphylum Crustacea, Class Malacostraca, and Order Decapoda (Martin and Davis 2001). Other decapods include freshwater and marine shrimps, crabs, and lobsters. All members of the Order possess five pairs of jointed walking legs and the first three pairs of thoracic appendages modified as maxillipeds for feeding and other functions. Approximately 530 species of crayfishes are currently recognized worldwide and they are assigned to one of three families: Astacidae, Cambaridae, and Parastacidae. These families are mostly differentiated by the reproductive structures of the male. In both Astacidae and Cambaridae the males possess a long modified structure for sperm transfer called the gonopod or first pleopod (Fig. 1); males in the family Parastacidae lack a modified gonopod. In Cambaridae the gonopod changes shape as the male alternates between reproductively active (form I) and reproductively inactive (form II) stages; in Astacidae the gonopod shape does not change. The Astacidae is a small family composed of approximately 13 species and is distributed in Europe, western Asia, and the extreme northwestern United States and southwestern Canada. Only five species are known to occur in the United States, all of which occur west of the Continental Divide. The Parastacidae, or Southern Hemisphere crayfishes, are found in Australia, New Guinea, Madagascar, and temperate South America. Containing approximately 130 species, the family is relatively diverse. With over 380 species, the family Cambaridae contains over two-thirds of the world's crayfish species. Members of the family are found natively in North America and eastern Asia; however, it is in North America that the family has reached an unparalleled level of success. Ninety-nine percent of all cambarids occur in North America and of those, approximately 90% occur in the eastern United States, primarily in the southeast.

Home to 52 described and 1 undescribed species (Table 1) of cambarids in 6 genera, the state of Kentucky contains one of the most diverse crayfish faunas in the United States. The state's fauna is dominated by the genera *Cambarus* (22 described and 1 undescribed species) and *Orconectes* (23 species), while the genera *Cambarellus* (2 species), *Fallicambarus* (1 species), *Procambarus* (3 species), and *Barbicambarus* (1 species) are also present.

Physiography of Kentucky and Faunal Distributions

The crayfish faunal richness observed in Kentucky can most likely be explained by the physiographical diversity found within the state. Another contributing factor is that the state was spared from Pleistocene glacial advances. These advances drastically altered the landscapes and decimated aquatic faunas in areas north of the Ohio River in more recent geologic times. Kentucky occupies 40,598 square miles with elevations ranging from Big Black Mountain in Harlan County at 4,150 feet to the Mississippi River in Fulton County at 260 feet. Aquatic habitats range from high-gradient headwater creeks in the eastern portion of the state to lowland swamps and sloughs in the Mississippi River floodplain of extreme western Kentucky. Large rivers, subterranean creeks, and springs are also found in the state. Three distinctive physiographic provinces have long been recognized in Kentucky—the Coastal Plain, Interior Low Plateaus, and Appalachian Plateaus provinces—with each being subdivided into several sections and in some cases subsections (Fig. 2) (Mengel, 1965; Burr and Warren 1986). Strong correlations

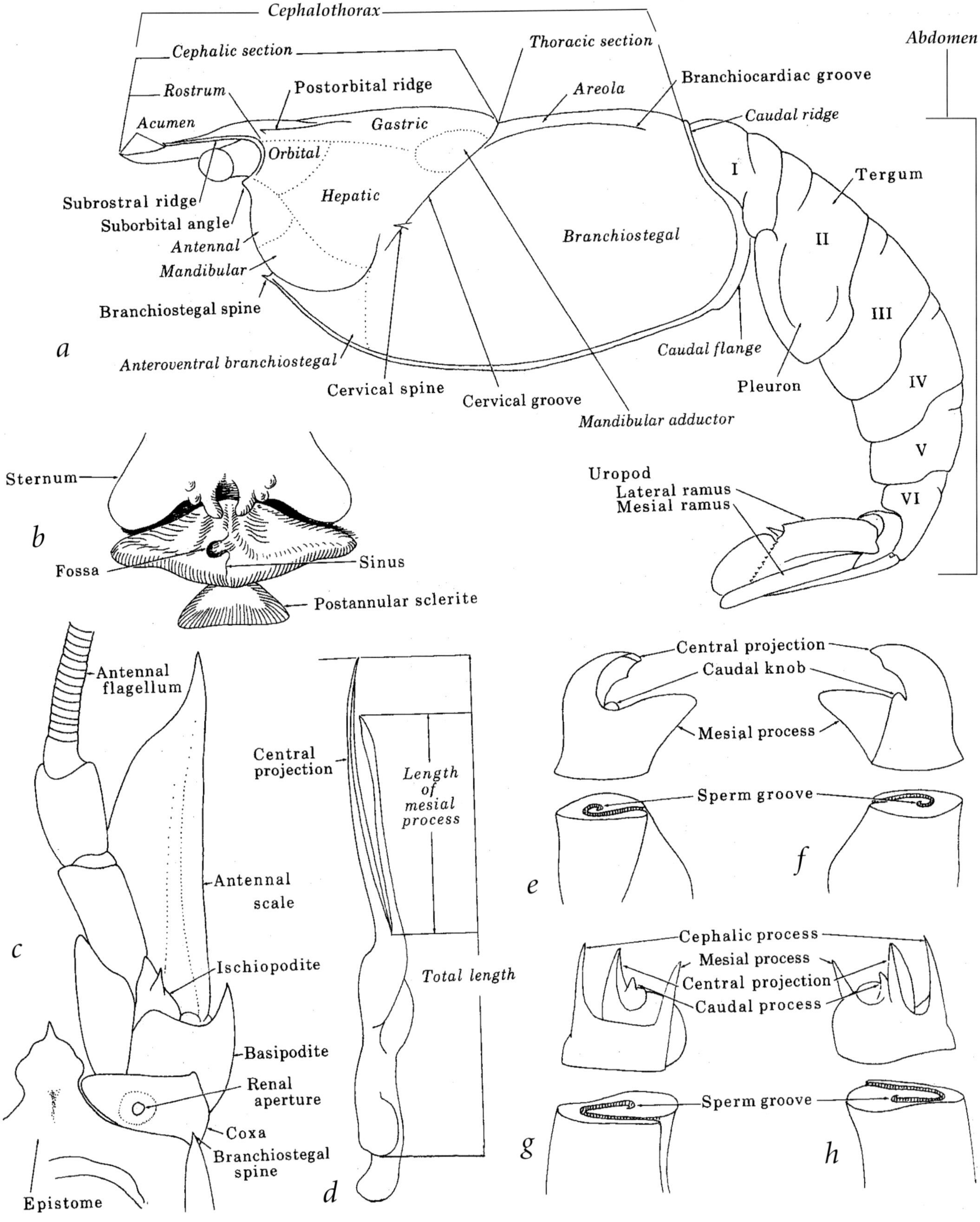

Fig. 1. Morphological structures of cambarid crayfish; a, lateral view of carapace and abdomen; b, annulus ventralis; c, ventral view of antenna; d, mesial view of *Orconectes* form I gonopod; e, mesial view of *Cambarus* form I gonopod; f, lateral view of *Cambarus* form I gonopod; g, mesial view of *Procambarus* form I gonopod; h, lateral view of *Procambarus* form I gonopod. Modified from Hobbs (1981).

between biota and physiographic provinces in Kentucky have been reported for other taxonomic groups such as birds (Mengel 1965) and fishes (Burr and Warren 1986). The crayfish information presented in this book matches those correlations.

Coastal Plain Province. The Coastal Plain Province generally includes that part of Kentucky west of the Tennessee River and is the northern extension of the much larger Gulf Coastal Plain Province. This extension, sometimes referred to as the Mississippi Embayment, represents the limit of the Cretaceous seas and their adjacent plains. The area is mostly characterized as a gently rolling plain of low hills and flat-topped, gravel-capped ridges with local relief rarely exceeding 100 feet. Elevations above sea level in the area range from 260 feet at the Mississippi River on the western border of Kentucky to 640 feet at the Tennessee Valley Divide at the Kentucky-Tennessee state line. Sediments in the Coastal Plain Province are dominated by unconsolidated gravels, sands, and clays, which are subject to rapid erosion (Mengel 1965). Gravel sediments are frequently encountered as the substrate in the creeks draining into the Ohio and Tennessee rivers (Fig. 3) in the central and eastern portions of the region while unconsolidated sand and clay constitute the predominant substrate in the slow-flowing lowland creeks that drain into the Mississippi River in the western edge of the province (Fig. 4). The western edge of the province also contains numerous clay and silt bottomed lowland ponds, ditches, and oxbow lakes (Fig. 5). Many of the ponds and ditches are ephemeral, drying up in later summer and fall months. Burr and Warren (1986) subdivided the Coastal Plain Province into three subsections: the Mississippi Alluvial Plain in the far west, the Eastern Gulf Coastal Plain in the central region, and the Tennessee River Plain in the east (Fig. 2). Major river drainages in the Coastal Plain Province include the Tennessee, Clarks, Ohio, and Mississippi River drainages. Smaller drainages include the Mayfield and Obion Creek drainages and Bayou du Chien (Fig. 6).

The Coastal Plain Province of Kentucky is home to 14 species of crayfishes representing 5 of the 6 total genera known from the state. Six of the species are near the northern limits of their total global ranges and occur in lowland, floodplain habitats. These include *Cambarellus puer*, *C. shufeldtii*, *Orconectes lancifer*, *O. palmeri palmeri*, *Cambarus ludovicianus*, and *Procambarus clarkii*. All of these six species are found nowhere else in Kentucky. Three species of *Orconectes* (i.e., *O. placidus*, *O. burri*, and *O. tricuspis*) occur in more permanent creeks with firmer gravel and clay substrates in the central and

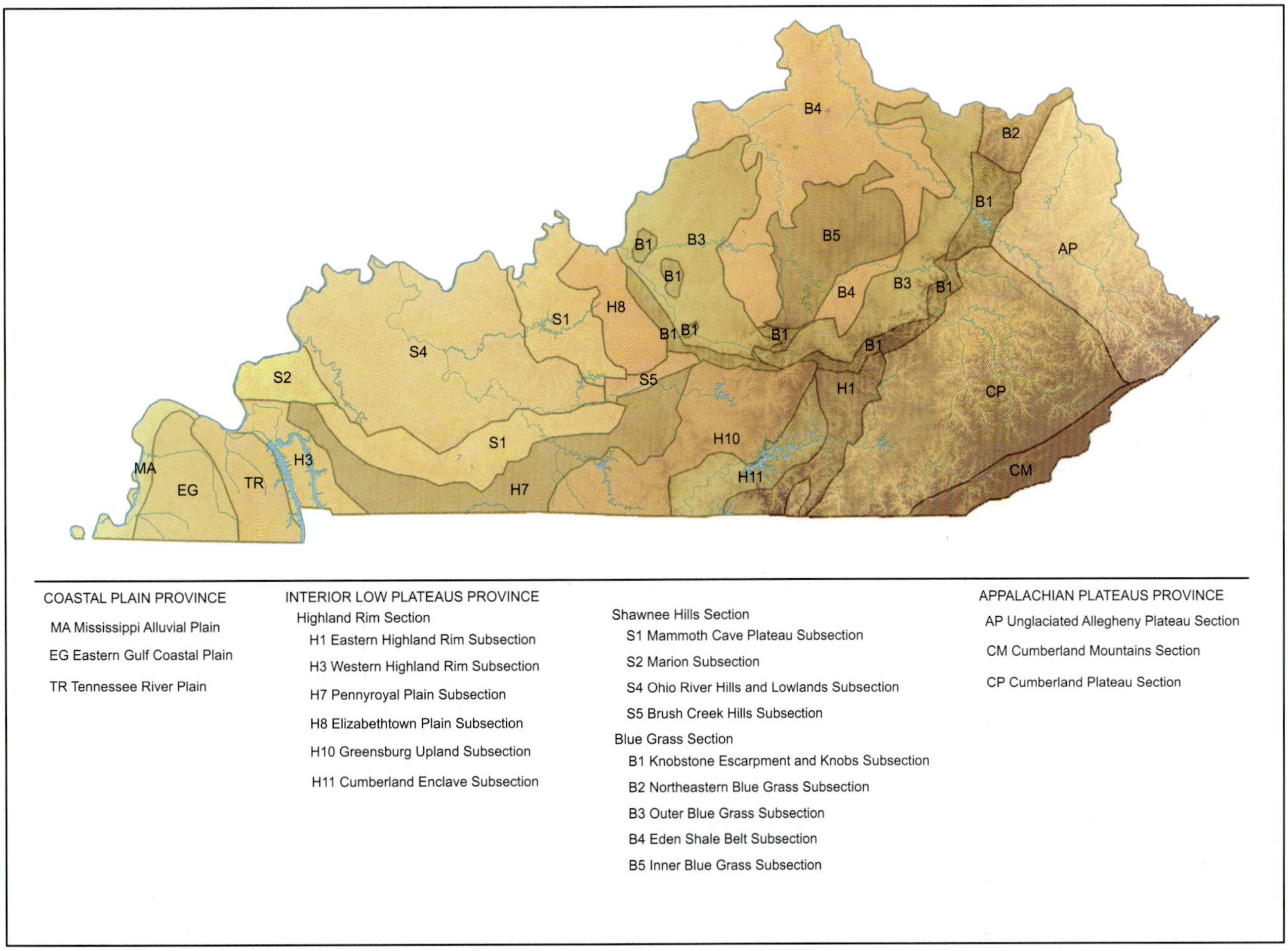

Fig. 2. The physiographic provinces of Kentucky following Mengel (1965) and Burr and Warren (1986).

Fig. 3. Wildcat Creek, Calloway County, a typical gravel-bottomed creek of the Tennessee River drainage in the Coastal Plain Province.

Fig. 4. Brush Creek, Hickman County, a typical lowland creek of the Coastal Plain Province.

Table 1. Checklist and current taxonomic arrangement of the crayfishes (Family Cambaridae) of Kentucky.

- Subfamily Cambarellinae Laguarda 1961
  - Genus *Cambarellus* Ortmann 1905
    - Subgenus *Dirigicambarus* Fitzpatrick 1983
      - *Cambarellus (D.) shufeldtii* (Faxon 1884)
    - Subgenus *Pandicambarus* Fitzpatrick 1983
      - *Cambarellus (P.) puer* Hobbs 1945
- Subfamily Cambarinae Hobbs 1942
  - Genus *Barbicambarus* Bouchard 1972
    - *Barbicambarus cornutus* (Faxon 1884)
  - Genus *Cambarus* Erichson 1846
    - Subgenus *Cambarus* Erichson 1846
      - *Cambarus (C.) bartonii cavatus* Hay 1902
      - *Cambarus (C.) ortmanni* Williamson 1907
      - *Cambarus (C.) sciotensis* Rhoades 1944
    - Subgenus *Depressicambarus* Hobbs 1969
      - *Cambarus (D.) deweesae* Bouchard and Etnier 1979
      - *Cambarus (D.) graysoni* Faxon 1914
      - *Cambarus (D.) sphenoides* Hobbs 1968
      - *Cambarus (D.) striatus* Hay 1902
    - Subgenus *Erebicambarus* Hobbs 1969
      - *Cambarus (E.) rusticiformis* Rhoades 1944
      - *Cambarus (E.) tenebrosus* Hay 1902
    - Subgenus *Glareocola* Bouchard and Bouchard 1995
      - *Cambarus (G.) friaufi* Hobbs 1953
    - Subgenus *Jugicambarus* Hobbs 1969
      - *Cambarus (J.) batchi* Schuster 1973
      - *Cambarus (J.) bouchardi* Hobbs 1970
      - *Cambarus (J.) distans* Rhoades 1944
      - *Cambarus (J.) dubius* Faxon 1884
      - *Cambarus (J.) parvoculus* Hobbs and Shoup 1947
    - Subgenus *Lacunicambarus* Hobbs 1969
      - *Cambarus (L.) diogenes* Girard 1852
      - *Cambarus (L.) ludovicianus* Faxon 1884
    - Subgenus *Puncticambarus* Hobbs 1969
      - *Cambarus (P.) buntingi* Bouchard 1973
      - *Cambarus (P.) cumberlandensis* Hobbs and Bouchard 1973
      - *Cambarus (P.) robustus* Girard 1852
      - *Cambarus (P.) veteranus* Faxon 1914
    - Subgenus *Tubericambarus* Jezerinac 1993
      - *Cambarus (T.) thomai* Jezerinac 1993
      - *Cambarus* (T.) sp. A
  - Genus *Fallicambarus* Hobbs 1969
    - Subgenus *Creaserinus* Hobbs 1973
      - *Fallicambarus (C.) fodiens* (Cottle 1863)
  - Genus *Orconectes* Cope 1872
    - Subgenus *Buannulifictus* Fitzpatrick 1987
      - *Orconectes (B.) palmeri palmeri* (Faxon 1884)
    - Subgenus *Crockerinus* Fitzpatrick 1987
      - *Orconectes (C.) bisectus* Rhoades 1944
      - *Orconectes (C.) burri* Taylor and Sabaj 1998
      - *Orconectes (C.) jeffersoni* Rhoades 1944
      - *Orconectes (C.) margorectus* Taylor 2002
      - *Orconectes (C.) rafinesquei* Rhoades 1944)
      - *Orconectes (C.) sanbornii* (Faxon 1884)
      - *Orconectes (C.) tricuspis* Rhoades 1944
    - Subgenus *Gremicambarus* Fitzpatrick 1987
      - *Orconectes (G.) compressus* (Faxon 1884)
    - Subgenus Orconectes Cope 1872
      - *Orconectes (O.) australis packardi* Rhoades 1944
      - *Orconectes (O.) inermis inermis* Cope 1872
      - *Orconectes (O.) pellucidus* (Tellkampf 1844)

Subgenus *Procericambarus* Fitzpatrick 1987

Orconectes (P.) barrenensis Rhoades 1944

Orconectes (P.) cristavarius Taylor 2000

Orconectes (P.) durelli Bouchard and Bouchard 1995

Orconectes (P.) juvenilis (Hagen 1870)

Orconectes (P.) placidus (Hagen 1870)

Orconectes (P.) putnami (Faxon 1884)

Orconectes (P.) ronaldi Taylor 2000

Orconectes (P.) rusticus (Girard 1852)

Subgenus *Rhoadesius* Fitzpatrick 1987

Orconectes (R.) kentuckiensis Rhoades 1944

Subgenus *Tragulicambarus* Fitzpatrick 1987

Orconectes (T.) lancifer (Hagen 1870)

Subgenus *Trisellescens* Bouchard and Bouchard 1995

Orconectes (T.) immunis (Hagen 1870)

Genus Procambarus Ortmann 1905

Subgenus *Ortmannicus* Fowler 1912

Procambarus (O.) acutus (Girard 1852)

Procambarus (O.) viaeviridis (Faxon 1914)

Subgenus *Scapulicambarus* Hobbs 1972

Procambarus (S.) clarkii (Girard 1852)

eastern regions of the province. The remaining five species (i.e., *Cambarus diogenes*, *Fallicambarus fodiens*, *Orconectes immunis*, *Procambarus acutus*, and *P. viaeviridis*) have very large total ranges in North America and most occur commonly in several habitat types in western Kentucky.

Interior Low Plateaus Province. The Interior Low Plateaus Province encompasses a wide area, occupying approximately two-thirds of the total Kentucky land mass. It ranges from the eastern shore of the Tennessee River to the westernmost edges of the Allegheny and Cumberland Plateaus in eastern Kentucky (Fig. 2). The province contains a wide variety of habitats for aquatic organisms as evidenced by the fact that it has been divided into three sections, each with numerous subsections. Burr and Warren (1986) characterized the province as a low plateau based predominately on Mississippian-age rocks containing a complex of uplands and dissected plains. It contains a vast karst region in its northwest portion composed of sinkhole plains, springs (Fig. 7), subterranean drainages (Fig. 8), and caverns. The province also has highly developed alluvial plains along the margins of the major rivers that dissect it, such as the Cumberland, Kentucky, Green, and Salt rivers (Fig. 6). These alluvial plains contain numerous swamps and oxbow lakes. Creeks in the province are generally of a higher gradient than those in the Coastal Plain and usually have unconsolidated gravel or limestone cobble substrates (Fig. 9). Elevations within the province range from 400 feet along the major rivers that bound the province (Ohio River to the north and Tennessee River to the west) to 1,100 feet in more eastern and southern regions.

The three sections of the Interior Low Plateaus Province are the Highland Rim, the Shawnee Hills, and the Blue Grass (Fig. 2). The Highland Rim Section forms a horseshoe-shaped region in the south-central part of Kentucky. Burr and Warren (1986) describe the section as an extensive limestone upland with karst topology in the eastern, western, and southern portions of the section, with the remaining nonkarst regions being a well-dissected plateau with long, narrow ridges with steep slopes extending down to narrow valley bottoms. Major river drainages in the section include the Cumberland, Tennessee, Green, and Barren (Fig. 6). The Highland Rim Section contains six subsections: Eastern Highland Rim, Western Highland Rim, Pennyroyal Plain, Elizabethtown Plain, Greensburg Upland, and Cumberland Enclave.

Also known as the Western Kentucky Coal Field, the Shawnee Hills Section is mostly an undulating to hilly region in the northwestern portion of the state (Fig. 2). Like the Highland Rim Section, the Shawnee Hills Section also contains well-developed karst topology and broad alluvial plains and associated wetlands along its major rivers. The section contains the well-known Mammoth Cave system and the Dripping Springs Escarpment to the south and east. The single major river drainage in the section is the Green, whose major tributaries are the Mud, Nolin, Pond, and Rough rivers. The Tradewater River drainage, an Ohio River tributary, also encompasses a large portion of the

Fig. 5. An unnamed cypress swamp found in the western edge of the Coastal Plain Province.

section. The Shawnee Hills Section has been subdivided into four subsections: the Mammoth Cave Plateau, Marion, Ohio River Hills and Lowlands, and Brush Creek Hills.

The Blue Grass Section is an upland region composed of a structural dome of Ordovician-aged limestone rimmed by an irregular circular belt of Ordovician and Silurian limestones. This belt displays the hilliest topology of the region and is recognized as having distinct topographic zones, which have been described as five subsections. These subsections are the Knobstone Escarpment and Knobs, Northeastern Blue Grass, Outer Blue Grass, Eden Shale Belt, and Inner Blue Grass. The Blue Grass Section is drained entirely by Ohio River tributaries including the lower Licking, lower Kentucky, and Salt River drainages (Fig. 6).

The crayfish fauna of the Interior Low Plateaus Province is the most diverse in the state and consists of 40 described and 1 undescribed species. Given its upland nature with firm gravel and cobble-bottomed creeks, the province is dominated by the genera *Cambarus* with 18 species (17 described and 1 undescribed) and *Orconectes* with 20 species. The genera *Fallicambarus*, *Procambarus*, and *Barbicambarus* are each represented in the province by single species. Seven species are also endemic to the province in Kentucky: *Cambarus batchi*, *Orconectes bisectus*, *O. jeffersoni*, *O. margorectus*, *O. rafinesquei*, *O. juvenilis*, and *O. ronaldi*.

Appalachian Plateaus Province. The Appalachian Plateaus Province of Kentucky encompasses the eastern fourth of the state (Fig. 2). The province contains the most rugged and highest topology in Kentucky. It consists of irregular hills, steep ridges, and mountains composed primarily of shales, conglomerates, siltstones, and coal-capped sandstones of Pennsylvanian age. These high hills and mountains are dissected by moderate to high gradient creeks and rivers with narrow floodplains and rocky and sandy substrates (Fig. 10). Elevations in the province range from 1,200 feet along its western edge to 4,150 feet in the extreme southeast. Major river drainages in the province include the Big Sandy, Little Sandy, upper Licking, upper Kentucky, and upper Cumberland (Fig. 6). Three sections within the Appalachian Plateaus Province have been recognized: the Unglaciated Allegheny Plateau Section in the northern third of the province, Cumberland Plateau Section comprising the southern two-thirds of the province, and Cumberland Mountains Section. The Cumberland Mountains of Kentucky form a narrow southwesterly trending strip along the extreme southeastern border of the state (Fig. 2).

While rugged, the overall topology of the Appalachian Plateaus Province is rather uniform. This factor most likely accounts for its low crayfish diversity. The province is home to nine species of crayfishes and is dominated by the genus *Cambarus* (eight species). One species of *Orconectes*, *O. cristavarius*, is widespread in the province.

Given the physiographic diversity of Kentucky and number of recognized sections (6) and subsections (21), for the most part we have chosen not to discuss each section and subsection individually. We instead refer the reader to Burr and Warren (1986) for more detailed discussions of the topology and geology of Kentucky.

In recent years the concept of ecological regions, or ecoregions, within physiographic regions has become an important tool in water quality resource management. Ecoregions are based on regional physiography, land usage, vegetation, soil types, and geology. Ecoregions have been developed into several levels of detail. Omernik (1987) mapped the Level III ecoregions for the lower 48 states, and that level has been used by Kentucky state agencies since its publication. Recently, Woods et al. (2002) published Level IV ecoregions for the state (Fig. 11), and these are currently being used to make management decisions. The concept assumes that creeks of the same size are faunistically similar to each other within a particular ecoregion, and consequently differ from creeks in other ecoregions. This has allowed managers to calibrate water quality metrics for each ecoregion. Therefore, the calibrations for metrics will be better for Level IV than for Level III ecoregions. In Kentucky there are 7 Level III and 25 Level IV ecoregions (Fig. 11). The Level III ecoregions very closely follow the physiographic regions discussed above.

HYDROGRAPHIC FEATURES

Aquatic and semi-aquatic ecosystems in Kentucky drain into 1 of 10 major drainages. An 11th drainage, the Obion, is represented in Kentucky only by Terrapin Creek and several other small creeks in an approximate 4.6-square-mile area of southern Graves and Calloway counties in the Coastal Plain Province. The 10 major drainages are as follows: Mississippi River and minor tributaries, Tennessee River drainage, Cumberland River drainage, Tradewater River drainage, Ohio River and minor tributaries, Green River drainage, Salt River drainage, Kentucky River drainage, Licking River drainage, and Big Sandy River drainage (Fig. 6). Table 2 lists crayfish species known to occur in each of those 10 drainages. As with topology and geology above, we refer the reader to Burr and Warren (1986) for detailed descriptions of each major river drainage.

STRUCTURES AND IDENTIFICATION

Crayfishes have well-developed nervous, sensory, circulatory, and digestive systems; however, since all structures used for identification are external, we will focus only on those. The crayfish body is divided into two main regions, the cephalothorax and the abdomen

Fig. 6. The river drainages of Kentucky, from Burr and Warren (1986).

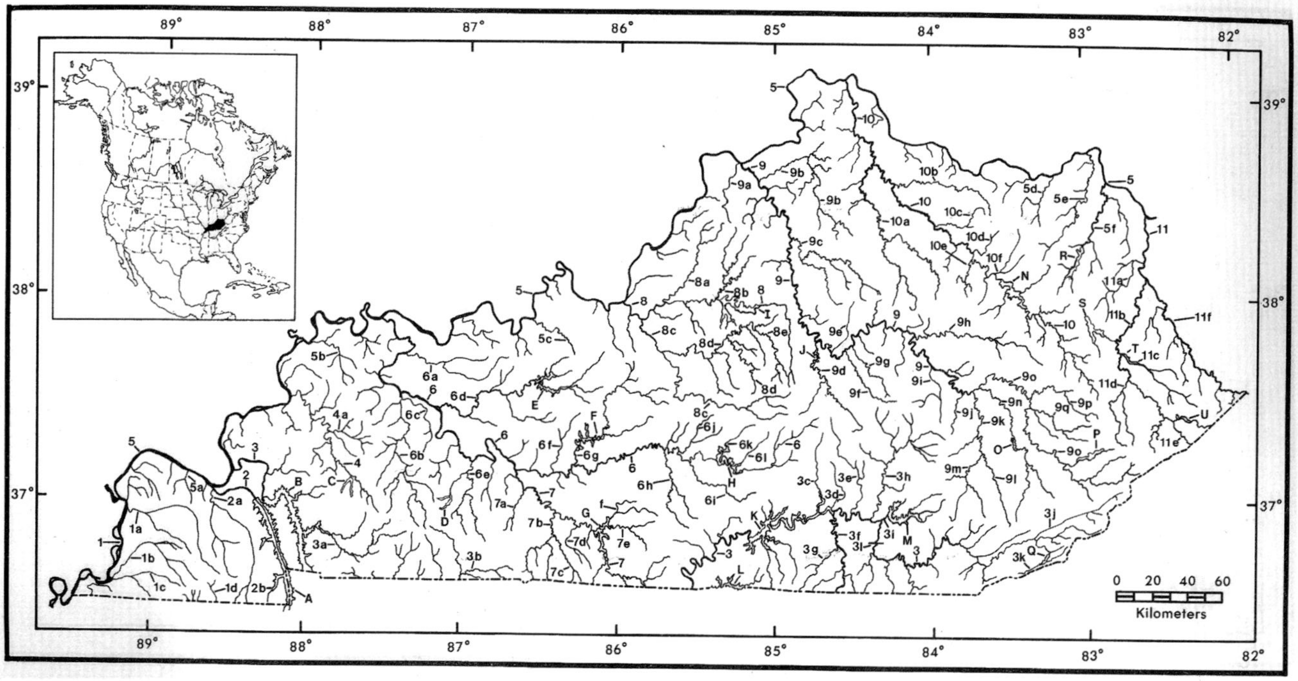

RIVER SYSTEMS

1. Mississippi River
   a. Mayfield Creek
   b. Obion Creek
   c. Bayou du Chien
   d. Terrapin Creek (Obion River)
2. Tennessee River
   a. Clarks River
   b. Blood River
3. Cumberland River
   a. Little River
   b. Red River
   c. Fishing Creek
   d. Pitman Creek
   e. Buck Creek
   f. Big South Fork
   g. Little South Fork
   h. Rockcastle River
   i. Laurel River
   j. Poor Fork
   k. Martins Fork
   l. Cumberland Falls
4. Tradewater River
   a. Clear Creek
5. Ohio River
   a. Massac Creek
   b. Highland Creek
   c. Sinking Creek
   d. Kinniconick Creek
   e. Tygarts Creek
   f. Little Sandy River
6. Green River
   a. Panther Creek
   b. Pond River
   c. Cypress Creek
   d. Rough River
   e. Mud River
   f. Bear Creek
   g. Nolin River
   h. Little Barren River
   i. Russell Creek
   j. Pitman Creek
   k. Robinson Creek
   l. Casey Creek
7. Barren River
   a. Gasper River
   b. Drakes Creek
   c. Trammel Fork
   d. Bays Fork
   e. Skeggs Creek
   f. Beaver Creek
8. Salt River
   a. Floyds Fork
   b. Brashears Creek
   c. Rolling Fork
   d. Beech Fork
   e. Chaplin River
9. Kentucky River
   a. Little Kentucky River
   b. Eagle Creek
   c. Elkhorn Creek
   d. Dix River
   e. Hickman Creek
   f. Paint Lick Creek
   g. Silver Creek
   h. Red River
   i. Station Camp Creek
   j. Sturgeon Creek
   k. South Fork
   l. Red Bird River
   m. Goose Creek
   n. Middle Fork
   o. North Fork
   p. Quicksand Creek
   q. Troublesome Creek
10. Licking River
    a. South Fork
    b. North Fork
    c. Fleming Creek
    d. Fox Creek
    e. Slate Creek
    f. Triplett Creek
11. Big Sandy River
    a. Blaine Creek
    b. Levisa Fork
    c. Johns Creek
    d. Beaver Creek
    e. Russell Fork
    f. Tug Fork

RESERVOIRS

A. Kentucky Lake
B. Lake Barkley
C. Lake Beshear
D. Lake Malone
E. Rough River Lake
F. Nolin Lake
G. Barren River Lake
H. Green River Lake
I. Taylorsville Lake
J. Herrington Lake
K. Lake Cumberland
L. Dale Hollow Lake
M. Laurel River Lake
N. Cave Run Lake
O. Buckhorn Lake
P. Carr Fork Lake
Q. Martins Fork Lake
R. Grayson Lake
S. Paintsville Lake
T. Dewey Lake
U. Fishtrap Lake

Table 2. Crayfish distributions in Kentucky by major river drainage. 1) Mississippi River; 2) Tennessee River; 3) Cumberland River; 4) Tradewater River; 5) Ohio River; 6) Green River; 7) Salt River; 8) Kentucky River; 9) Licking River; 10) Big Sandy River.

Taxa	1	2	3	4	5	6	7	8	9	10
Barbicambarus										
cornutus						X				
Cambarellus										
puer	X									
shufeldtii	X				X					
Cambarus										
bartonii cavatus					X		X	X	X	X
batchi								X		
bouchardi			X							
buntingi			X							
cumberlandensis			X			X				
deweesae			X			X	X	X		
diogenes	X	X	X	X	X	X				
distans			X					X		
dubius			X					X	X	X
friaufi			X			X				
graysoni			X			X	X	X		
ludovicianus	X									
ortmanni					X	X	X	X	X	X
parvoculus			X					X		
robustus								X	X	X
rusticiformis			X		X	X				
sciotensis					X					X
sphenoides			X					X		
striatus						X		X		
tenebrosus			X	X	X	X	X	X	X	
thomai			X		X	X	X	X	X	X
veteranus										X
sp. A		X	X	X	X	X				
Fallicambarus										
fodiens	X	X		X	X	X				
Orconectes										
australis packardi			X							
barrenensis						X				
bisectus					X					
burri		X								
compressus			X			X				
cristavarius			X		X			X	X	X
durelli		X	X			X				
immunis	X	X	X	X	X	X				
inermis inermis					X	X				
jeffersoni					X					
juvenilis			X		X		X	X		
kentuckiensis			X	X	X					
lancifer	X				X					
margorectus			X		X					
palmeri palmeri	X									
pellucidus			X			X				
placidus		X	X			X				
putnami			X		X	X				
rafinesquei					X	X				
ronaldi						X				
rusticus			X		X	X	X	X	X	
sanbornii					X					
tricuspis		X	X			X				
Procambarus										
acutus	X	X	X		X	X		X	X	
clarkii	X	X			X					
viaeviridis	X	X								
TOTALS	11	11	29	6	25	26	8	17	9	8

Fig. 7. An unnamed spring in Breckinridge County.

(Fig. 1), and is covered by an exoskeleton consisting of chiton. Chiton is a compound of complex polysaccharides hardened with inorganic salts.

The cephalothorax is formed by the fusion of the head and thoracic sections and is covered by the carapace. The head contains a pair of stalked compound eyes, a median rostrum between the eyes, and two pairs of antennae. The shorter pair of antennae is known as antennules while the longer pair is the antenna (Fig. 12); both pairs serve as touch and smell sensory structures. The antennules each have two branches and the antennae have flattened bladelike structures called antennal scales at their bases (Fig. 1). A pair of hardened mandibles, two pairs of maxillae, and three pairs of maxillipeds are located on the underside to the head region. The mandibles are used to chew food items while the maxillae and maxillipeds are used to manipulate food, clean the antennae, and bail water through the gill chambers. The thoracic section or region is a large vaulted region that covers the gills, digestive structures, and heart and is separated from the head section or region by the cervical groove (Fig. 1). The dorsal surface of the thoracic section contains two grooves that run backwards from the cervical groove (Fig. 1). Each groove at first curves gradually inward before continuing rearward parallel to or touching the other. This region is known as the areola (Figs. 1 and 12) and the space between the two grooves at its narrowest point is known as the areola width. Areola width is highly variable among crayfish species. The underside of the thoracic region contains 5 pairs of walking legs, also known as pereiopods (Fig. 12). The first pair of pereiopods terminate in enlarged pincers or claws, called chelae. The chelae usually have one to three rows of swollen bumps called tubercles along the their inner margins; the number of tubercles is variable among crayfish species, especially in the genus *Cambarus*. Chelae are used for capturing food, defense, and social interactions. The second and third pairs of pereiopods also have chelae, however these chelae are not enlarged. Each leg is composed of

Fig. 8. A creek passage in L & N Cave, Barren County. Photo by S. Taylor.

seven segments; moving away from the body these include the coxa, basis, ischium, merus, carpus, propodus, and dactyl. In male crayfish the openings of the testes are found at the bases of the coxa of the fifth pair of pereiopods, while in the female the openings of the ovaries are found at the bases of the coxa of the third pair of pereiopods (Fig. 13). Females also have a sperm storage organ called an annulus ventralis located between the bases of the fourth or fifth pair of pereiopods (Figs. 1 and 13). The annulus ventralis is usually elevated and subcircular or subrhomboidal in shape and can be directed to either the right or left in the same species. Male crayfishes also have hooks on the ischia of the second and third, just third, or third and fourth pairs of pereiopods (Fig. 12, couplet 5 in key). These hooks are larger and more developed in reproductively active males (see below).

The abdomen of crayfishes is segmented and flexible and includes the large abdominal flexor muscles that are used for swimming. The abdomen is composed of seven segments with the first through fifth segments each having a pair of appendages known as swimmerets, or pleopods, on their undersides. In males, the first and second pairs of pleopods have been highly modified for sperm transfer. They are thickened and elongated and lie flat against the underside of the crayfish directed anteriorly between the pereiopods (Figs. 1 and 12). The morphology of the first pair of

Fig. 9. Falling Timber Creek, Barren County, a typical upland creek found in the Interior Low Plateaus Province. Photo by C. Lukhaup.

Fig. 10. Big South Fork of the Cumberland River, McCreary County, a typical creek found in the Appalachian Plateaus Province.

pleopods, or gonopods, in males radically differs between form I and form II individuals and those differences are discussed below in the Biology of Crayfishes section. The gonopods of males are also very important for identification purposes because the shape and number of the terminal elements are quite variable among North American species. For Kentucky species, the number of terminal elements ranges from two to four. The seventh segment of the abdomen is also highly modified. It consists of five flattened structures including a median telson and lateral uropods each composed of two flattened plates (Figs. 1 and 12). The modified seventh segment is commonly referred to as the tail fan.

BIOLOGY OF CRAYFISHES

Detailed life-history information is sorely lacking for most Kentucky crayfish species. With this in mind, most of the burrowing behavior, growth, reproduction, longevity, and foraging behavior information presented below represent a summation of generalized life-history data gathered from species found across North America.

Burrowing. As is the case elsewhere in North America, crayfishes in Kentucky occupy a wide range of habitats. They occur in rivers, creeks, swamps, sloughs, subterranean creeks, springs, ephemeral ponds, reservoirs, ditches, and wet fields. The primary life-history trait of crayfishes that determines the habitats in which they can live is their propensity for burrowing. Crayfishes have been classified into three categories of burrowers: primary, secondary, and tertiary. Crayfishes respire with gills and to extract oxygen from the air their gills must remain moist. Burrowing crayfishes will usually dig vertically into the substrate to reach the water table, thus allowing them to colonize habitats with semi-permanent standing water such as low-lying ditches (Fig. 14), shallow ponds, wet fields (Fig. 15), and prairies, or to survive times of drought in more permanent aquatic habitats. Primary burrowers are those species that excavate elaborate burrows, spending almost their entire lives in them. These burrows can be up to nine feet in depth and be composed of multiple chambers, tunnels, and surface openings (Fig. 16a). Primary burrowers will emerge from their burrows to forage for food or look for mates during wet spring or fall months and on warm, humid nights. Secondary burrowers are those species that spend a significant portion of their lives in burrows but will also frequent creeks and ponds throughout the year. The burrows of secondary burrowers are usually less complex than those of primary burrowers and differ from those in that they

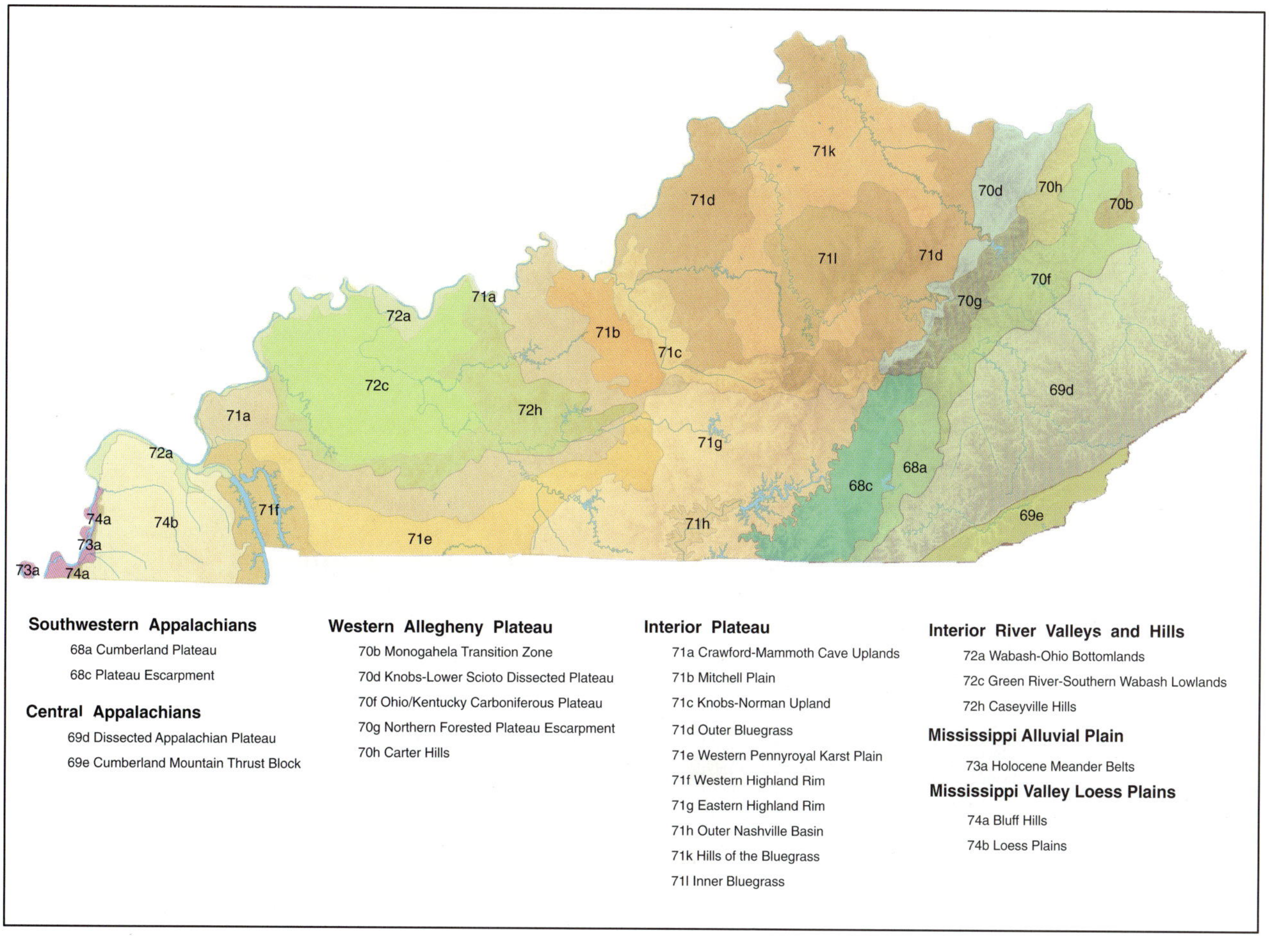

Fig. 11. The ecoregions of Kentucky following Woods et al. (2002).

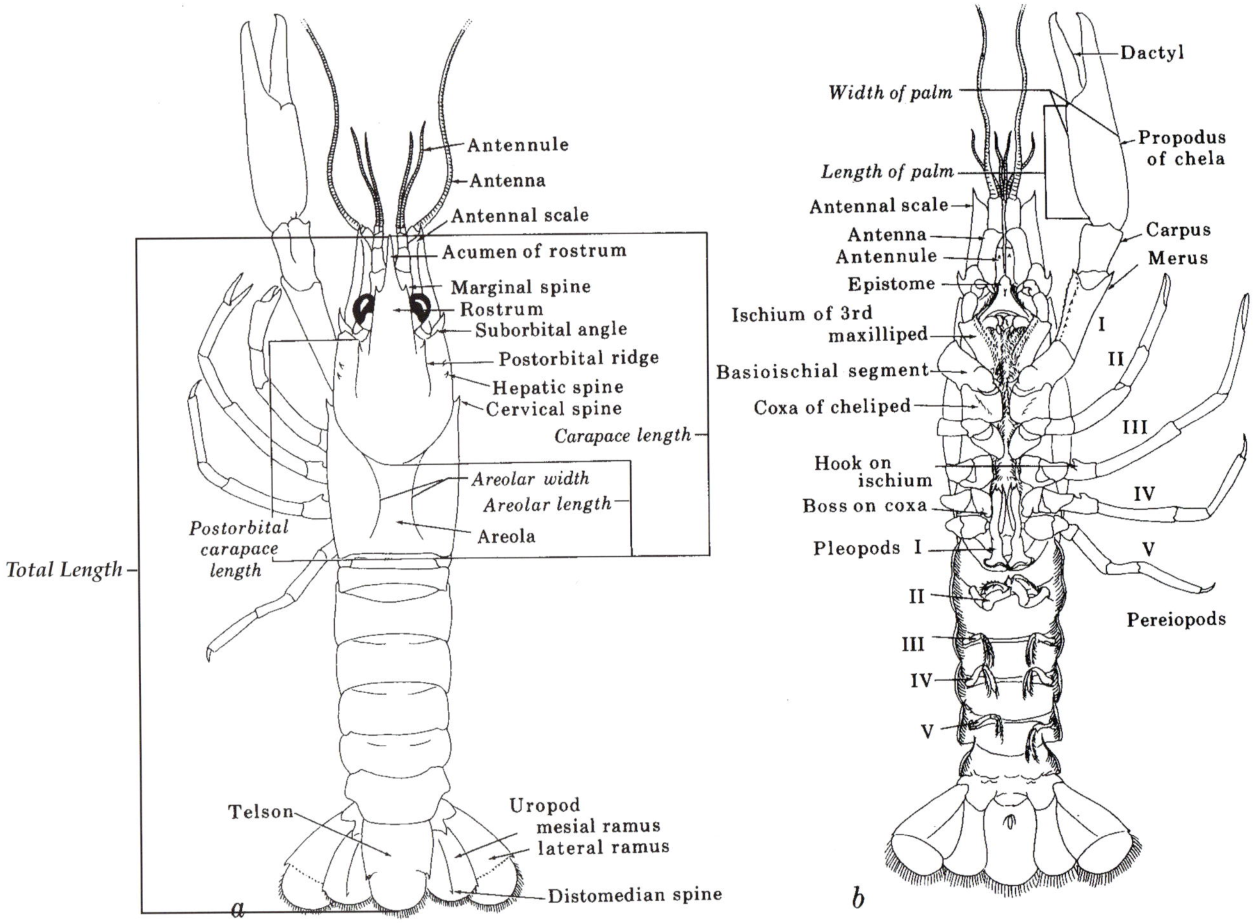

Fig. 12. Morphological structures of cambarid crayfish; a, dorsal view; b, ventral view. Modified from Hobbs (1981).

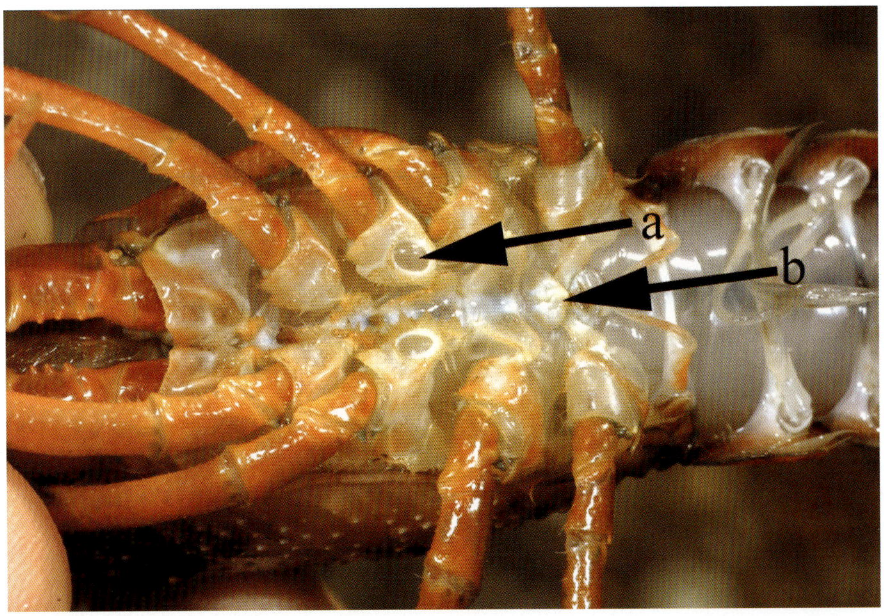

Fig. 13. Ventral view of a female *Procambarus* specimen showing the openings of the genital pores at the base of the third pair of pereiopods (a) and the annulus ventralis (b). Photo by C. Lukhaup.

usually have a connection or tunnel leading to a permanent water body (Fig. 16d). Tertiary burrowers are those species that only occasionally retreat into simple burrows during times of drought or when females tend eggs. Tertiary burrowers usually inhabit permanent flowing rivers and creeks and build shallow vertical shafts into the substrates to reach water tables in the dry months of late summer or during other times of extreme drought (Fig. 16e). The tertiary burrower category can be considered something of a "catch all" category since almost all nonprimary or secondary burrowing crayfishes will construct some form of a simple burrow to escape desiccation during periods of drought.

As crayfish excavate their burrows they will push the loose soil or mud up and out of the burrow opening. This activity will usually lead to the formation of a chimney rising several inches above the ground (Fig. 17). Chimneys of secondary burrowers can be frequently seen lining creek and river margins while those of primary burrowers can be found in open pastures, low-lying ditches, ephemeral ponds, and even plowed agricultural fields. Burrowing crayfishes are also known to plug the openings of their burrows with mud during dry or cold months and remain inactive for weeks or even months.

Life Cycle, Growth, and Reproduction. The process of molting plays a significant role in the life history of a crayfish. The lives of crayfishes consist of a continuous process of shedding, or molting, their old exoskeleton, and growing and hardening the new exoskeleton. In the days leading up to molting, crayfishes begin to break down the calcium salts in their exoskeleton and store a small amount of them in small whitish disklike structures called gastroliths. Gastroliths are found in the stomachs of crayfishes and provide the important initial source of calcium used to harden the new exoskeleton. Crayfishes in the molting process are very vulnerable to predation because they are soft and unable to move effectively. During this time they will usually find refuge under rocks, in woody debris, or in burrows.

Crayfishes begin life as fertilized eggs attached to the underside of the female's abdomen (Fig. 18). While considerable variation exists in the timing of reproductive activities of crayfishes, in Kentucky females are usually found with eggs attached to their abdomens from early March to late May. These females are referred to as being ovigerous or "in berry" and will carry their eggs for several days to several weeks. Crayfishes have direct development and the newly hatched young resemble smaller forms of the adults at hatching. After hatching from the egg, juvenile crayfishes will stay attached to the underside of the female's abdomen for several days (Fig. 19), usually going through two or three molts as body size increases. After detaching from the female, juvenile crayfishes begin life as free-living individuals foraging for food, finding shelter, and growing to adult size.

Growth is most rapid in the first summer of life; crayfishes can molt as many as 14 times during this period and triple or quadruple in body size. As food

Fig. 14. A roadside ditch in McLean County, a habitat type frequently occupied by primary burrowing species of crayfishes.

Fig. 15. A temporary pond along a roadside in Livingston County, a habitat type frequently occupied by primary burrowing species of crayfishes.

becomes less abundant and water temperatures decline during the winter months, growth and molting will typically cease. As water temperatures warm in the spring, entire populations may simultaneously molt in a given reach of creek or pond. Adults of most species will usually molt only twice each year during the remainder of their lives, once in the spring and then again in the fall.

All Kentucky crayfishes are members of the family Cambaridae, and as such, the males exhibit a cycle of dimorphism associated with reproduction. Males are classified as either form I or form II. Form I males are reproductively active and differ from form II males in having longer, more pointed terminal elements of the gonopod and larger hooks on the ischia of the pereiopods. The tips of the terminal elements of the gonopod have a yellowish, translucent appearance and are described as being corneous. Usually in the late spring and early summer and again in the fall, form I males will molt into the reproductively inactive form II stage. Males of most species can reach sexual maturity (form I) in their first year of life. Recent work has also shown that female crayfishes may alternate between reproductively active and inactive states (Wetzel 2002). Reproductively active "form I" females are usually found in fall and spring months and differ from form II females in having wider abdominal segments and swollen glair glands (described below).

Copulation in cambarid crayfishes is believed to occur mostly in the fall months from September to November. Sperm is transferred from openings at the base of the male's last pair of pereiopods to the female via the male's gonopods. The sperm is inserted into the annulus ventralis located between the last two pairs of pereiopods (Fig. 1) of the female where it is stored overwinter. In an effort to prevent other males from mating with the same female, male crayfishes can deposit a gummy substance known as a sperm plug over the opening of the annulus ventralis. However, recent evidence has shown that female crayfishes will mate with multiple males, implying that males can remove the sperm plug deposited by other males (Walker et al. 2002). In the late winter or early spring months (usually February or March), females will move to a secluded location and will release their unfertilized eggs from openings at the bases of their third pair of pereiopods (Fig. 13). She will simultaneously release the sperm stored in her annulus ventralis, thereby fertilizing the eggs. Prior to egg-laying, the pleopods are coated with a sticky substance, known as glair, excreted from glands on the underside of the abdomen. The fertilized eggs are moved to the underside of her abdomen where they become attached to the pleopods. Females carrying eggs will protect them by curling her tail fan under her abdomen. The total number of eggs fertilized and carried by a female is dependent on species and the body size of the female; larger females generally carry more eggs. The

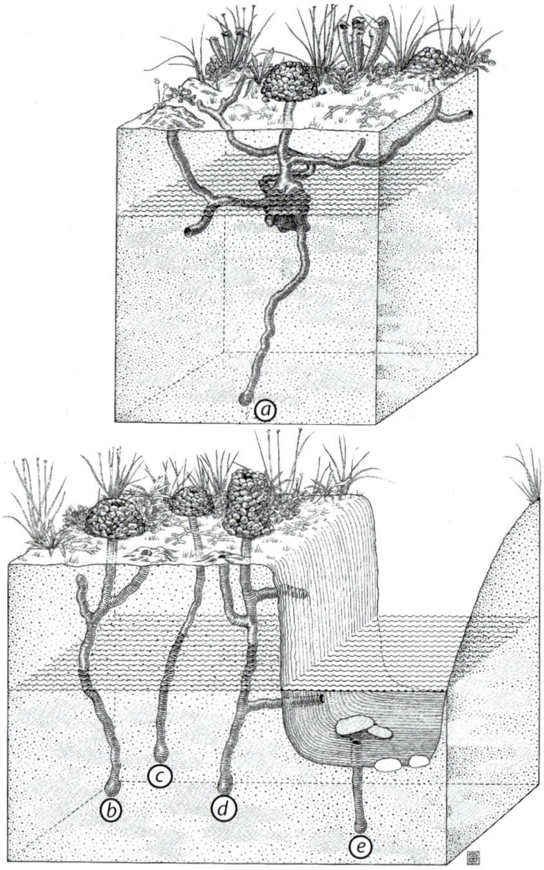

Fig. 16. Morphology of crayfish burrows: a, that of a primary burrowing species; b, c, d, those of a secondary burrowing species; e, that of a tertiary burrowing species. From Hobbs (1981).

Fig. 17. An example of a mud chimney found at the opening of a crayfish burrow. Photo by T. Anton.

Fig. 18. **An ovigerous or "in berry" female *Procambarus clarkii*. Photo by C. Lukhaup.**

number of eggs found on a female can range from less than 50 to 400. In most Kentucky species, copulation and egg-laying occur during the fall and spring, respectively. However, some species may also copulate in early spring months with females using the sperm to fertilize eggs later that same spring. In more southern latitudes of North America, copulation commonly occurs in the spring with eggs being extruded in the fall.

Copulation is usually initiated by the male after coming in contact with a female. There is some courtship behavior, which can initially appear to be hostile, that results in species and sex recognition. This initially starts with antennal flicking, followed by the male grabbing the female with his chelae. She eventually capitulates and allows the male to turn her onto her back. Once this is done he will crawl on top of her and hold her chelae and pereiopods with his chelae. At this point she is completely subdued, and the pair become locked together as the male fits the hooks on the ischia of his pereiopods into corresponding depressions on the underside of the female (Fig. 20). The copulating pair can remain locked together for an extended period of time, often an hour or more, during which the male will transfer sperm via his gonopods into the female's annulus ventralis.

Longevity. It is estimated that most Kentucky crayfishes live from two to three years. In general, species that grow to larger sizes such as *Barbicambarus cornutus* and *Cambarus cumberlandensis* may live for five or six years while smaller species, such as those in the genus *Cambarellus*, may live only one year. Evidence suggests that some populations of cave-dwelling species in the genus *Orconectes* may live for at least 20 years (Cooper 1975). Kentucky is home to three

Fig. 19. A female *Fallicambarus fodiens* carrying young.

subterranean species of *Orconectes*; however, it is unknown if individuals in these populations can live that long.

Diet. Historically treated as opportunistic omnivores, crayfishes have been shown to eat a variety of food items, both living and dead. These include aquatic and terrestrial vegetation, plant detritus, insects, snails, and aquatic crustaceans (e.g., isopods, amphipods, and copepods). Some evidence has suggested that crayfishes are primarily obligate carnivores, but in the absence of animal material, they become facultative omnivores and scavengers (Momot 1995). Crayfishes play an important role in ecosystem function by consuming detritus and vegetation and converting that material into animal protein. That

Fig. 20. A copulating pair of crayfish. Photo by C. Lukhaup.

protein is then made available to a wide range of other aquatic and terrestrial organisms such as predaceous insects, fishes, birds, and mammals. Butler et al. (2003) reported that over 240 different species of animals are known to feed on crayfishes. Crayfishes are also known to occasionally consume carrion and assist with its breakdown in aquatic ecosystems.

HISTORY, COLLECTING, AND CONSERVATION OF CRAYFISHES IN KENTUCKY

History. Kentucky's crayfish fauna has received early but unfortunately sporadic attention. The uniqueness of Mammoth Cave in west-central Kentucky and its associated fauna attracted the attention of early naturalists. The first published record of a crayfish inhabiting Kentucky waters came from Theodor Tellkampf, who described *Astacus pellucidus* (= *Orconectes pellucidus*) from that cave in 1844 (Tellkampf 1844). While several species were described in the succeeding 87 years, the first work to present distributional information for multiple species known from the state appeared in 1931. In his monograph of crayfishes from the southern Appalachians and Cumberland Plateau, Ortmann (1931) provided identification keys and distributional data for a crayfish fauna that included several species that occurred in eastern and central Kentucky. At this time a total of 15 species were recorded from the state.

The first and only comprehensive effort to report exclusively on the crayfishes of Kentucky was that of Rhoades (1944). Rhoades gave the results of a preliminary survey of crayfishes conducted in each of Kentucky's 120 counties and listed 27 taxa (species and subspecies) from the state. He described six species and four subspecies new to science and provided notes on distribution, ecology, and morphology for each taxon. Horton Hobbs, Jr. (1914–1994) was undoubtedly the most productive and respected North American astacologist. For over 50 years Hobbs published extensively on the crayfishes of the eastern United States; however, his attention was never directed towards Kentucky and its fauna. In 1989 Hobbs published his monumental checklist of American crayfishes (Hobbs 1989) in which he listed known taxa from every country, province, and state in North, Central, and South America. In that publication new distributional records and species descriptions had increased Kentucky's crayfish list to 45 taxa.

Other than the field efforts of Rhoades (1944) and those of the authors, which began in earnest in 1993, crayfish collecting in Kentucky during the 20th century mostly occurred after 1960 (Fig. 21). Figure 21 shows all collections housed at major museums (listed below), including those made by us, and indicates that almost all of the records plotted on our distribution maps have been made in recent times. Collection numbers from the 1930s and 1940s are somewhat low since most of Rhoades' collections cannot be located; it is unknown if they were ever deposited in a museum. Crayfish collections made by those other than Rhoades and ourselves mostly fell under three categories: 1) ancillary specimens collected by biologists working with other aquatic organisms such as fishes, mussels, or insects; 2) specimens collected as parts of taxonomic studies of crayfish groups that had members occurring in Kentucky; or 3) specimens collected by various state agencies during general aquatic faunal surveys. Noteworthy collecting efforts were conducted by researchers from several institutions including the

Academy of Natural Sciences of Philadelphia (ANSP), Eastern Kentucky University (EKU), Illinois Natural History Survey (INHS), Ohio State University (OSU), Southern Illinois University at Carbondale (SIUC), and the University of Tennessee (UT). The specimens collected by these researchers have invariably ended up in museums with large holdings of crayfishes such as the Branley A. Branson Museum of Zoology at EKU, The Museum of Biological Diversity at OSU (OSM), the INHS Crustacean Collection, the ANSP Crustacean Collection, and the United States National Museum of Natural History Smithsonian Institution (USNM) Crustacean Collection.

Collecting Crayfishes. Crayfishes can easily be collected from, and sometimes observed in, their habitats. However, given the variation in habitats in which they occur, a variety of methods need to be used. A nylon minnow seine with no larger than 1/8-inch mesh and measuring approximately 10 feet in length and 4 feet in depth can be used effectively in a variety of habitats. The seine should have small lead weights along the bottom edge and be attached between brail poles. Seines are most effective in shallow riffles and pools of creeks and in lakes and ponds lacking lush growths of aquatic vegetation. In riffles the seine should be set perpendicular to the current with the lead line firmly held against the bottom; the substrate immediately upstream of the seine should then be vigorously disturbed by kicking and/or turning over rocks by hand. Pools of creeks or ponds can be seined by rapidly pulling the seine through the water, being careful to keep the lead line against the bottom. In one continuous motion, the seine should then be pulled out of the water onto a gradually sloping bank. In ponds or pools with abundant submergent vegetation, hand-held dip-nets often prove to be more effective. An alternative method of collecting crayfishes in pools and ponds is with baited traps. Numerous trap designs are available; however, those similar to a minnow trap made with wire or hard plastic mesh with an opening at the end of a funnel are the most effective. Traps can be baited with dog food or any type of animal flesh or tissue such as beef liver and left submerged for one to several days.

Burrowing crayfishes present an altogether different situation for collecting. While many adult and juvenile individuals can be collected in standing water with a minnow seine, their presence in standing water is highly seasonal. Extracting crayfishes from their burrows can be a difficult and time-consuming endeavor. The most successful method is to use hand shovels or trowels to physically enlarge the opening and dig out the burrow. Crayfishes will usually retreat to the deepest chamber or tunnel of the burrow complex as one begins to excavate the burrow, hence the difficulty in extracting them. As many burrows can be unoccupied, it is best to look for and dig out those burrows with signs of fresh mud that has been pushed

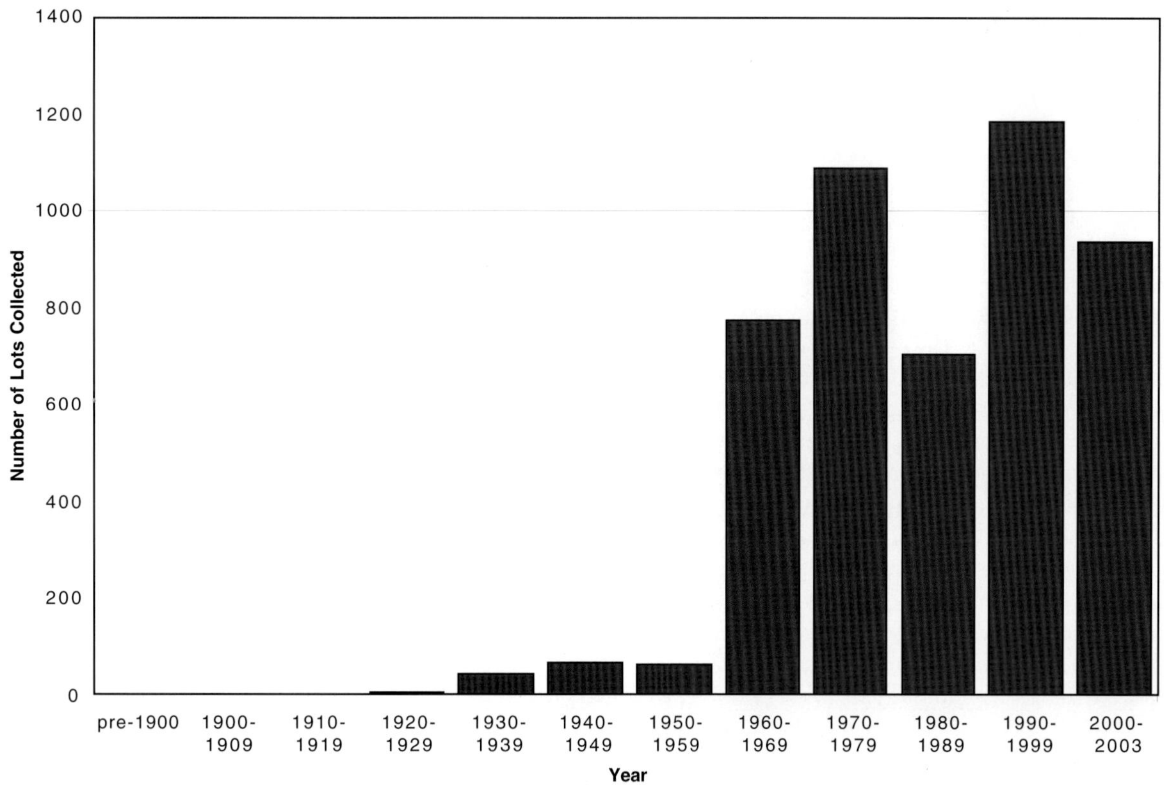

Fig. 21. Number of crayfish lots collected per decade in Kentucky that are housed in major museum collections and plotted on distribution maps. Lot = a unit containing at least one vouchered specimen of a species collected at a given location.

out of the burrow opening. If the burrow is relatively shallow and the water table only a few inches below the surface, one can enlarge the opening by hand and agitate the water. If the collector remains quiet, he or she will occasionally be rewarded by a crayfish coming to the water's surface to investigate the disturbance. The crayfish's presence will usually be given away as it waves it antennae near the water's surface. The collector must then quickly pin the crayfish against the burrow wall and extract it. Finally, adult burrowing crayfishes will sometimes come out of their burrows on warm and humid or rainy nights in search of food or mates and can be collected by hand using a flashlight.

Conservation of Crayfishes. Crayfishes are susceptible to the same threats facing other aquatic organisms, mainly habitat loss/alteration and the introduction of non-native species. However, crayfishes are especially vulnerable to these threats given the limited native ranges of numerous species. A recent review of the conservation status of North American crayfishes listed limited range as the primary factor responsible for the finding that almost 50% of all species were in need of conservation attention (Taylor et al. 1996). Several species occurring in Kentucky are found only in a single creek or river drainage in a one- or two-county area. Habitat alteration in the form of creek channelization with the removal of rocky substrates or woody debris, wetland drainage, or impoundment could drastically affect the total population size of some of these species.

The introduction of non-native crayfishes has perhaps had the greatest negative impact on crayfishes and deserves the most attention. When introduced into new areas, whether continents, countries, or drainages within the same state, non-native crayfishes often rapidly increase in population size and can aggressively displace native species. There are several examples in the United States of native species being totally eliminated from lakes or stretches of rivers after a non-native species was introduced (Lodge et al. 2000). As of the time of printing of this book, Kentucky has been spared from the establishment of crayfish species that are not native to the state. For Kentucky to remain free of non-natives, vigilance must be exercised by individuals who come in contact with live crayfishes. This would include educators who use live crayfishes for laboratory teaching, aquarium hobbyists, and fisherman who use them for bait. Under no circumstances should any crayfish be released into waters from which they were not collected.

Taylor et al. (1996) listed four Kentucky species as endangered, three as threatened, and two as special concern. The Kentucky State Nature Preserves Commission listed five species as endangered, three as threatened, and five as special concern (KSNPC 2000). Listing by either entity does not afford a species legal projection from direct taking or destruction of habitat.

METHODS

Data presented in this book come from three sources: collections of crayfishes and notes on life history made by the authors, crayfish specimens housed in major museums, and life-history data presented in peer-reviewed literature. Between the years 1993 and 2003 approximately 320 different sites in Kentucky were sampled for crayfishes by the authors. These field efforts resulted in the collection of 663 lots of crayfishes. A lot is defined as a unit containing at least one vouchered specimen of a species collected at a given location on a given date. While the number of specimens per lot range from 1 to 25, a conservative estimate of 5 specimens per lot brings the total number of specimens collected and examined by the authors to 3,315. Crayfishes were collected using a variety of methods; however, standard minnow seines and hand-held dip-nets were the predominate methods employed. Burrowing crayfishes were usually collected by digging out burrows with shovels and/or handheld trowels. Specimens were preserved in the field with 70% ethanol and deposited into the crustacean collections at either EKU or INHS.

Additional Kentucky records were obtained from the holdings of EKU, INHS, OSM, Museum of Comparative Zoology at Harvard University (MCZ), and USNM. A total of 2,053 lots at EKU and 570 lots at INHS representing Kentucky collections made prior to 1993 and 113 lots at OSM and 50 at MCZ were examined and plotted. Approximately 1,200 lots of Kentucky crayfishes are held at USNM. About 480 of those lots were examined and plotted. The remaining 720 lots were plotted on distribution maps in this book, given that they were identified by taxonomic experts, mainly Horton H. Hobbs, Jr. In total, approximately 4,650 Kentucky crayfish records were used for distributional data.

Unless otherwise credited, morphological illustrations were drawn by the senior author with the assistance of a camera lucida mounted on a stereo dissecting microscope.

OTHER SOURCES OF INFORMATION

The World Wide Web contains a plethora of information on crayfish biology and photographs. Readers are encouraged to visit the following websites for additional information:

http://www.ncwildlife.org/pg07_WildlifeSpeciesCon/nccrayfishes/nc_crayfishes.html

http://nas.er.usgs.gov/crustaceans/crayfish.htm

http://www.bio.ilstu.edu/Perry/crayfish/thhuxley/crayfish.htm

http://mdc.mo.gov/nathis/arthopo/crayfish/

http://www.crayfishworld.com/

http://crayfish.byu.edu/IAA/index.htm

HOW TO USE THIS BOOK

The remainder of this book is made up of a dichotomous identification key and individual accounts for all 53 crayfish taxa (described and undescribed species) known to occur in Kentucky. Terms used in the identification key and species accounts are also discussed and illustrated in the preceding pages and in the Glossary.

Identification of Specimens

The dichotomous key in this book is based on morphological characters found on form I males. As such, the most accurate method of identifying crayfishes with this book is to examine either a live or dead form I male under a dissecting microscope or with a hand lens and progressively move through the couplets of the key. After arriving at a species name, the reader should then turn to the page indicated and confirm that the distributional data and illustrations presented for that species match the specimen and location from which it was collected. If a form I male is unavailable, the reader may be able to infer the specimens' identification by studying the color plates and illustrations of non-form I characters of species known to occur in the region from which the specimen was collected. However, it should be noted that due to their constant cycle of molting, the intensity of colors shown by crayfishes can be highly variable. Freshly molted individuals have much brighter colors, especially reds and blues, than individuals that have not molted in several weeks or months. These older intermolt individuals can also be covered with thin layers of dark brown or black-stained debris. Because of this variation, the reader should be cautious in arriving at an identification of a specimen based solely on color.

Several members of the genus *Cambarus* are diagnosed using chelae morphology. Crayfishes can regenerate chelae that are lost and since these regenerated chelae differ in size, shape, and tubercle pattern from original (nonregenerated chelae), only original chelae should be used for identification purposes. Regenerated chelae will usually be longer and thinner than original chelae and will have tubercles of equal size along the opposable margins of the fingers of the chelae (Fig. 22). Finally, some characters such as rostrum shape and areola width can change as some individuals of certain species mature. For example, juveniles of *Orconectes immunis* may have small spines on the lateral margins of the rostrum and slightly wider areolas while adults lack rostral spines and have narrower areolas.

Given the variation in life color, life-cycle (form I vs. form II), ontogeny, and their ability to regenerate, the identification of crayfishes can be difficult. As with all of life's endeavors, the more experience one has the easier the endeavors become. Even with experience,

Fig. 22. A *Cambarus cumberlandensis* specimen with a regenerated and an original chela. Regenerated chelae, such as the one at the top, have shorter palm regions, longer fingers, and tubercles along the opposable margins of the fingers that are all of the same size.

certain species groups can remain problematic. For example, because of close evolutionary relationships the following species pairs can be difficult to discern: *Orconectes cristavarius*/*O. putnami*, *Cambarus* buntingi/ *C. cumberlandensis*, and *C. distans*/*C. parvoculus*.

Species Accounts

Common names used in the species accounts were taken from published sources (Williams et al. 1989; Pflieger 1996; Thoma and Jezerinac 2000) if available or are proposed by us. Each account contains detailed descriptions of the morphological characters, life colors, and sizes for that species. Because different characters are used to identify species within genera (i.e., suborbital angle in *Cambarus*), the CHARACTERS sections do differ between species accounts. Maps showing the distribution of each species in Kentucky and its native North America range and a color plate or plates are also included. As several sites within close proximity to one another are occasionally sampled, dots on the Kentucky distribution maps may represent more than one locality. The MAXIMUM SIZE section contains the maximum known size of the species in total length either for Kentucky specimens examined by us or reported in the literature for out-of-state specimens if larger. The MOST LIKE section calls attention to other crayfish species found in Kentucky that would most likely be confused with the species under consideration. Notes on the life history of the species and the types of habitats in which it would be found are presented. The CRAYFISH ASSOCIATES section lists those species of crayfishes also collected from habitats in Kentucky that contained the species under consideration. Finally, it is noted under CONSERVATION STATUS if a species was recognized as in need of conservation attention by either the American Fisheries Society Endangered Species Subcommittee for Crayfishes (Taylor et al. 1996) and/or the Kentucky State Nature Preserves Commission (KSNPC 2000).

Key to Crayfishes of Kentucky

Based on form I males. First number in parentheses indicates referring couplet.

1a Without eyes; body color white, no pigment; restricted to caves 2

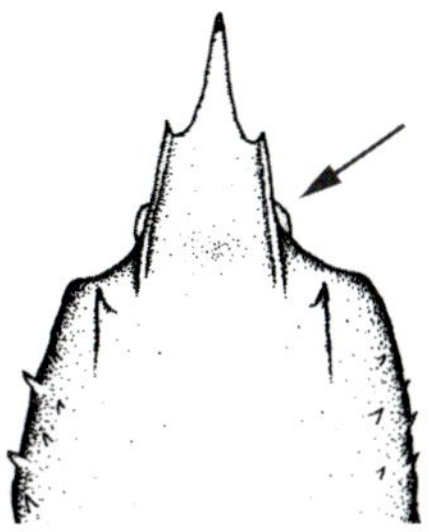

1b With eyes; body pigmented; not restricted to caves 4

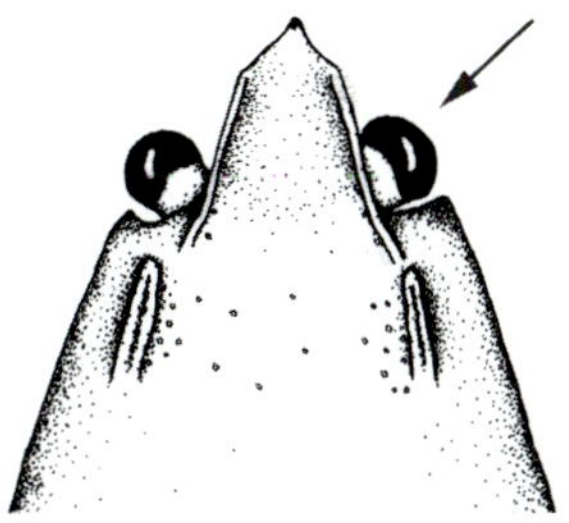

2a(1) Mesial process of gonopod extending distinctly farther distally than central projection (Hart to Trigg counties, Mammoth Cave system). . . . . *Orconectes pellucidus* (p. 178)

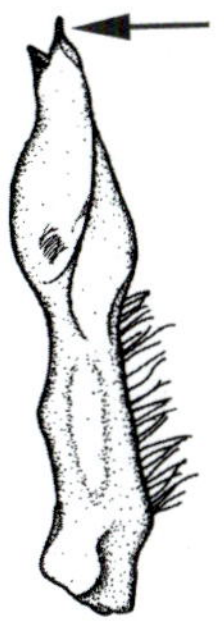

2b Mesial process of gonopod extending only slightly, if at all, farther distally than central projection 3

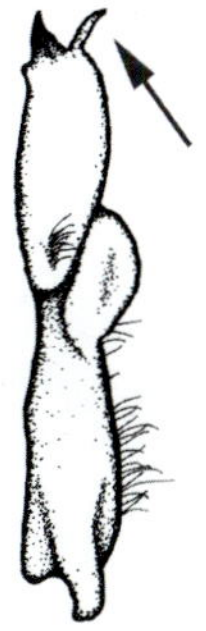

3a(2) Cephalodistal portion of gonopod with rounded or weak angular shoulder at base of central projection (caves in Cumberland River drainage, southeastern Kentucky). . . . *Orconectes australis packardi* (p. 135)

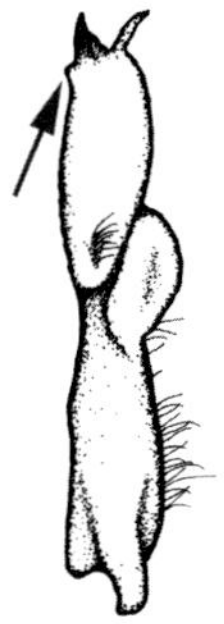

3b Cephalodistal portion of gonopod without shoulder at base of central projection (caves from Hart County north to Ohio River) . . . . *Orconectes inermis inermis* (p. 158)

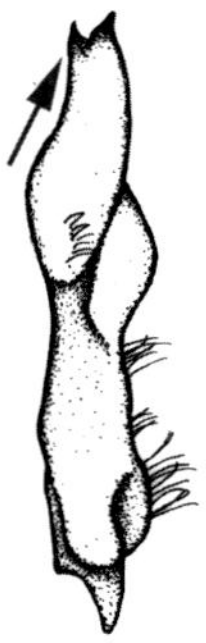

4a(1) Lateral margin of antenna bearded with long, dense setae (upper Green River drainage) . . . . *Barbicambarus cornutus* (p. 59)

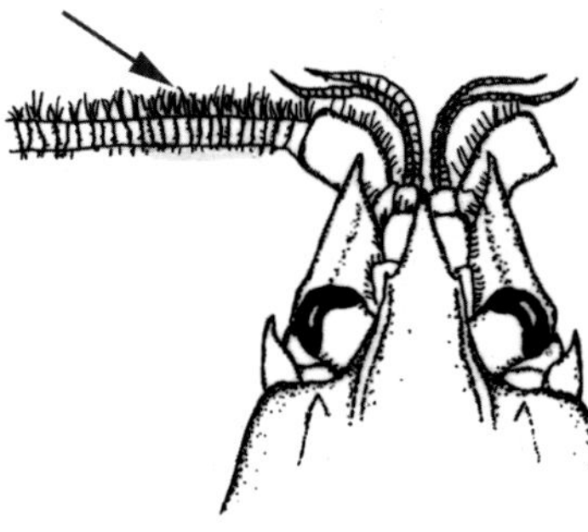

(modified from Hobbs 1972)

4b Lateral margin of antenna with setae, but not long or dense 5

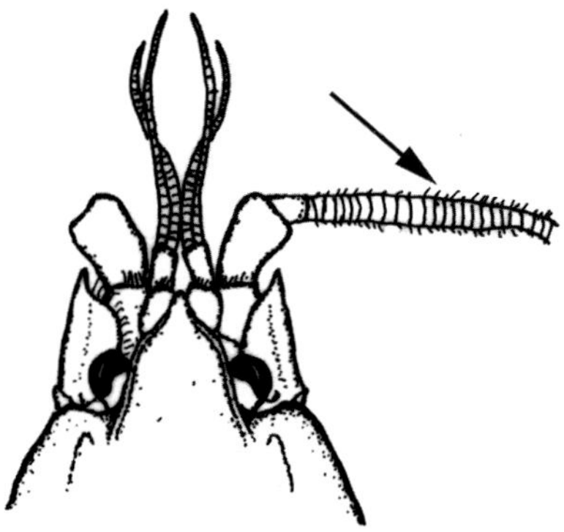

(modified from Hobbs 1972)

5a(4) Males with hooks on ischia of second and third pereiopods; adult size small, never exceeding 40 mm TL 6

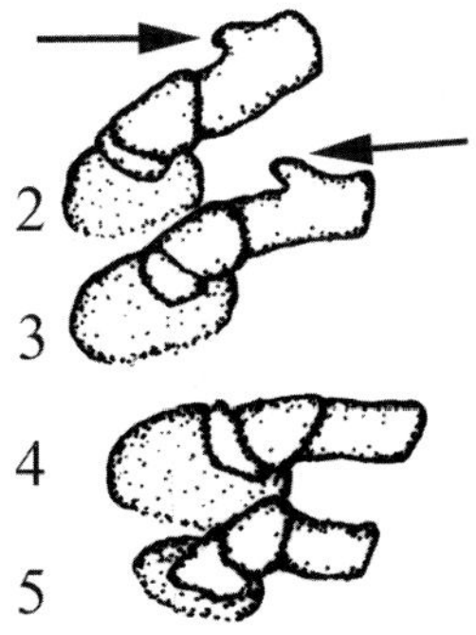

5b Males with hooks on ischia of third or third and fourth pereiopods, never with hooks on second pereiopod 7

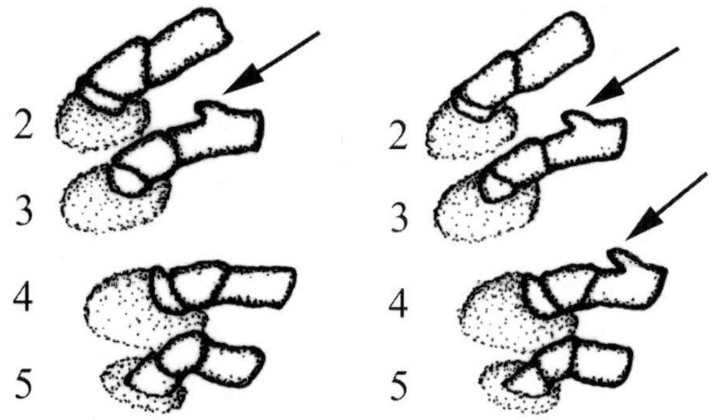

6a(5) Terminal elements of gonopods curved (Mayfield Creek, Mississippi River drainage) . . . . *Cambarellus puer* (p. 52)

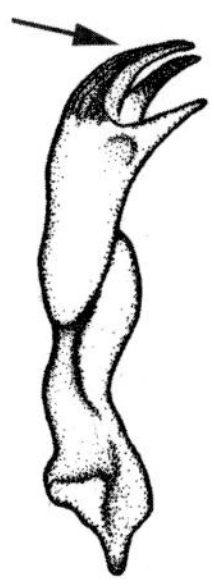

6b Terminal elements of gonopods straight or nearly so (Mississippi River drainage) . . . . *Cambarellus shufeldtii* (p. 55)

7a(5) Areola obliterated 8

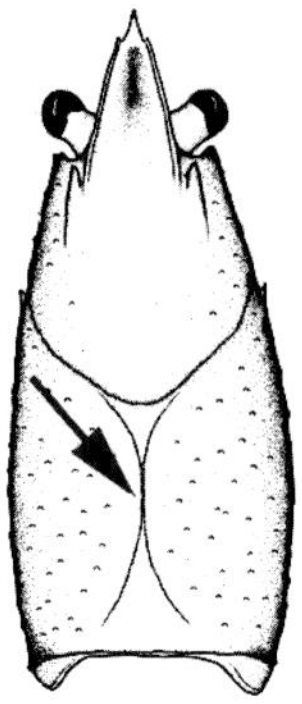

7b Areola open, space very narrow to wide 17

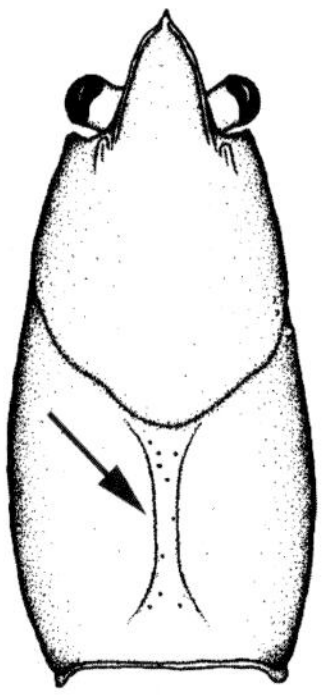

8a(7) Gonopods with more than two terminal elements, tubercles entirely covering chelae and carapace (lower Tennessee, lower Ohio and Mississippi River drainages) . . . . *Procambarus clarkii* (p. 206)

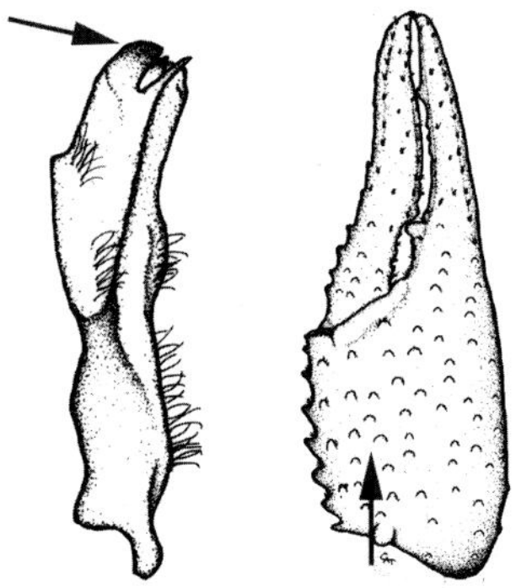

8b Gonopods with two terminal elements, tubercles covering only portions of chelae and carapace 9

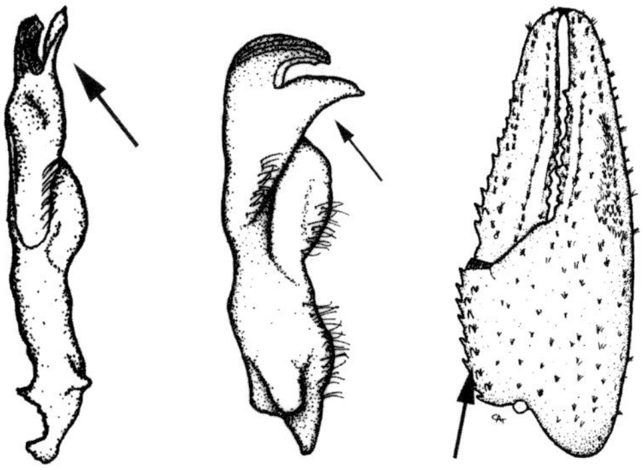

9a(8) Rostral margins terminating in spines or tubercles 10

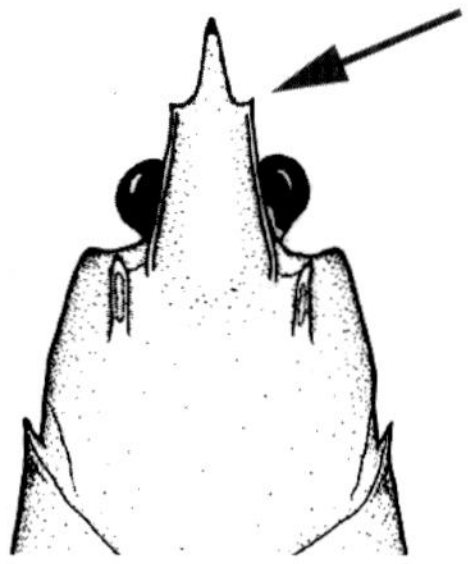

9b Rostral margins broadly rounded, not terminating in spines or tubercles 11

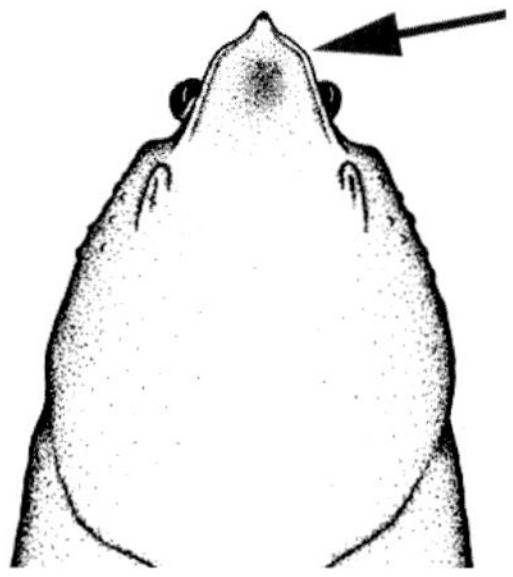

10a(9) Terminal elements of gonopods very short; length of dactyl of chelae less than length of mesial margin of palm (Mississippi and Ohio River drainages). . . . . *Orconectes lancifer* (p. 170)

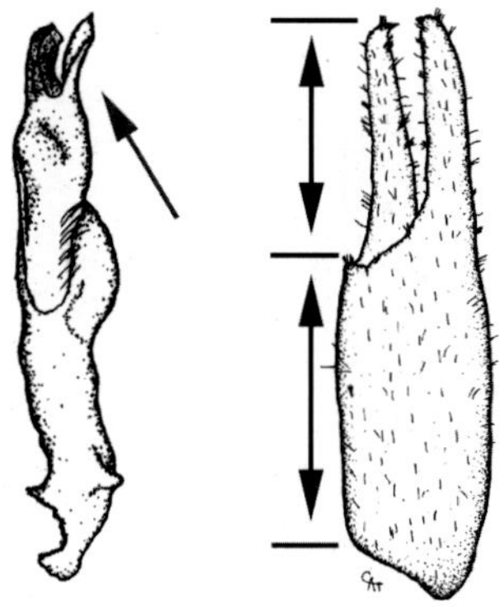

10b Terminal elements long and curved; length of dactyl of chelae distinctly longer than length of mesial margin of palm (Mississippi River drainages) . . . . *Orconectes palmeri palmeri* (p. 175)

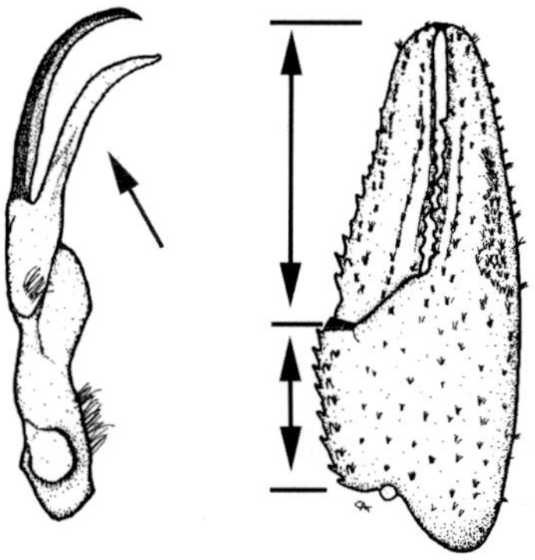

11a(9) Suborbital angle absent, area broadly rounded 12

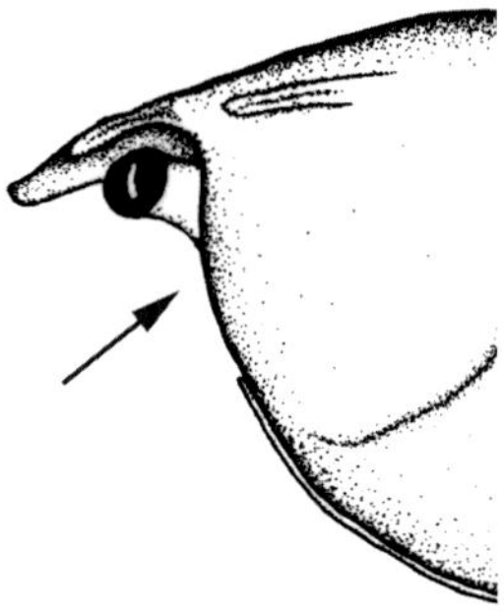

11b Suborbital angle distinct 14

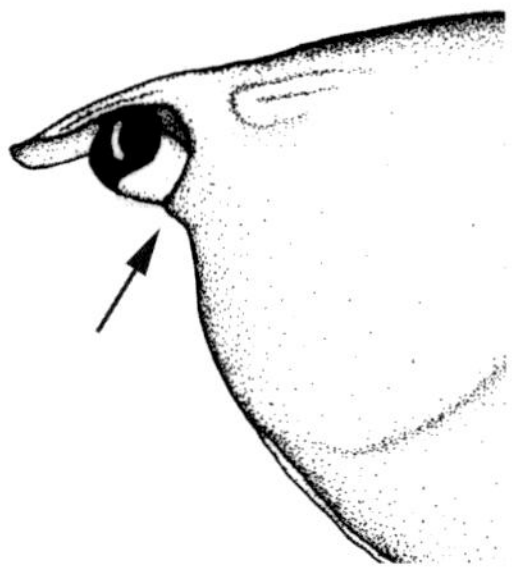

12a(11) Dactyl of chela with distinct basal notch; chelae flattened laterally (Salt River drainage west) . . . . *Fallicambarus fodiens* (p. 131)

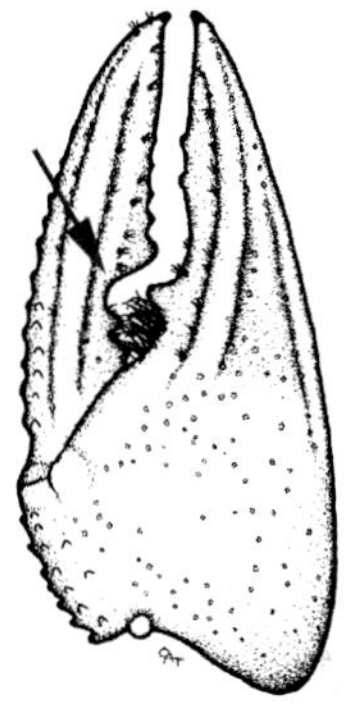

12b Dactyl of chela without basal notch; chelae inflated not flattened 13

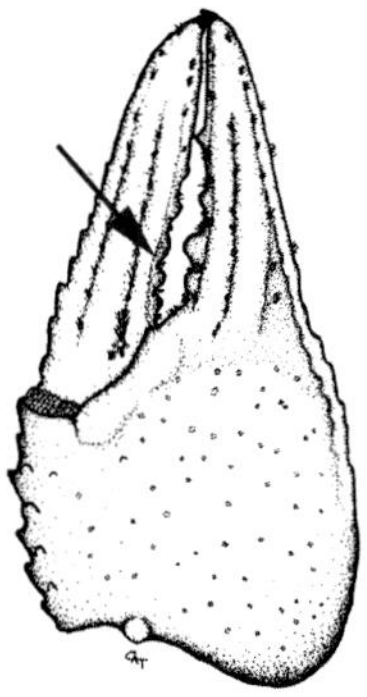

13a(12) Mesial margin of palm of chela with one row of tubercles; carapace smooth with few setal punctations; no subpalmar tubercles on chelae (restricted to the outer Blue Grass region around Madison County) .... *Cambarus batchi* (p. 66)

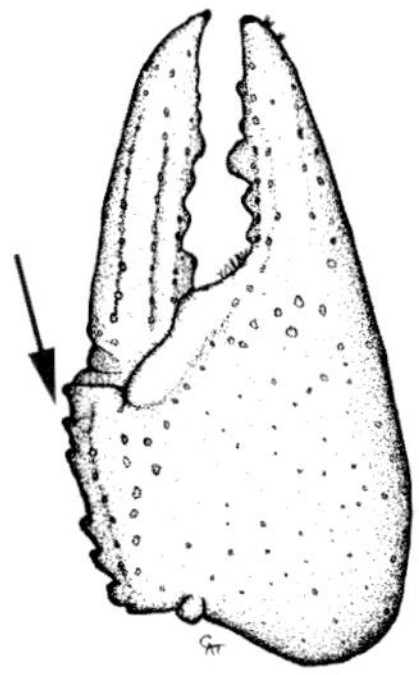

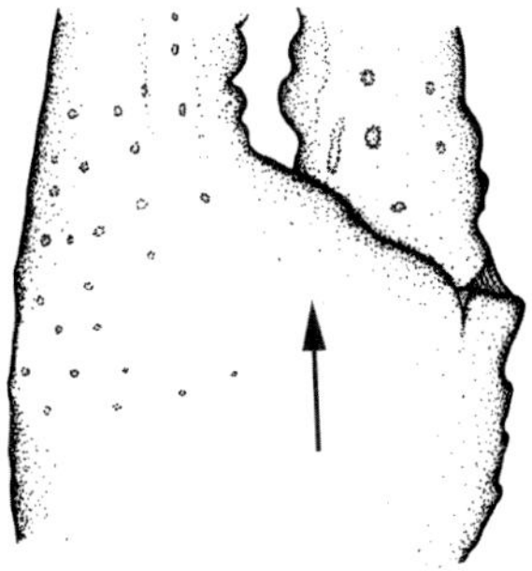

13b Mesial margin of palm of chela with two rows of tubercles; carapace covered with setal punctations giving rough texture; one to two subpalmar tubercles on chelae (widespread in central and eastern Kentucky) .... *Cambarus deweesae* (p. 77)

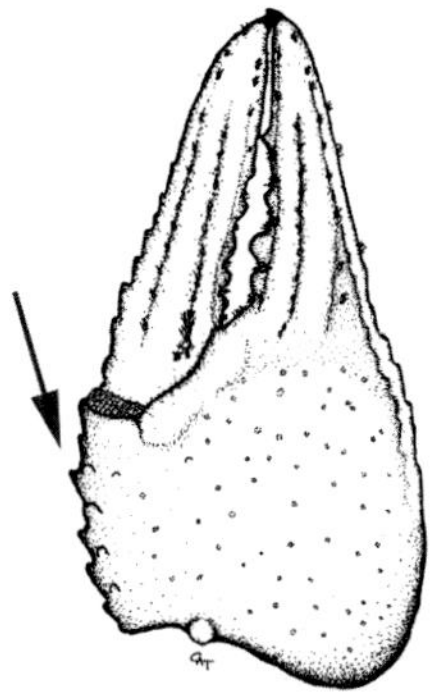

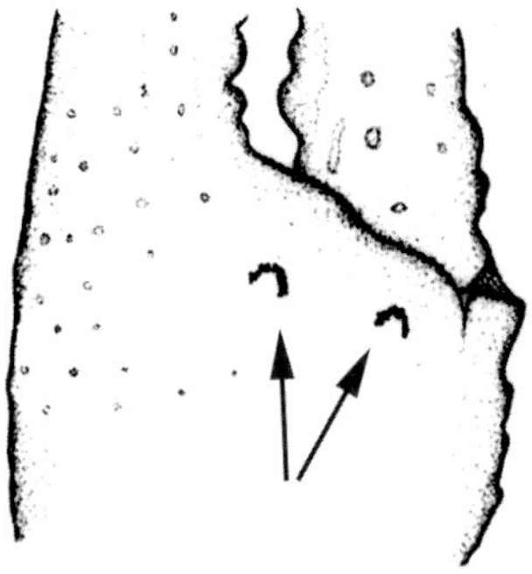

14a(11) Zero or one subpalmar tubercles on chelae; dactyl length less than 1.8 times palm length 15

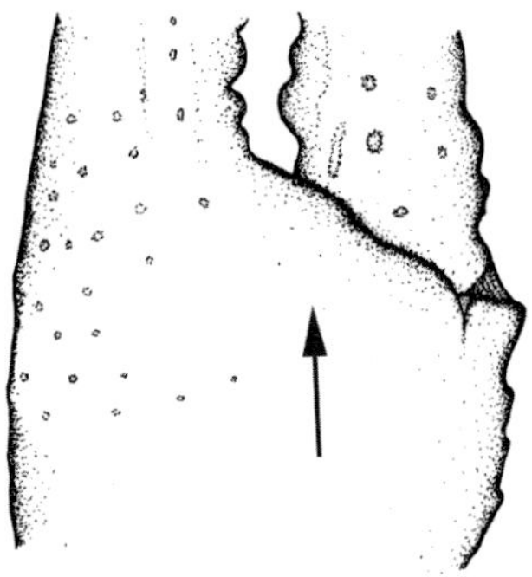

14b One or more subpalmar tubercles on chelae; dactyl length 1.8 times or greater than palm length 16

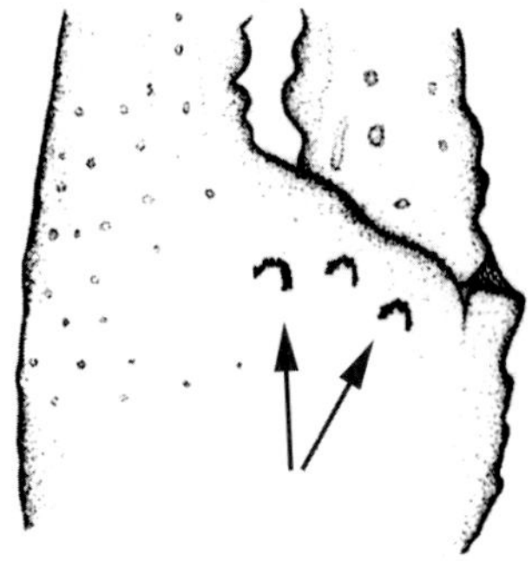

15a(14) Rostrum straight in lateral view, although somewhat angled downward (widespread in central and eastern Kentucky) . . . . *Cambarus thomai* (p. 121)

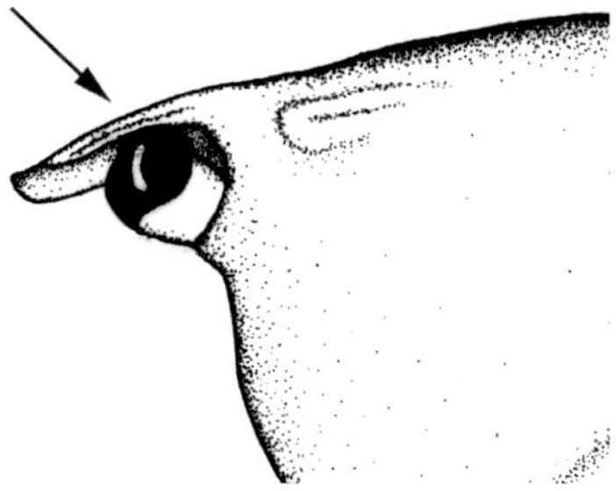

15b Rostrum broadly rounded in lateral view and strongly directed downward (widespread in central and western Kentucky). . . . . *Cambarus* sp. A (p. 126)

16a(14) Anteriolateral edge of pleuron of first abdominal segment straight; abdomen thin in lateral view; cephalic section of carapace 1.4 times or less the length of areola (Mississippi River drainage) . . . . *Cambarus ludovicianus* (p. 95)

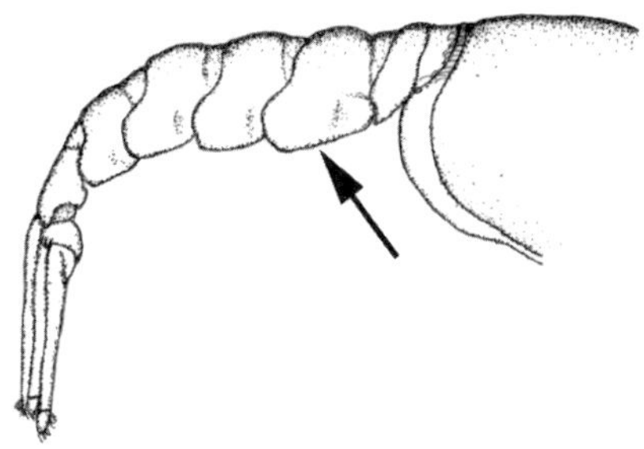

16b Anteriolateral edge of pleuron of first abdominal segment convex; abdomen thicker in lateral view; cephalic section of carapace more than 1.4 times length of areola (widespread in western Kentucky) . . . . *Cambarus diogenes* (p. 80)

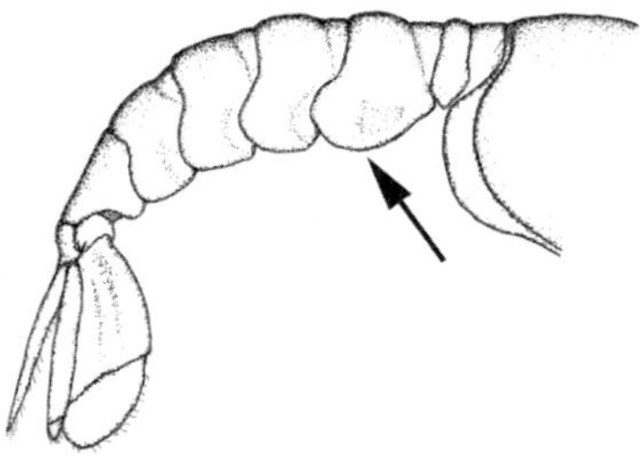

17a(7) Gonopods with more than two terminal elements 18

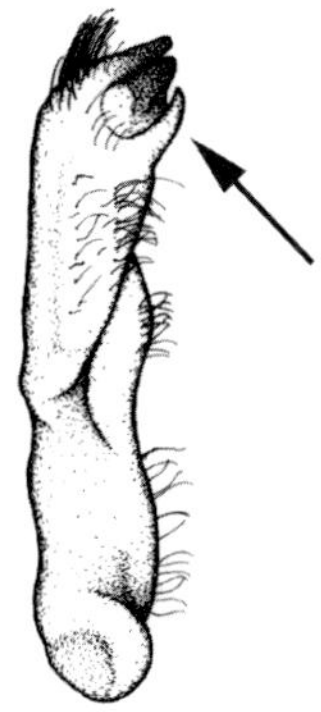

17b Gonopods with two terminal elements 19

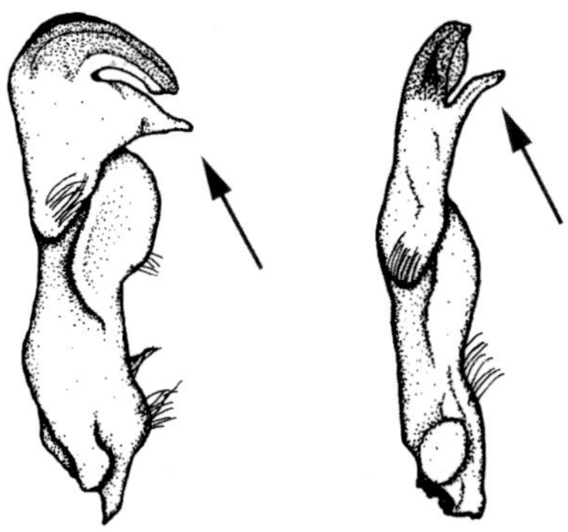

18a(17) Margins of rostrum converging anteriorly, terminating in small spines; all terminal elements of gonopods curved (western Kentucky, introduced populations elsewhere) . . . . *Procambarus acutus* (p. 203)

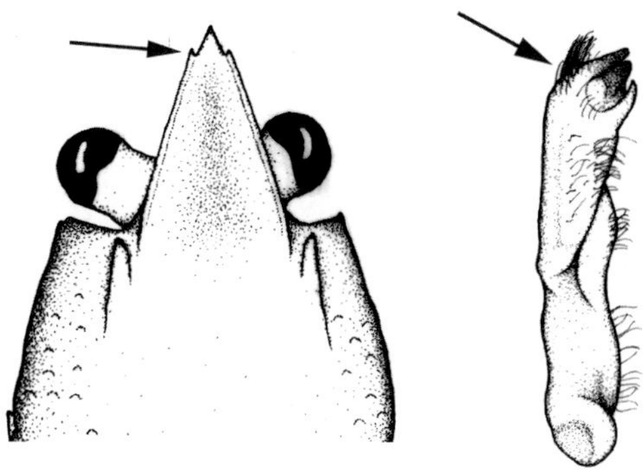

18b Margins of rostrum gradually converging, without spines; not all terminal elements of gonopods curved (lower Ohio, lower Tennessee, Mississippi River drainages) . . . . *Procambarus viaeviridis* (p. 209)

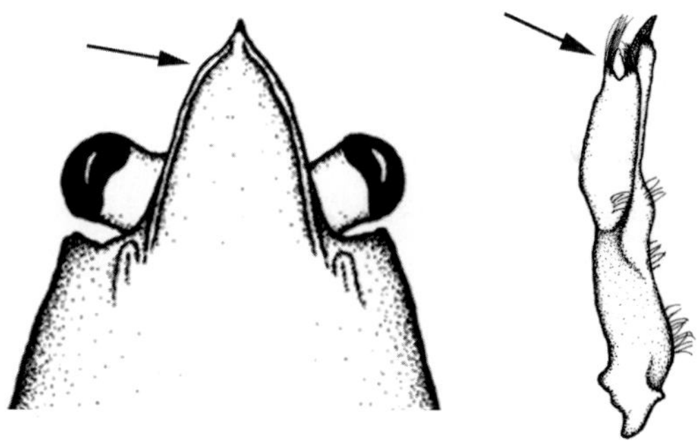

19a(17) Terminal elements of gonopods short, thick, and slightly curved 20

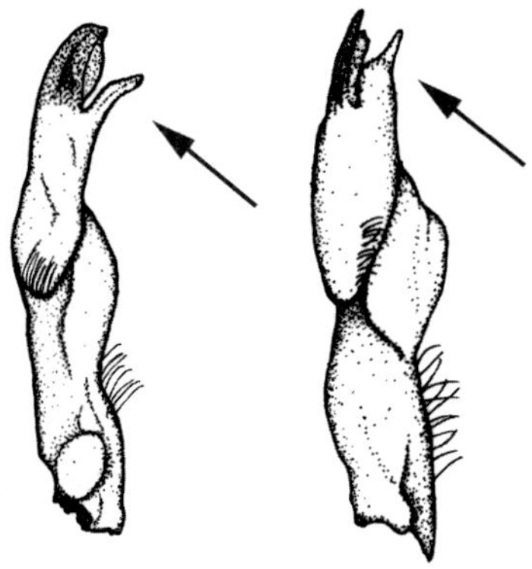

19b Terminal elements of gonopods not short, thick, and slightly curved 21

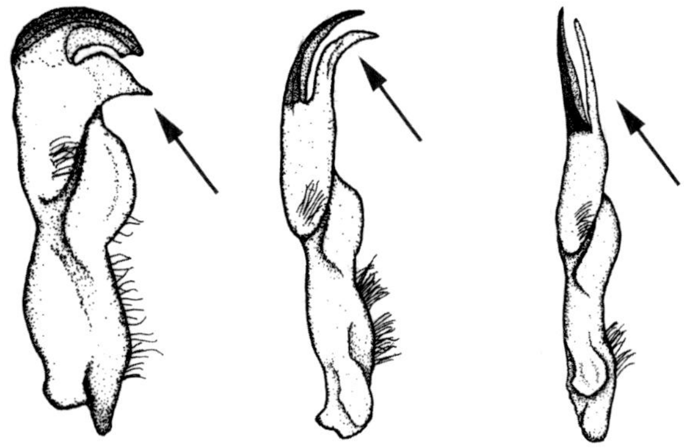

20a(19) Rostral margins terminating in spines; gonopod as shown (lower Ohio, lower Cumberland, and Tradewater River drainages) . . . . *Orconectes kentuckiensis* (p. 167)

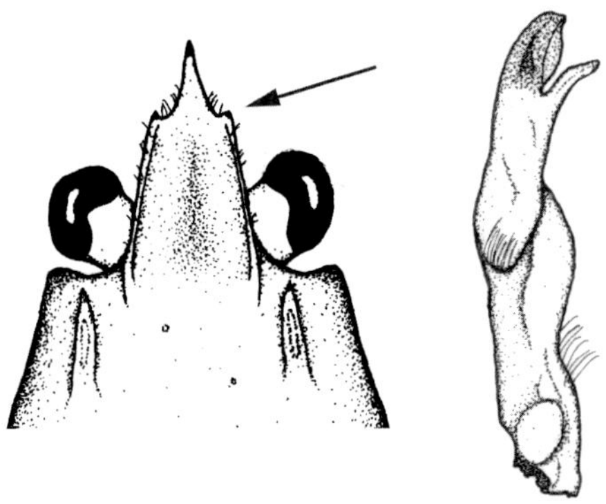

20b Rostral margins not terminating in spines; gonopod as shown (restricted to Roaring Paunch Creek, Big South Fork drainage) . . . . *Cambarus bouchardi* (p. 69)

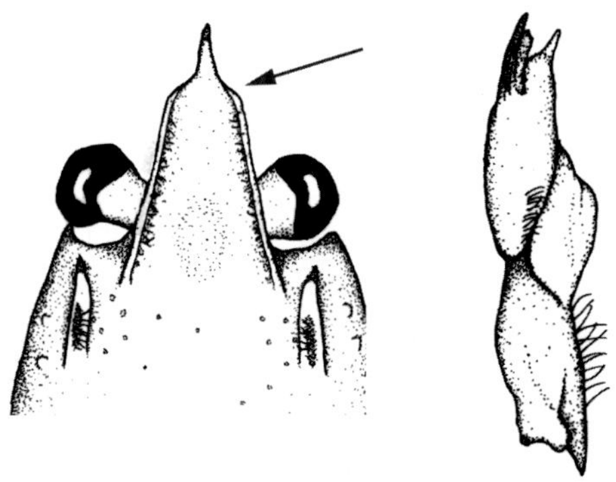

21a(19) Central projection not thick, rounded, or sickle shaped 22

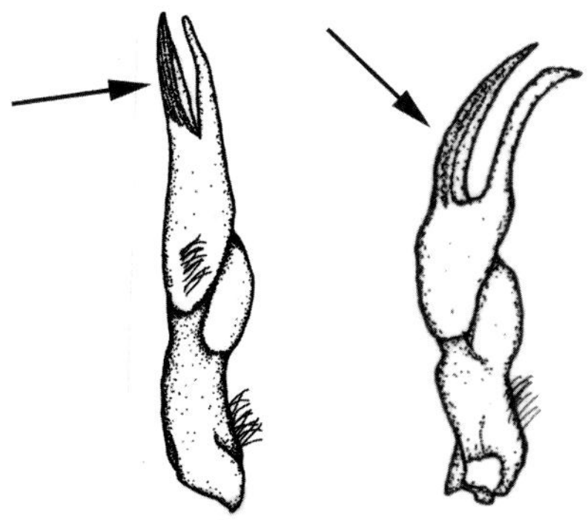

21b Central projection thick, rounded, and sickle shaped 39

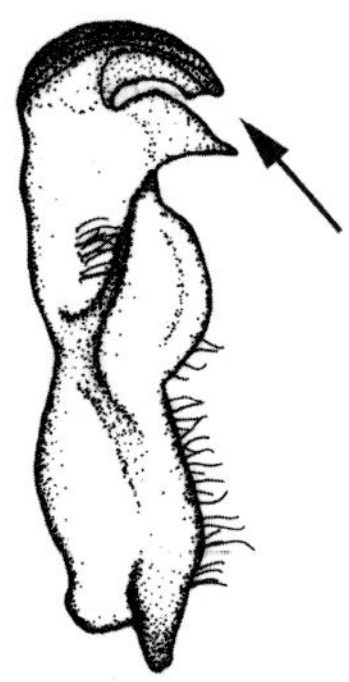

22a(21) Carapace strongly compressed laterally (Barren and lower Cumberland river drainages) . . . . *Orconectes compressus* (p. 146)

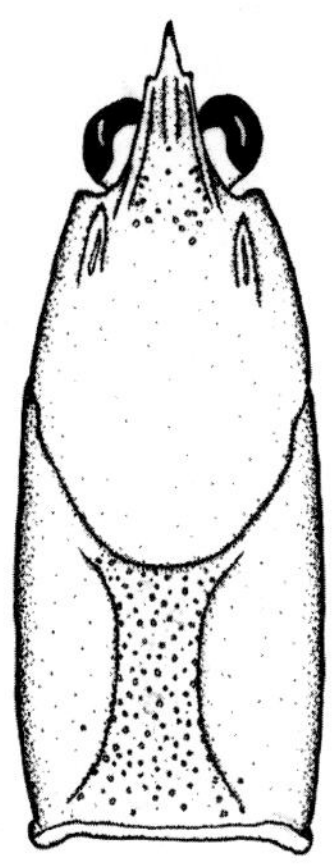

22b Carapace not strongly compressed laterally 23

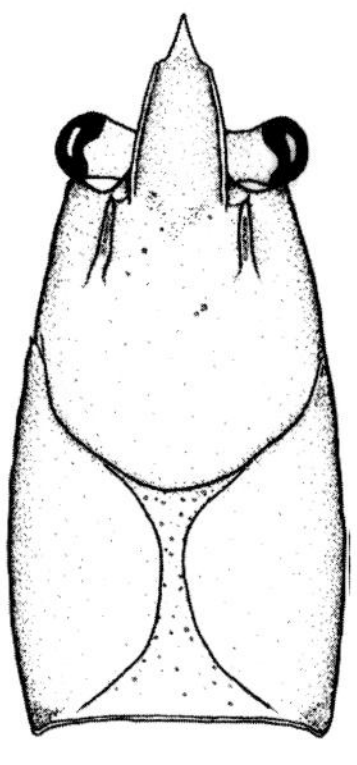

23a(22) Dactyl of chelae with distinct basal notch; terminal elements of gonopods thin and strongly curved (widespread in western Kentucky) . . . . *Orconectes immunis* (p. 155)

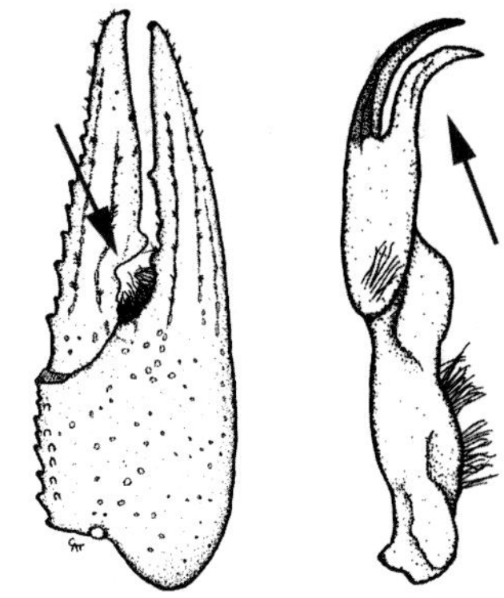

23b Dactyl of chelae without distinct basal notch; elements of gonopods straight or slightly curved 24

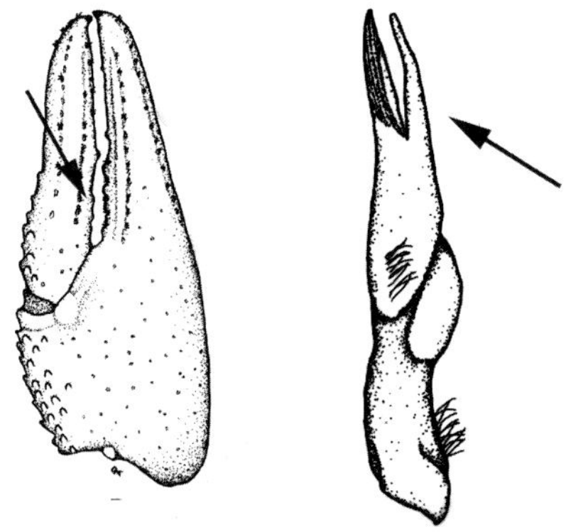

24a(23) Central projection of gonopods less than 35% of total gonopod length 25

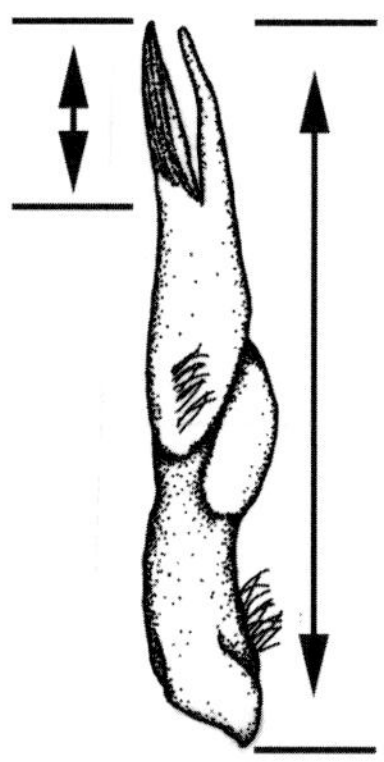

24b Central projection of gonopods greater than 35% of total gonopod length 31

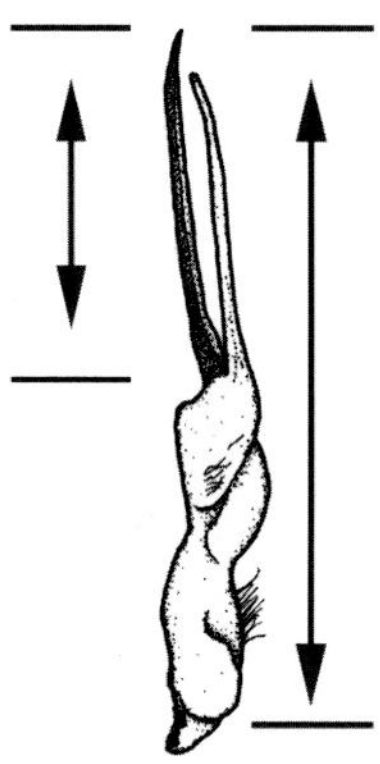

25a(24) Incisor region of mandible with a straight edge (extreme lower Cumberland and Ohio River drainages) . . . . *Orconectes margorectus* (p. 172)

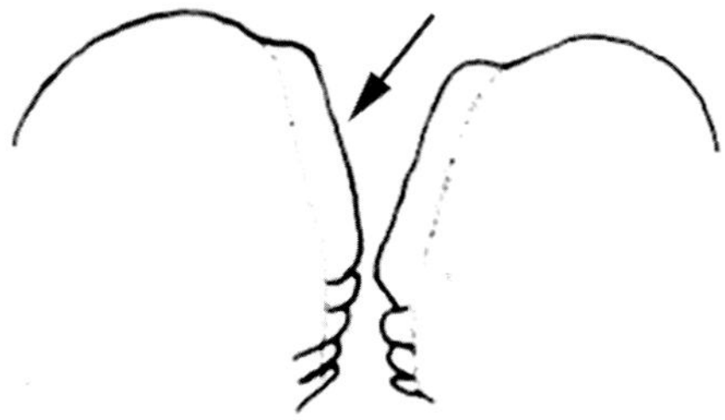

25b Incisor region of mandibles dentated, with distinct teeth 26

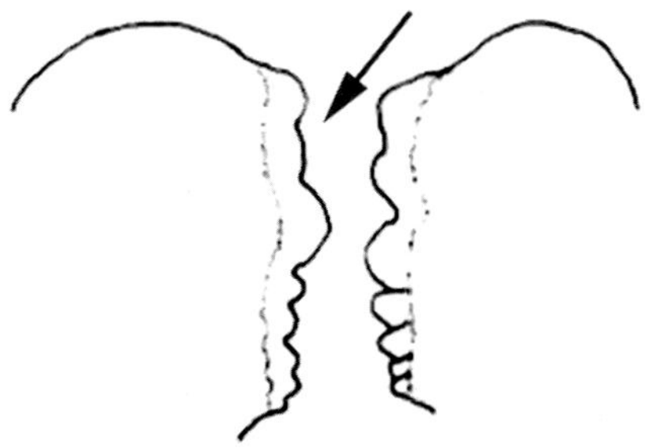

26a(25) Tip of central projection of gonopods strongly hooked, bent at 90° to the main shaft (restricted to Blood River drainage, western Kentucky) . . . . *Orconectes burri* (p. 143)

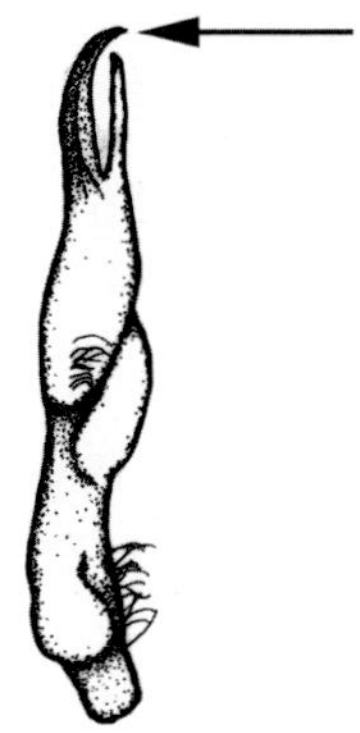

26b Tip of central projection of gonopods slightly curved to main shaft 27

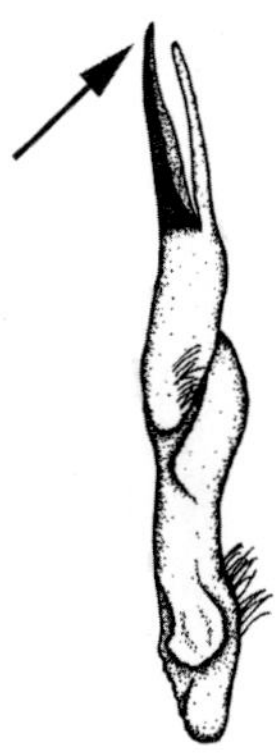

27a(26) Length of central projection of gonopod more than 25% of total gonopod length (restricted to Jefferson County vicinity) . . . . *Orconectes jeffersoni* (p. 161)

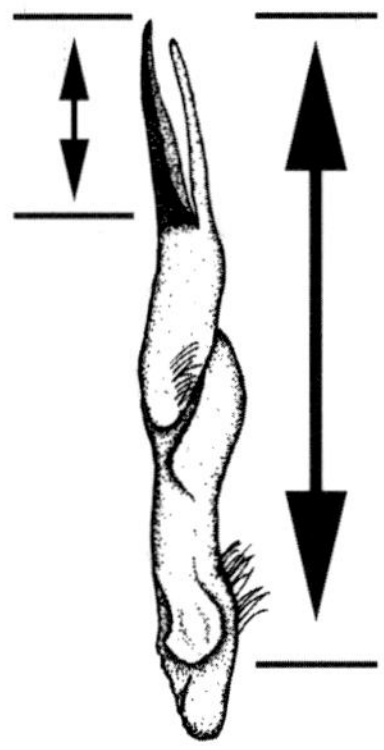

27b Length of central projection of gonopod less than 25% of total gonopod length 28

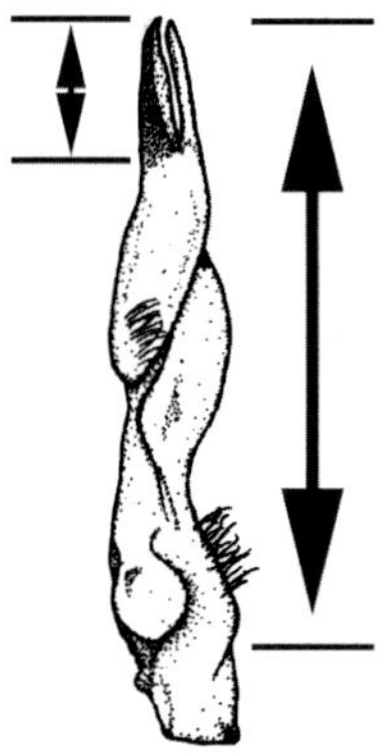

28a(27) Distal end of mesial process of gonopod dorsoventrally flattened 29

28b Distal end of mesial process of gonopod not dorsoventrally flattened 30

29a(28) Rostral carina absent; mesial process not curved cephalodistally; mesial process usually subequal in length to central projection (Tygarts Creek, and adjacent Ohio River drainages) . . . . . *Orconectes sanbornii* (p. 196)

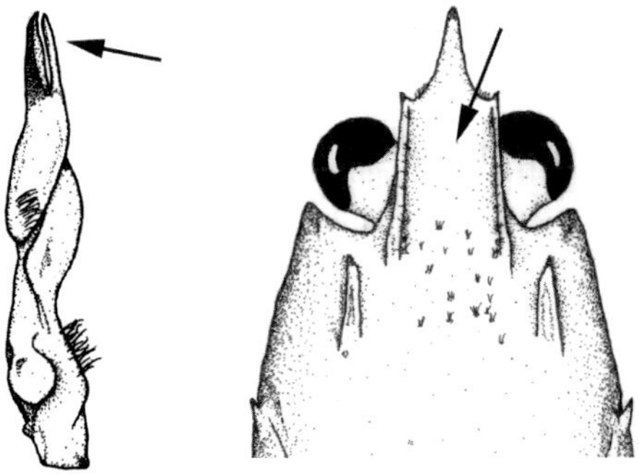

29b Rostral carina present; mesial process curved cephalodistally; mesial process usually equal in length to or extending past tip of central projection (Pond, lower Tennessee, and lower Cumberland River drainages) . . . . *Orconectes tricuspis* (p. 199)

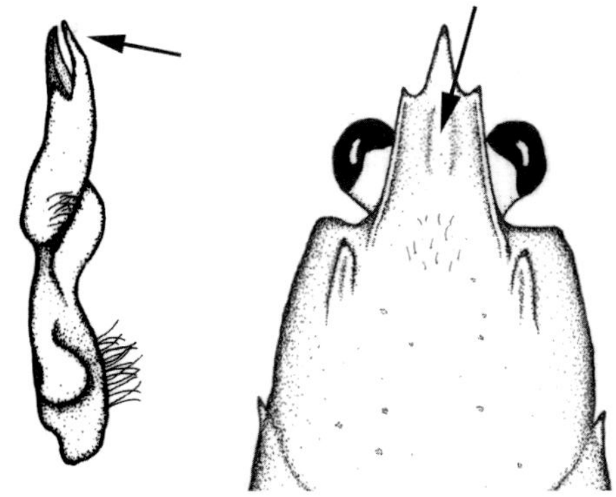

30a(28) Mesial processes nearly parallel; central projection tapering to sharp point (lower Green and lower Ohio River drainages) . . . . *Orconectes rafinesquei* (p. 187)

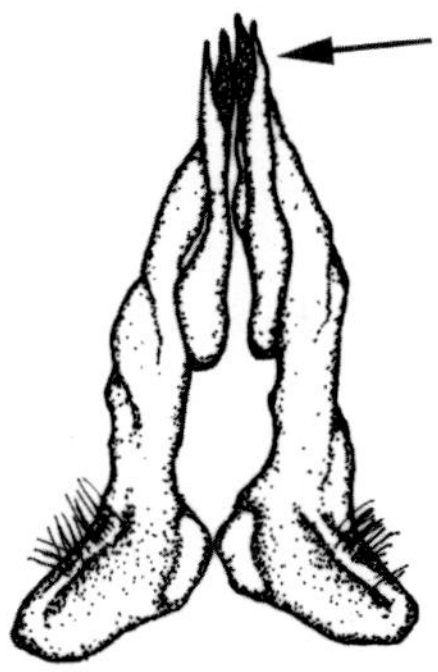

30b Mesial processes divergent laterally; central projection truncated at tip (restricted to Ohio River drainage in Crittenden County) . . . . *Orconectes bisectus* (p. 141)

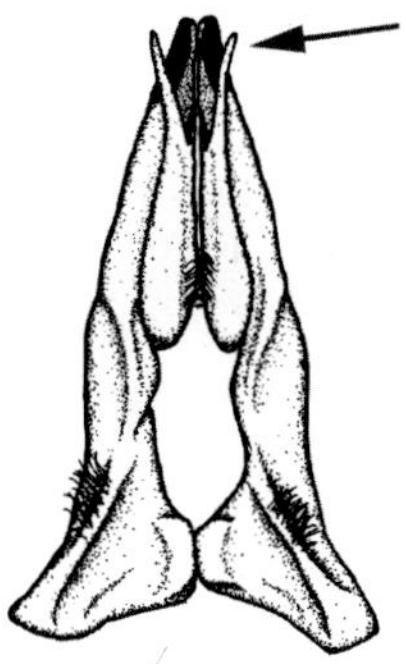

31a(24) No shoulder on cephalic margin of gonopods, cephalic margin nearly straight or broadly rounded . . . 32

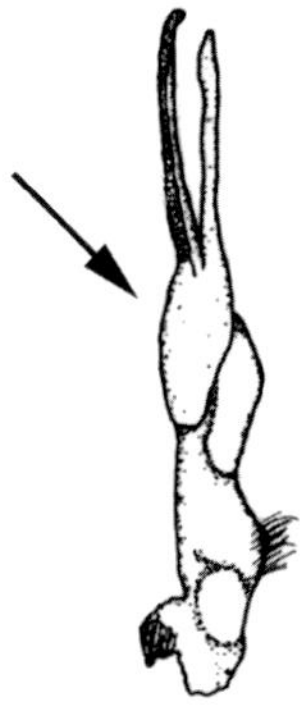

31b Angular shoulder present on cephalic margin of gonopod 34

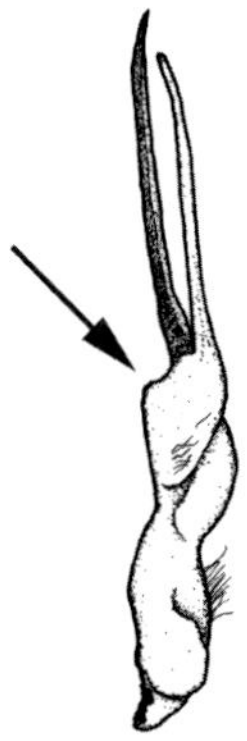

32a(31) Cervical spines absent; rostrum distinctly narrows anteriorly (Barren, Rough, and Nolin River drainages) . . . . *Orconectes barrenensis* (in part) (p. 138)

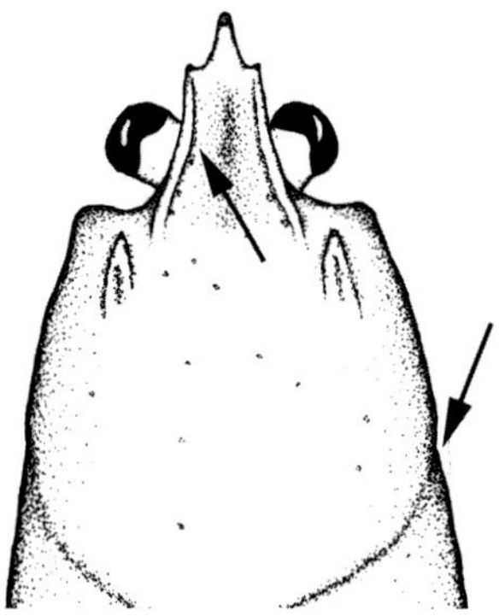

32b Cervical spines present; rostral margins nearly parallel 33

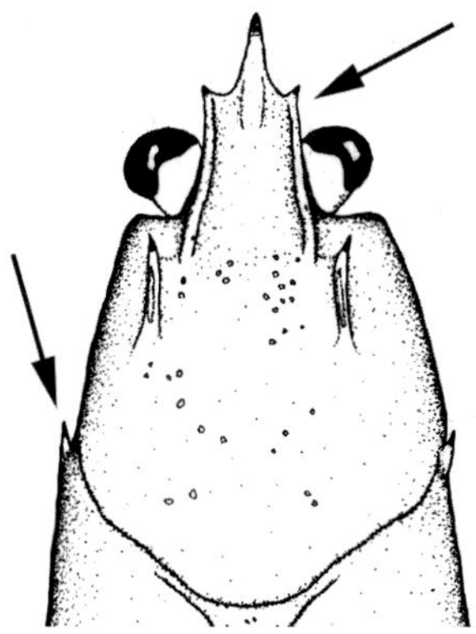

33a(32) Dactyl of chelae less than two times the length of mesial margin of palm of chelae; chelae stocky (lower Tennessee, Cumberland and Green River drainages) . . . . *Orconectes durelli* (p. 152)

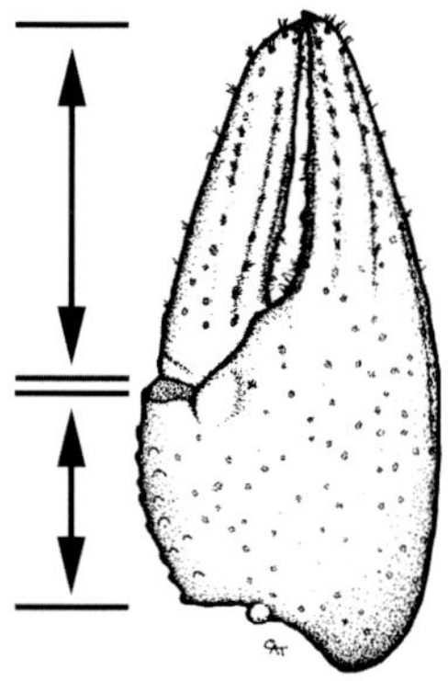

33b Dactyl of chelae greater than two times length of mesial margin of palm of chelae; chelae elongate (Cumberland, Tennessee, and lower Ohio River drainages) . . . . *Orconectes placidus* (p. 181)

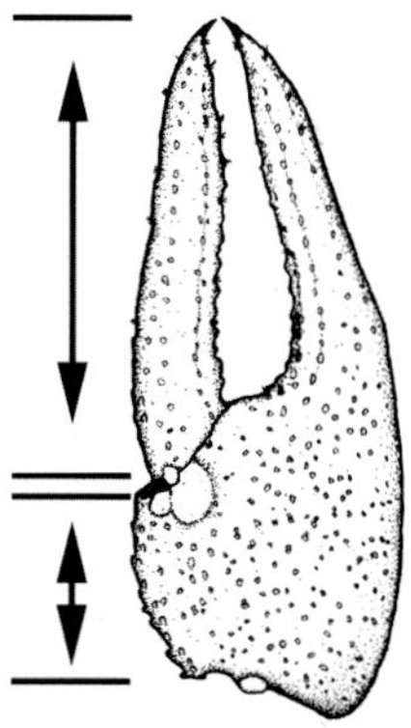

34a(31) Cervical spines absent; shoulder on cephalic margin of gonopods weak and broadly rounded (Barren, Rough, and Nolin river drainages) . . . . *Orconectes barrenensis* (in part) (p. 138)

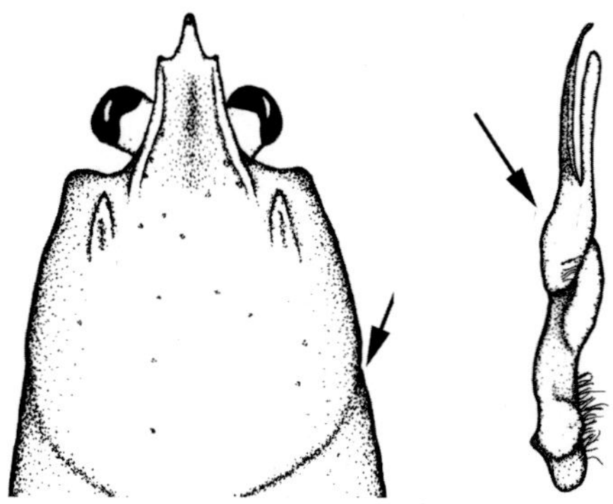

34b Cervical spines present; cephalic margin of gonopods with strong angular shoulder 35

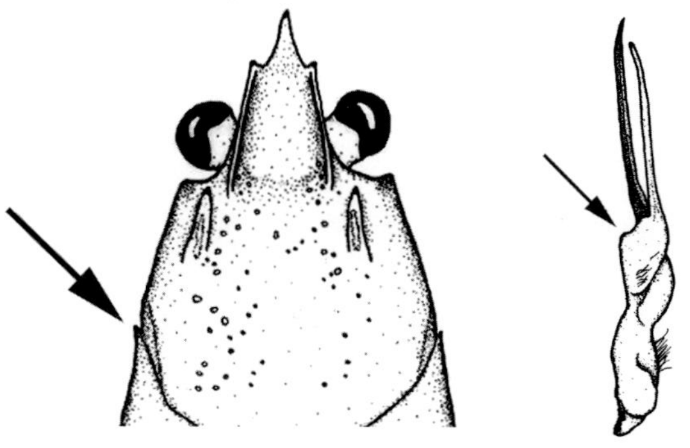

35a(34) Incisor region of mandible with a straight edge 36

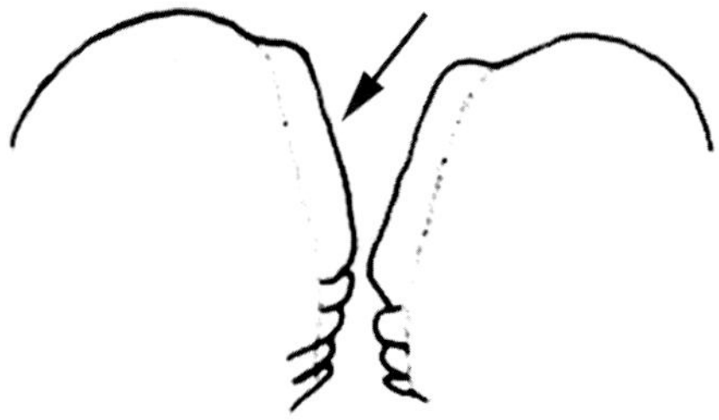

(modified from Bouchard 1978)

35b Incisor region of mandibles dentated, with distinct teeth 37

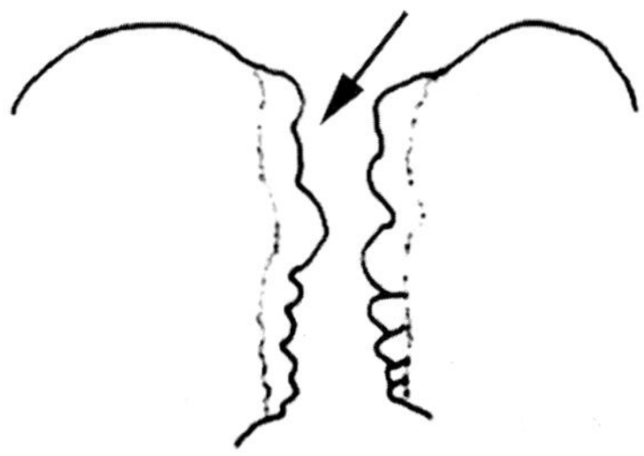

(modified from Bouchard 1978)

36a(35) Central projection at least 44% of total length of gonopods; gonopods usually reaching base of first pair of pereiopods when abdomen is flexed (lower Kentucky, upper Cumberland, Salt, and Ohio River drainages) . . . . *Orconectes juvenilis* (p. 164)

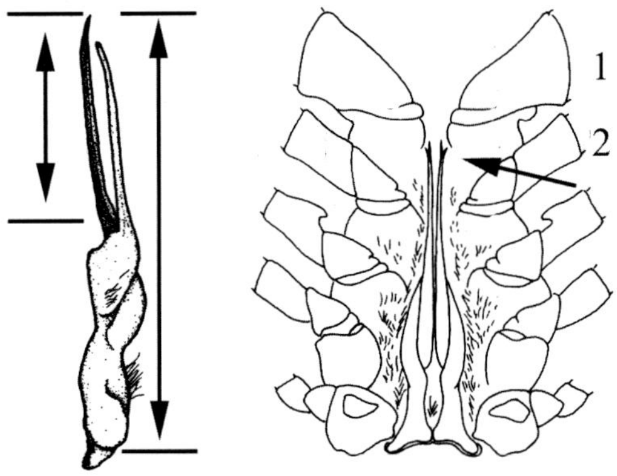

(from Hobbs 1972)

36b Central projection less than 44% of total length of gonopods; gonopods usually not reaching base of first pair of pereiopods when abdomen is flexed (lower Licking, middle Kentucky, Salt, Green River drainages; Ohio River proper) . . . . *Orconectes rusticus* (p. 192)

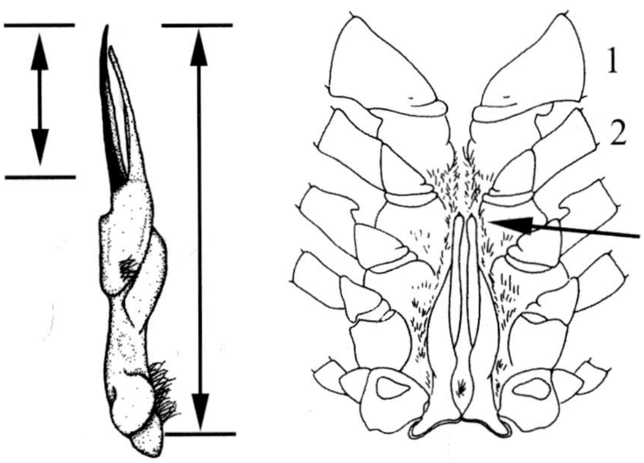

(from Hobbs 1972)

37a(35) Central projection of gonopods straight (restricted to Mud River and Muddy Creek drainages) . . . . *Orconectes ronaldi* (p. 190)

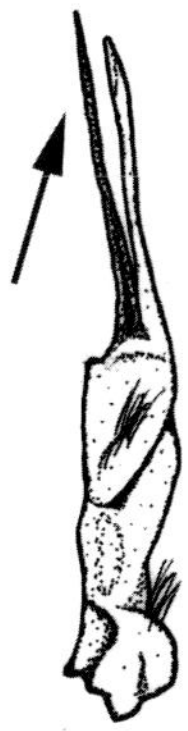

37b Distal end of central projection of gonopods slightly curved 38

38a(37) Central projection of gonopods less than 50% of total length of gonopod; total length of gonopods less than 45% of total carapace length (upper Cumberland, Kentucky and Licking River drainages; widespread throughout eastern Kentucky in Ohio River tributaries) . . . . *Orconectes cristavarius* (p. 149)

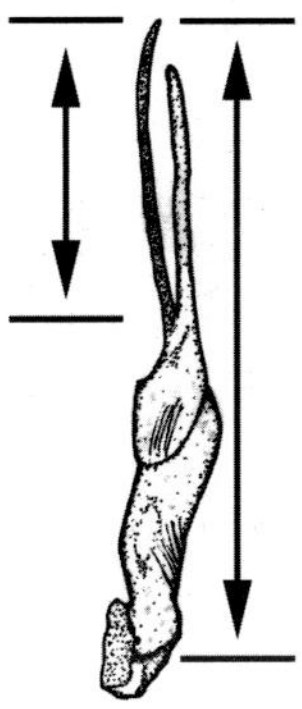

38b Central projection of gonopods 50% or greater than total length of gonopod; total length of gonopod 45% or greater than total carapace length (Green, upper Cumberland, and Ohio River drainages) . . *Orconectes putnami* (p. 184)

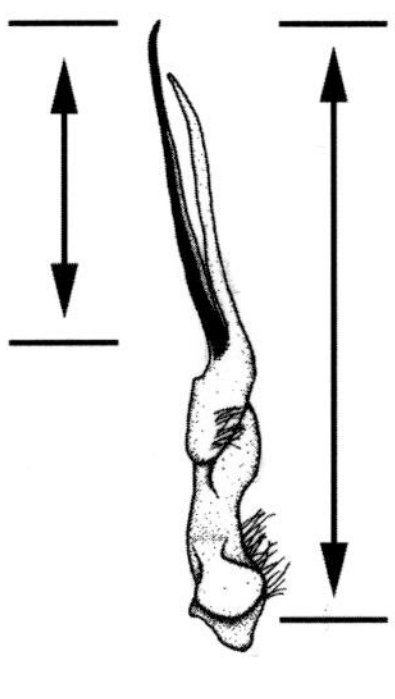

39a(21) Mesial margin of palm of chelae with row of eight or more tubercles 40

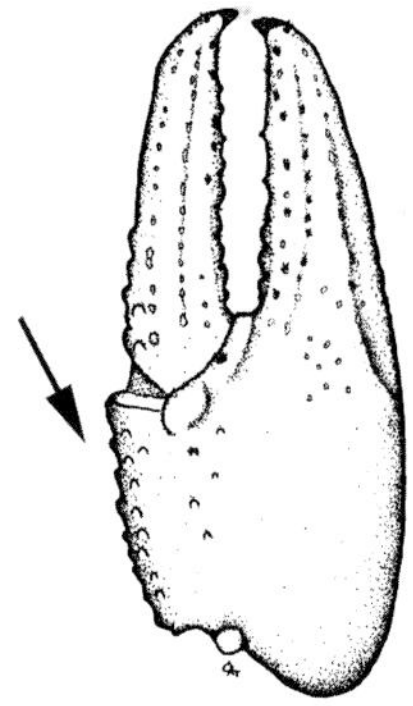

39b Mesial margin of palm of chelae with row of seven or fewer tubercles 45

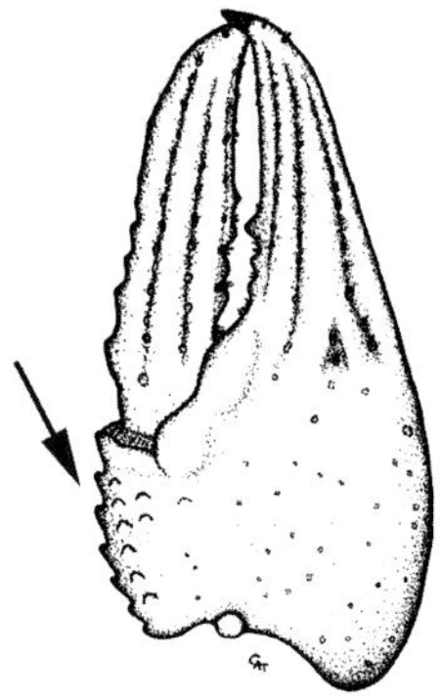

40a(39) Rostral margins of rostrum terminating in spines or tubercles 41

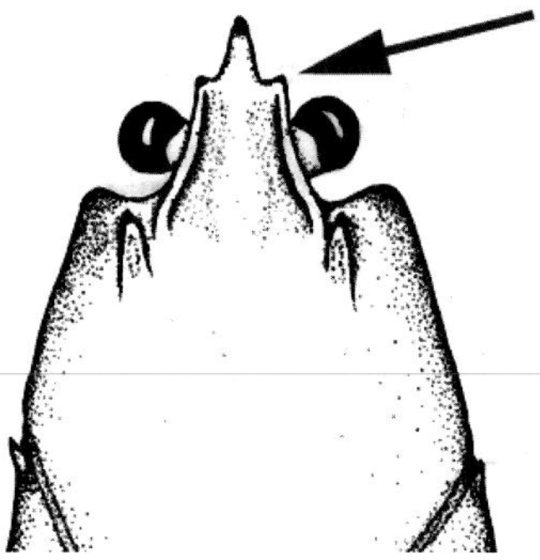

40b Rostral margins of rostrum not terminating in spines or tubercles 43

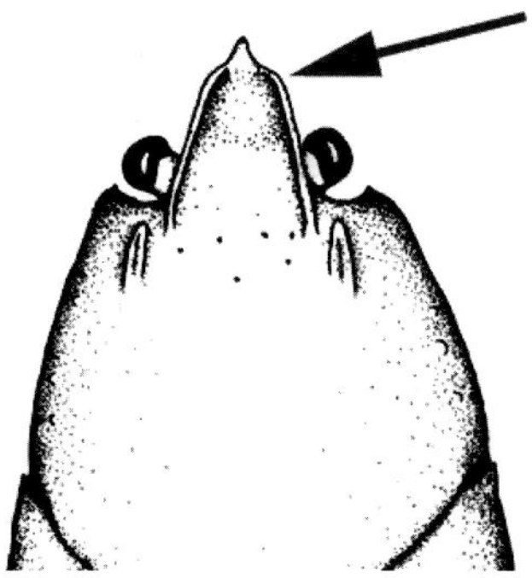

41a(40) Mesial margin of palm of chelae with one complete row of tubercles (Cumberland and Green River drainages, and lower Ohio River proper) . . . . *Cambarus rusticiformis* (p. 106)

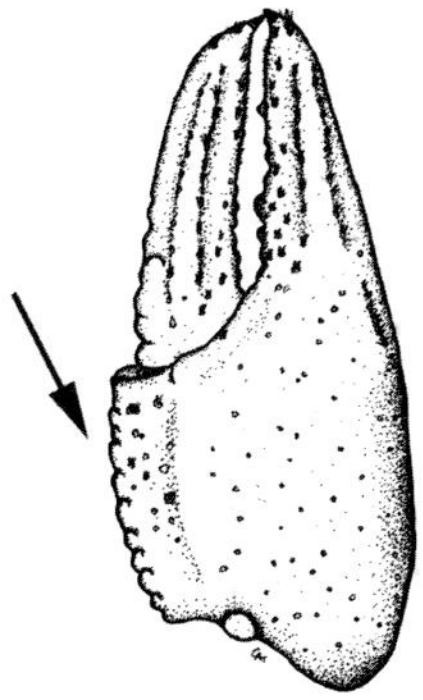

41b Mesial margin of palm of chelae with two complete rows of tubercles 42

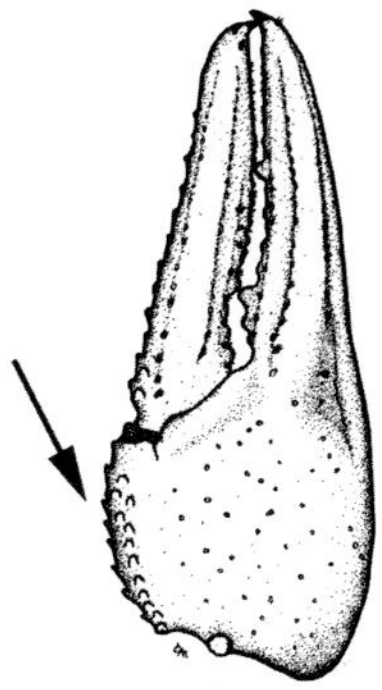

42a(41) Distal half of rostral margins usually straight and always terminating in strong spines; all spines on body well developed and sharply pointed (upper Cumberland and upper Green River drainages) . . . . *Cambarus cumberlandensis* (p. 74)

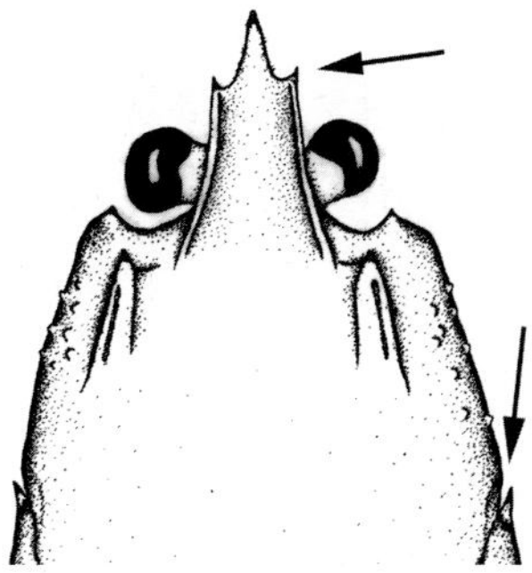

42b Distal half of rostral margins converging and always ending in short spines or tubercles; all spines on body not well developed (upper Cumberland River drainage above the falls) . . . . *Cambarus buntingi* (p. 71)

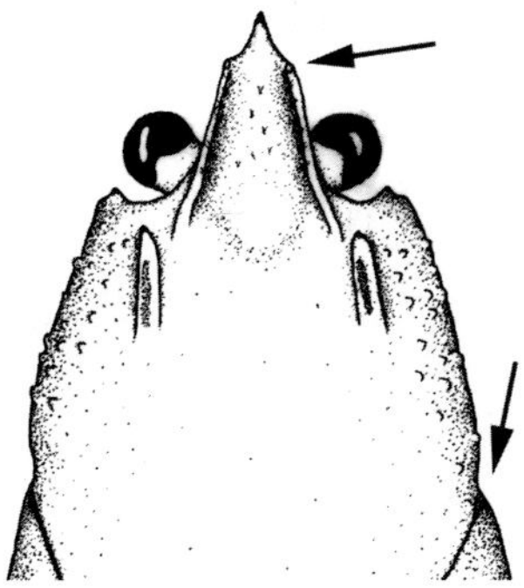

43a(40) Dorsoventral depression at lateral base of propodus 44

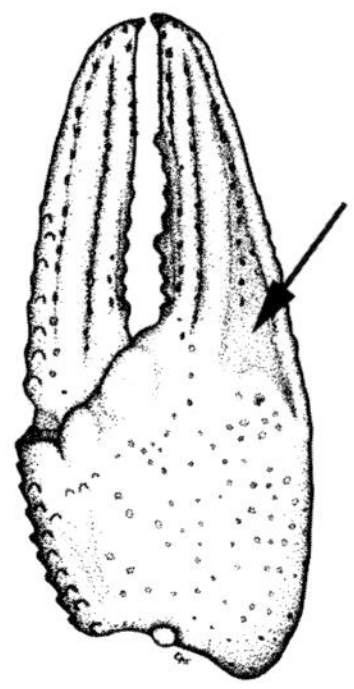

43b No dorsoventral depression at lateral base of propodus (widespread central and westsern Kentucky) .... *Cambarus tenebrosus* (p. 118)

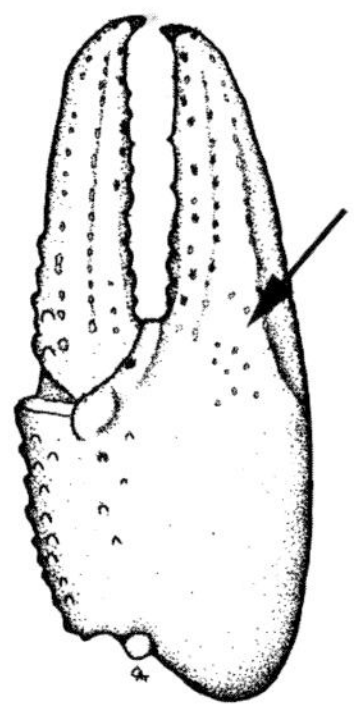

44a(43) Suborbital angle acute; mesial margin of palm of chelae usually with two rows of well-developed tubercles (widespread in eastern Kentucky) .... *Cambarus robustus* (p. 103)

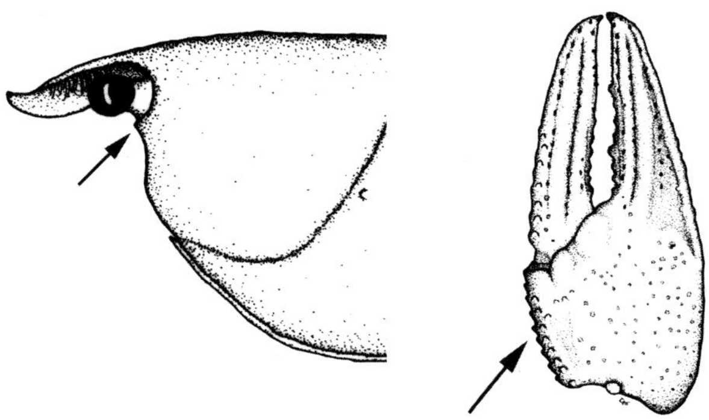

44b Suborbital angle absent or broadly rounded; tubercles along mesial margin of palm of chelae weakly developed and usually confined to one row (upper Big Sandy River drainage) .... *Cambarus veteranus* (p. 124)

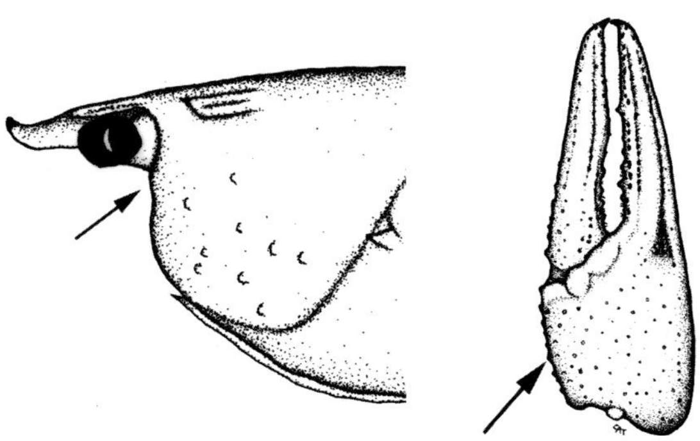

45a(39) Lobelike knob at base of central projection; body and chelae covered with long setae (upper Barren and lower Cumberland River drainages) . . . . *Cambarus friaufi* (p. 90)

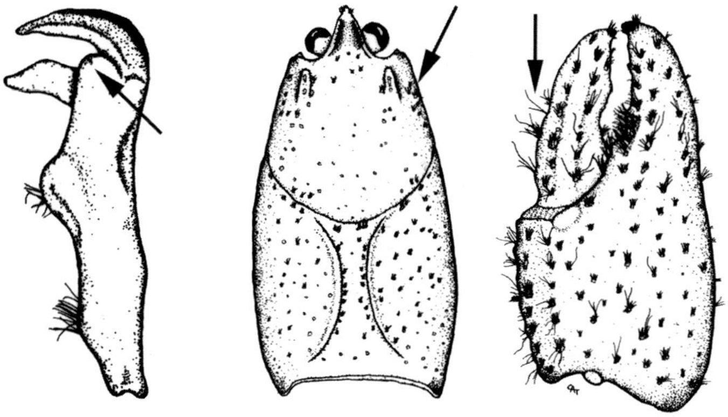

45b No lobelike knob at base of central projection; body and chelae not covered with long setae 46

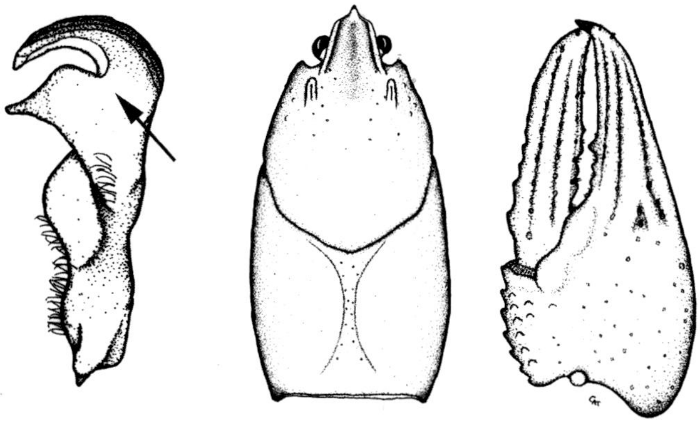

46a(45) Areola narrow with room for two or fewer punctations 47

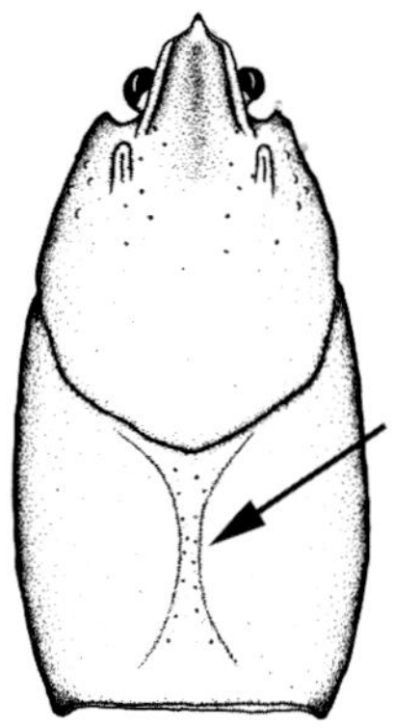

46b Areola wide with room for three or more punctations 50

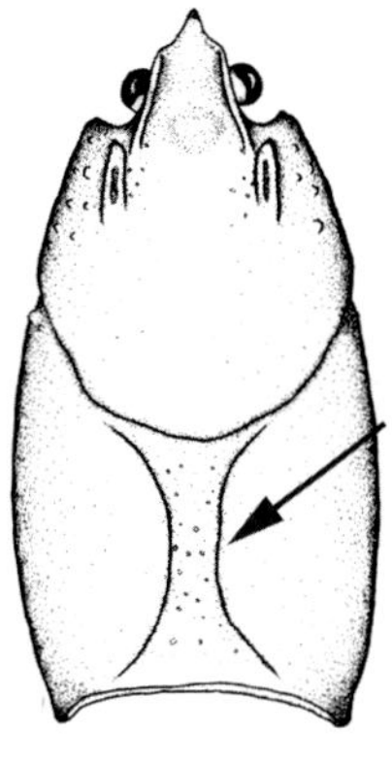

47a(46) Epistomal zygoma weakly arched (Kentucky, Salt, Cumberland, and Green River drainages) . . . . *Cambarus graysoni* (in part) (p. 92)

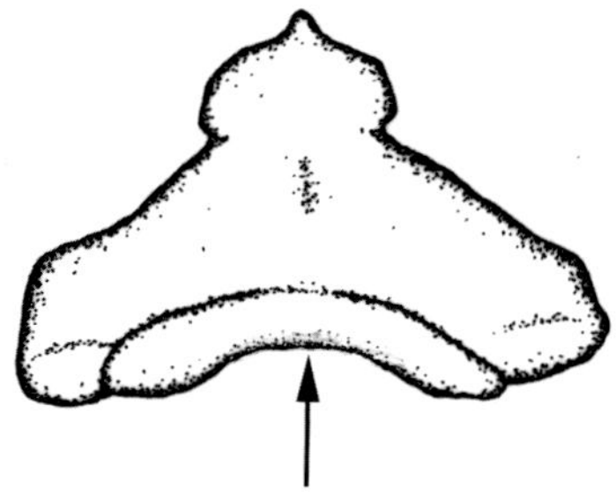

47b Epistomal zygoma strongly arched 48

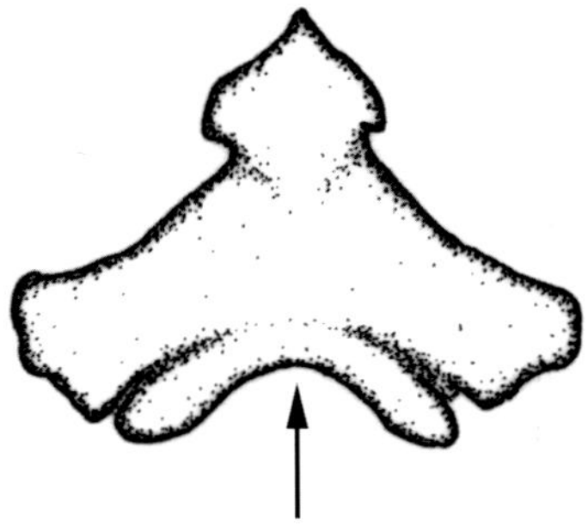

48a(47) Both fingers of chelae with well-developed longitudinal ridges 49

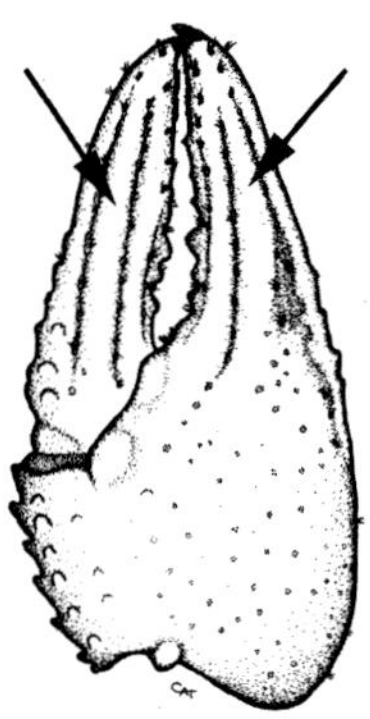

48b Fingers of chelae without well-developed longitudinal ridges (widespread in central and eastern Kentucky) . . . . *Cambarus ortmanni* (p. 98)

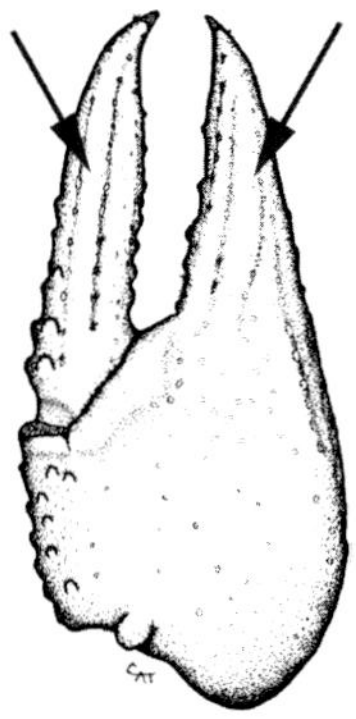

49a(48) Palm of chelae with two complete rows of tubercles; rows without gaps between tubercles; basalmost tubercles of mesial row not fused to form ridge (sporadic in Middle Kentucky and lower Green River drainages) . . . . *Cambarus striatus* (p. 115)

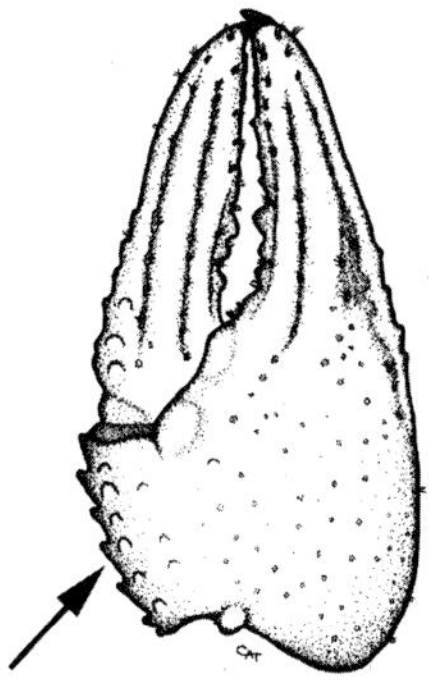

49b Palm of chelae with one complete row of tubercles along mesial margin with incomplete second row or no second row; basalmost tubercles of mesial row fused to from ridge (widespread in eastern Kentucky) . . . . *Cambarus dubius* (in part) (p. 86)

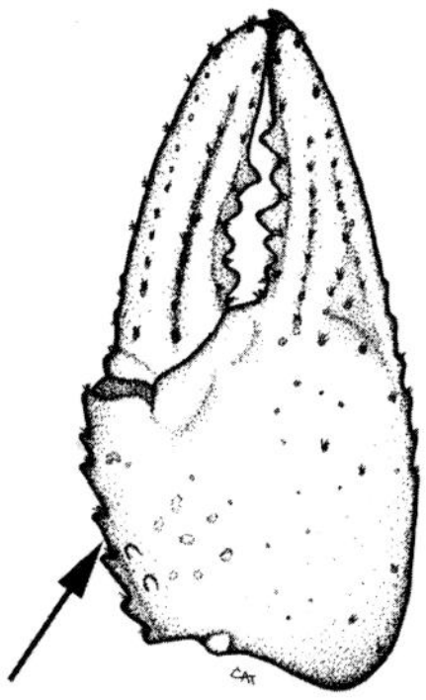

50a(46) Tubercles along mesial margin of palm of chelae low and rounded; fingers of chelae with weak longitudinal ridges (widespread in central and eastern Kentucky) . . . . *Cambarus bartonii cavatus* (p. 63)

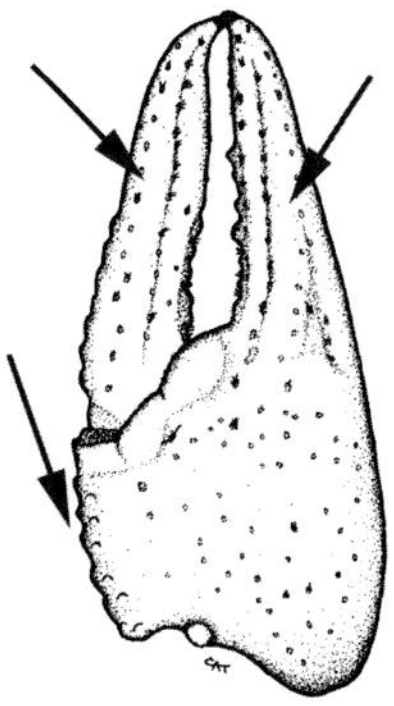

50b Tubercles along mesial margin of palm of chelae strong and sawtoothlike; fingers of chelae with well-developed longitudinal ridges 51

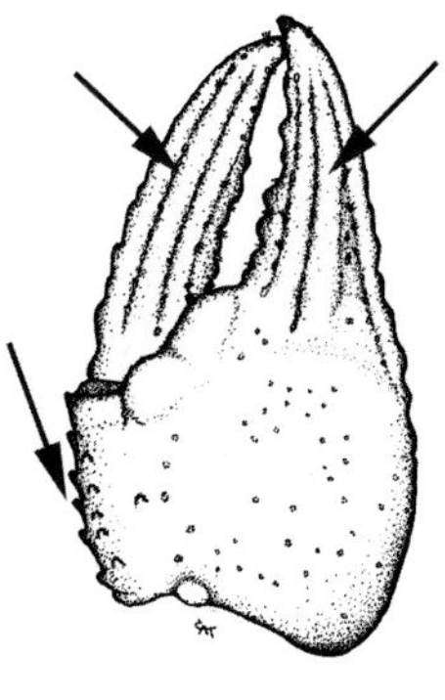

51a(50) Tubercles in second row on mesial margin of palm of chelae low and weakly developed (Big and Little Sandy River drainages) . . . . *Cambarus sciotensis* (p. 109)

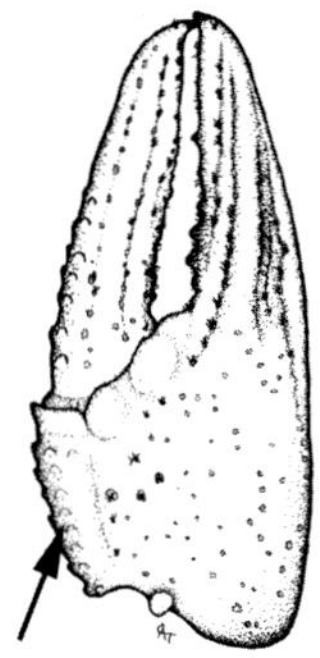

51b Tubercles in second row on mesial margin of palm of chelae well developed, strongly protruding above palm surface, or no second row present 52

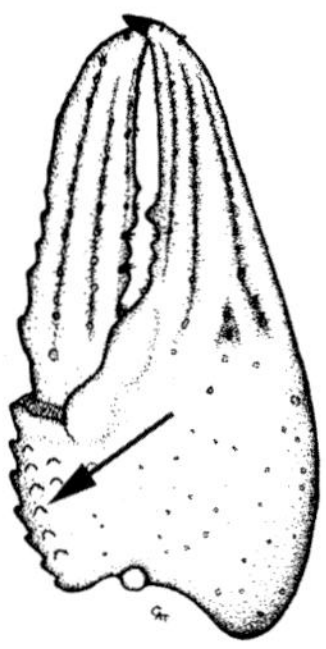

52a(51) Second row of tubercles on mesial margin of palm incomplete, or no second row present 53

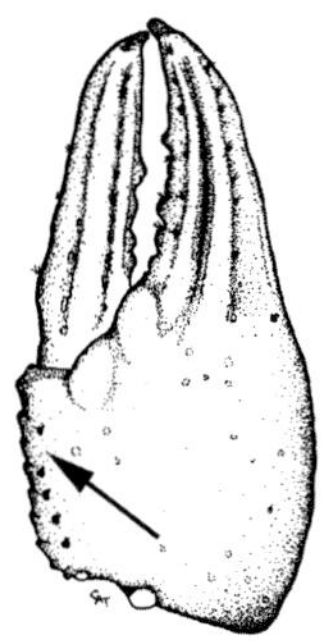

52b Second row of tubercles on mesial margin of palm complete, no gaps between tubercles 55

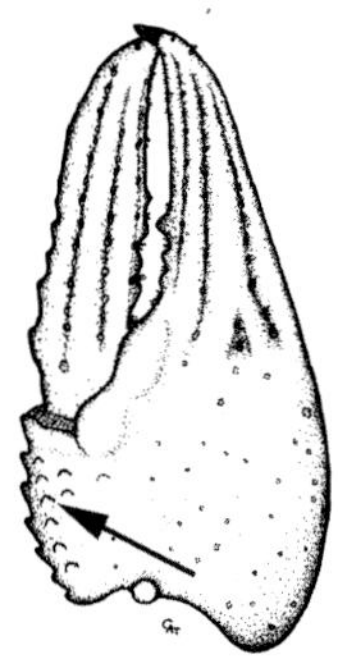

53a(52) Areola length at least nine times greater than areola width (widespread in eastern Kentucky) . . . . *Cambarus dubius* (in part) (p. 86)

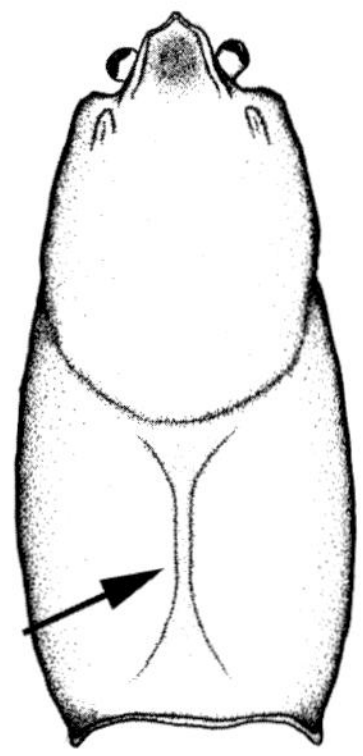

53b Areola length less than nine times greater than areola width 54

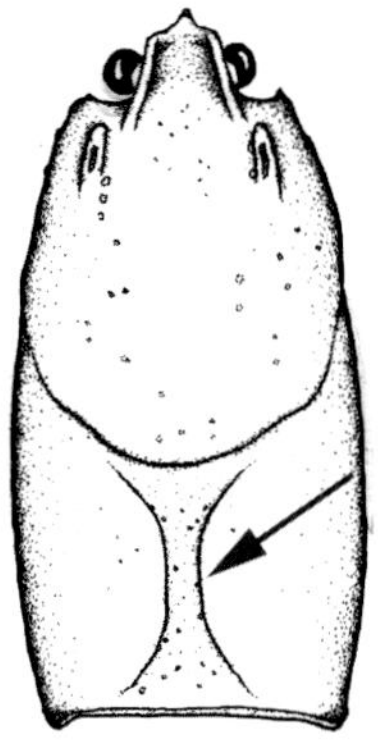

54a(53) Central projection of gonopod greatly recurved, much greater than 90° to main axis of shaft; rostrum short and wide, rostrum length less than 1.7 times rostrum width (sporadic in upper Cumberland and Kentucky River drainages) . . . . *Cambarus parvoculus* (p. 100)

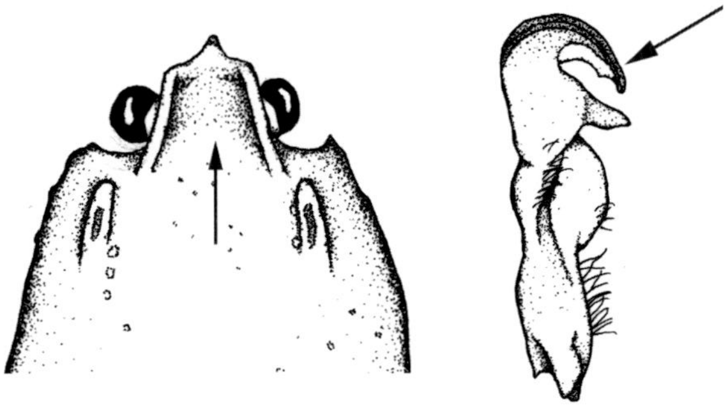

54b Central projection of gonopod not strongly recurved, approximately 90° to main axis of shaft; rostrum long, rostrum length at least 1.7 times greater than rostrum width (widespread in upper Cumberland and Kentucky River drainages) . . . . *Cambarus distans* (p. 83)

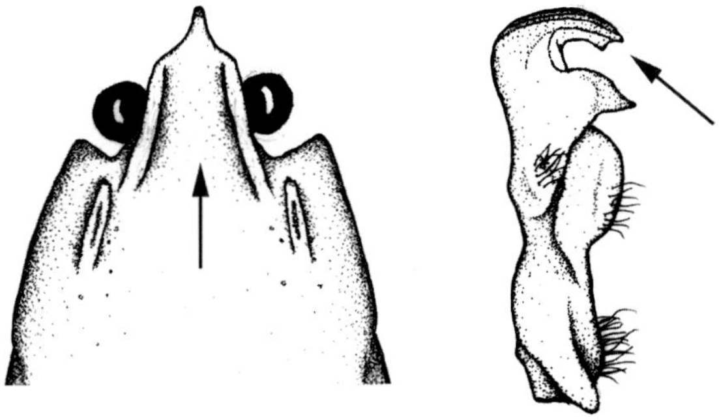

55a(52) Epistomal zygoma strongly arched (upper Cumberland and Kentucky River drainages) . . . .
. . . . *Cambarus sphenoides* (p. 112)

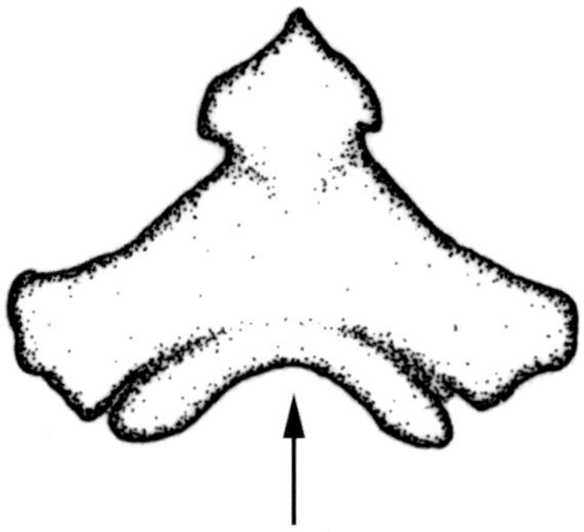

55b Epistomal zygoma weakly arched (Kentucky, Salt, Cumberland, and Green River drainages) . . . .
. . . . *Cambarus graysoni* (in part) (p. 92)

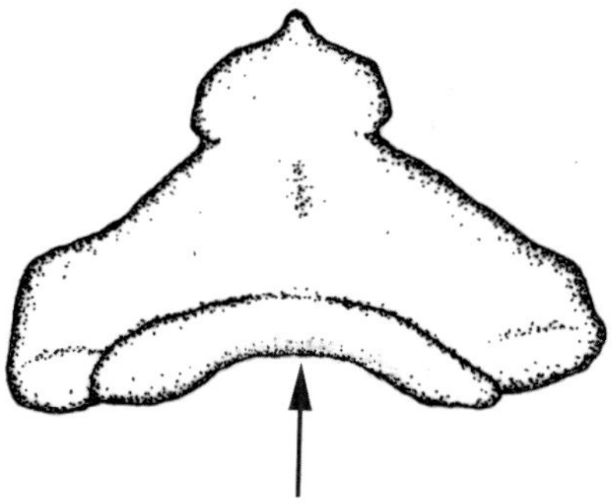

Species Accounts

Cambarellus puer **Photo by C. Lukhaup**

Cambarellus Ortmann

Members of the genus *Cambarellus* are known as the dwarf crayfishes. Species rarely exceed 30 mm in total length and occur almost exclusively in lowland swamps, sloughs, and ditches. Nine of the 17 *Cambarellus* species occur in central and eastern Mexico, the remaining members occur in the Gulf Coastal Plain of the United States from Texas to Florida. In addition to the small adult size, members of the genus are unique in that males have hooks on the ischia of the second and third pair of pereiopods. Males also have three terminal elements on the gonopod. Kentucky is home to two species of *Cambarellus*: *C. puer* and *C. shufeldtii*.

Cambarellus puer Hobbs
Swamp Dwarf Crayfish

Fig. 23 Photo by W. Pflieger

LIFE COLORS – (Fig. 23) Base color of dorsal and lateral surfaces of chelae, carapace, and abdomen variable, ranging from rust red to light brown to gray. Dorsum of carapace and abdomen with two longitudinal brown or black stripes or two rows of spots of the same color fused into uninterrupted bands. Tips of fingers of chelae lacking orange and/or black bands. Venter cream to whitish in color.

The occurrence of both stripes and spots in the species is similar to that reported for *C. shufeldtii*. Given the ecological and evolutionary closeness of the two species, the cause of the dimorphic color state in *C. puer* is likely due to the same actions discussed under LIFE COLORS for *C. shufeldtii*.

CHARACTERS – (Fig. 24) Body size small, adults rarely exceeding 28 mm in total length. Rostrum flat; margins converging distally, terminating in small spines at base of acumen. Acumen equal to or slightly longer than width of rostrum at marginal spines. Carapace laterally compressed and moderately arched dorsoventrally. Strong cervical spines present. Areola moderately wide to wide, 5.3 to 6.5 times longer than wide at its narrowest point, with room for five to seven shallow punctations. Antennal scale widest at midpoint. Chelae small and with short fingers. Fingers and palm with rounded mesial and lateral margins, single setae covering dorsal surfaces. Chelae lacking longitudinal ridges on fingers and rows of tubercles on mesial margin of palm. Dactyl of equal length or shorter than mesial margin of palm. Males with hooks on basis of second and third pereiopods.

Gonopod of form I male terminating in three caudodistally recurved elements of moderate and equal length: corneous, pointed central projection; pointed, noncorneous or corneous caudal process; and pointed, noncorneous mesial process. Annulus ventralis of female movable, subcircular with flattened or shallowly notched caudal edge; annulus with strongly elevated central region; lacking a fossa; occasionally with short, laterally directed sinus.

MAXIMUM SIZE – The maximum known size for the species is 32.0 mm total length.

MOST LIKE – In possessing three terminal elements on the form I male gonopod and hooks on the basis of the second and third pairs of pereiopods, *Cambarellus puer* can be confused only with the other member of the genus *Cambarellus* found in Kentucky, *C. shufeldtii*. *Cambarellus puer* differs from that species in having strongly curved elements of the male gonopod.

DISTRIBUTION AND HABITAT – (Fig. 25) *Cambarellus puer* occurs in the central Coastal Plain from extreme southern Illinois south to the coasts of eastern Texas and Louisiana (Hobbs 1989). Current records indicate that the species has a very limited distribution in Kentucky. It is known only from the Mayfield Creek drainage, a Mississippi River tributary, along the Graves-McCracken county line. Given the abundance of suitable habitat (see below) in western Kentucky, the limited distribution of *C. puer* in the state is curious. Potential explanations include incomplete sampling of lowland habitats or the interactions between congeners. Penn and Fitzpatrick (1962, 1963) reported *C. shufeldtii* to be dominant over *C. puer* in experimental trials and that *C. shufeldtii* had displaced *C. puer* at several Gulf Coast locations in recent years. Page (1985) suggested that this competitive exclusion between the two species might partially explain their distributions in southern Illinois. In western Kentucky *C. shufeldtii* is far more abundant and widespread in lowland habitats.

In Kentucky, *C. puer* is found in cypress swamps among live or dead vegetation. In southern Illinois Page (1985) reported collecting the species from mostly permanent water bodies

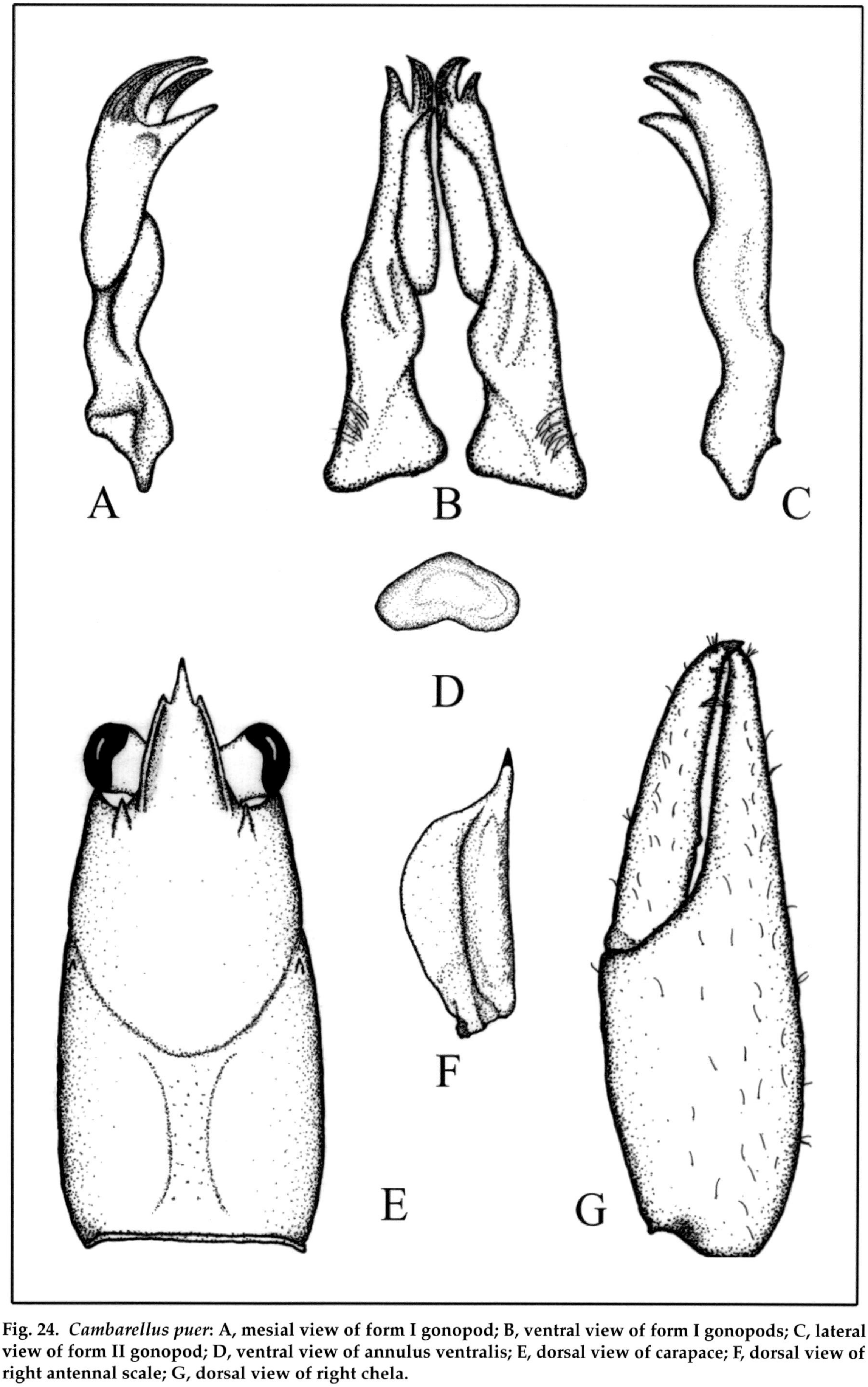

Fig. 24. *Cambarellus puer*: A, mesial view of form I gonopod; B, ventral view of form I gonopods; C, lateral view of form II gonopod; D, ventral view of annulus ventralis; E, dorsal view of carapace; F, dorsal view of right antennal scale; G, dorsal view of right chela.

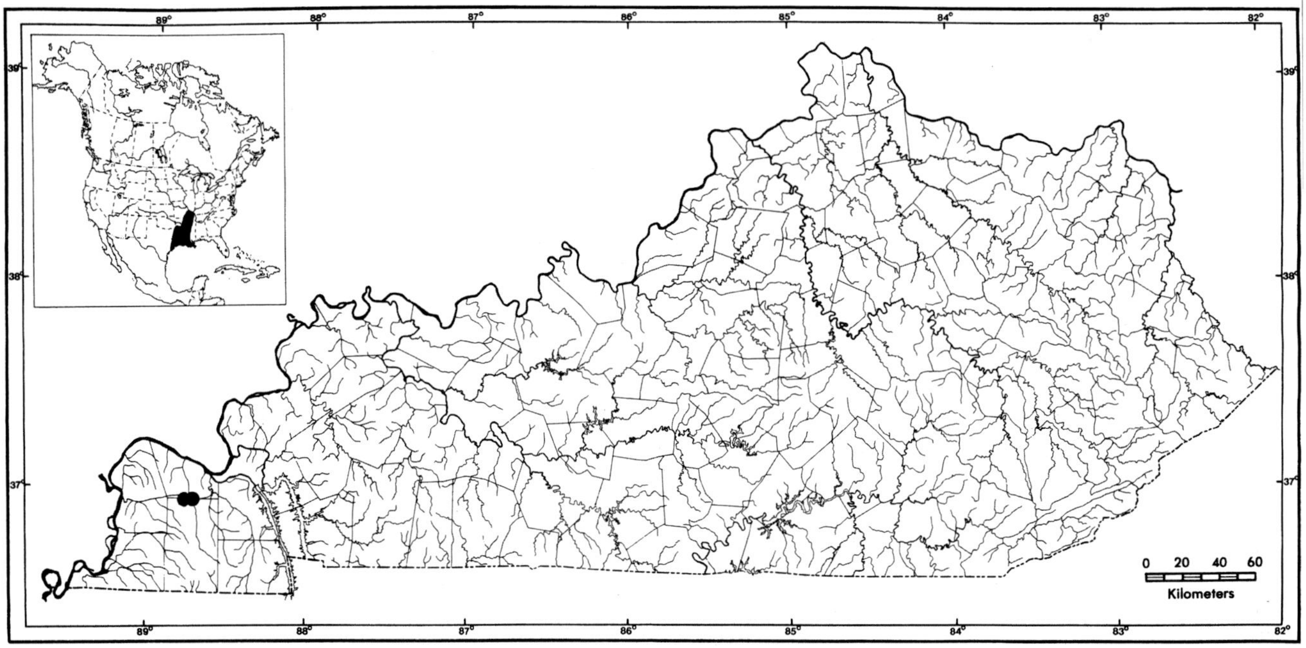

Fig. 25. Distribution of *Cambarellus puer* in Kentucky.

such as swamps, effluent creeks, and lowland areas that were formerly swamps before being drained. Pflieger (1996) reported the species from habitats that dried completely during summer and early fall months.

LIFE HISTORY – Given the small number of collections (< 5) of *C. puer* from Kentucky, very little is known about the species' life history in the state. Form I males have been collected in Kentucky in February and December, but no ovigerous females have been collected. Most of what is known about the species comes from data collected in other states, mostly Louisiana (Black 1966), Illinois (Page 1985), and Missouri (Pflieger 1996). The geographic proximity of southern Illinois and southeastern Missouri populations to Kentucky populations precludes significant variation in life history parameters between those populations. In Illinois and Missouri, ovigerous females have been collected from February to May and ranged in length from 22 to 28 mm; the number of eggs carried per female ranged from 43 to 100, with egg diameter averaging 1.0 to 1.1 mm. In Louisiana, Black (1966) reported two periods of reproductive activity, a more active period in late winter and early spring and a secondary period of activity in mid-summer. Black also reported that males required 13 or 14 months to reach sexual maturity, as such, most males did not breed during their first breeding season after hatching. Longevity in the species was estimated to be 15 to 18 months (Black 1966).

CRAYFISH ASSOCIATES – The following species have been collected from habitats containing *Cambarellus puer*: *Cambarus diogenes*, *Procambarus acutus*.

CONSERVATION STATUS – KSNPC Endangered.

Cambarellus shufeldtii (Faxon)
Cajun Dwarf Crayfish

Fig. 26 Photo by W. Pflieger

LIFE COLORS – (Fig. 26) Base color of dorsal and lateral surfaces of chelae, carapace, and abdomen variable, ranging from rust red to light brown to gray. Dorsum of carapace and abdomen with two longitudinal brown or black stripes or two rows of spots of the same color fused into uninterrupted bands. Two blue stripes on tail fan. Tips of fingers of chelae lacking orange and/or black bands. Venter cream to whitish in color.

The occurrence of both stripes and spots in the species has been attributed to the presence of a single dimorphic allele. Using mating experiments and field observations, Volpe and Penn (1957) reported that the striped condition was dominant to the spotted condition. Pflieger (1996) reported that the character was selectively neutral and that mating between individuals possessing different color conditions was random.

CHARACTERS – (Fig. 27) Body size small, adults rarely exceeding 28 mm in total length. Rostrum flat, margins converging distally and terminating in small spines at base of acumen. Acumen equal to or slightly longer than width of rostrum at marginal spines. Carapace laterally compressed and moderately arched dorsoventrally. Strong cervical spines present. Areola moderately wide, 4.5 to 6.5 times longer than wide at its narrowest point, with room for four to six shallow punctations. Antennal scale widest at midpoint. Chelae small and with short fingers; fingers and palm with rounded mesial and lateral margins; single setae covering dorsal surfaces. Dactyl of chela of equal length or shorter than mesial margin of palm. Chelae lacking longitudinal ridges on fingers and rows of tubercles on mesial margin of palm. Males with hooks on basis of second and third pereiopods.

Gonopod of form I male terminating in three straight elements of moderate length: corneous, pointed central projection; shorter, pointed, noncorneous caudal process; and longer, pointed, noncorneous mesial process. Annulus ventralis of female movable and kidney-shaped, thicker portion on either left or right side; with a strongly elevated central region; lacking fossa; short sinus usually present.

MAXIMUM SIZE – The maximum known size for the species is 35.0 mm total length.

MOST LIKE – In possessing three terminal elements on the form I male gonopod and hooks on the basis of the second and third pairs of pereiopods, *Cambarellus shufeldtii* can only be confused with the other member of the genus *Cambarellus* found in Kentucky, *C. puer*. *Cambarellus shufeldtii* differs from that species in having straight terminal elements on the male gonopod.

DISTRIBUTION AND HABITAT – (Fig. 28) *Cambarellus shufeldtii* has an almost identical distribution to that of *C. puer*; it differs only in that *C. shufeldtii* ranges slightly farther west along the Gulf Coast of Texas (Hobbs 1989). In Kentucky the species occurs in the Mississippi River floodplain and drainage in Ballard, Carlisle, Fulton, and Hickman counties.

Cambarellus shufeldtii is found in Kentucky in floodplain ditches and ponds, sloughs, and cypress swamps. Habitats in which the species are found usually have abundant aquatic vegetation, both rooted and unrooted, and soft mud substrate.

LIFE HISTORY – Form I males have been collected in Kentucky in April, June, September, October, and November. Ovigerous females have been collected in April and September. These data suggest that the species has two periods of reproductive activity, one during the late spring and one in the fall. This pattern would resemble the reproductive life cycle reported for populations of the species in Louisiana (Lowe 1961). In Louisiana populations of ovigerous females were collected year-round but were more frequently encountered in the early spring and late summer

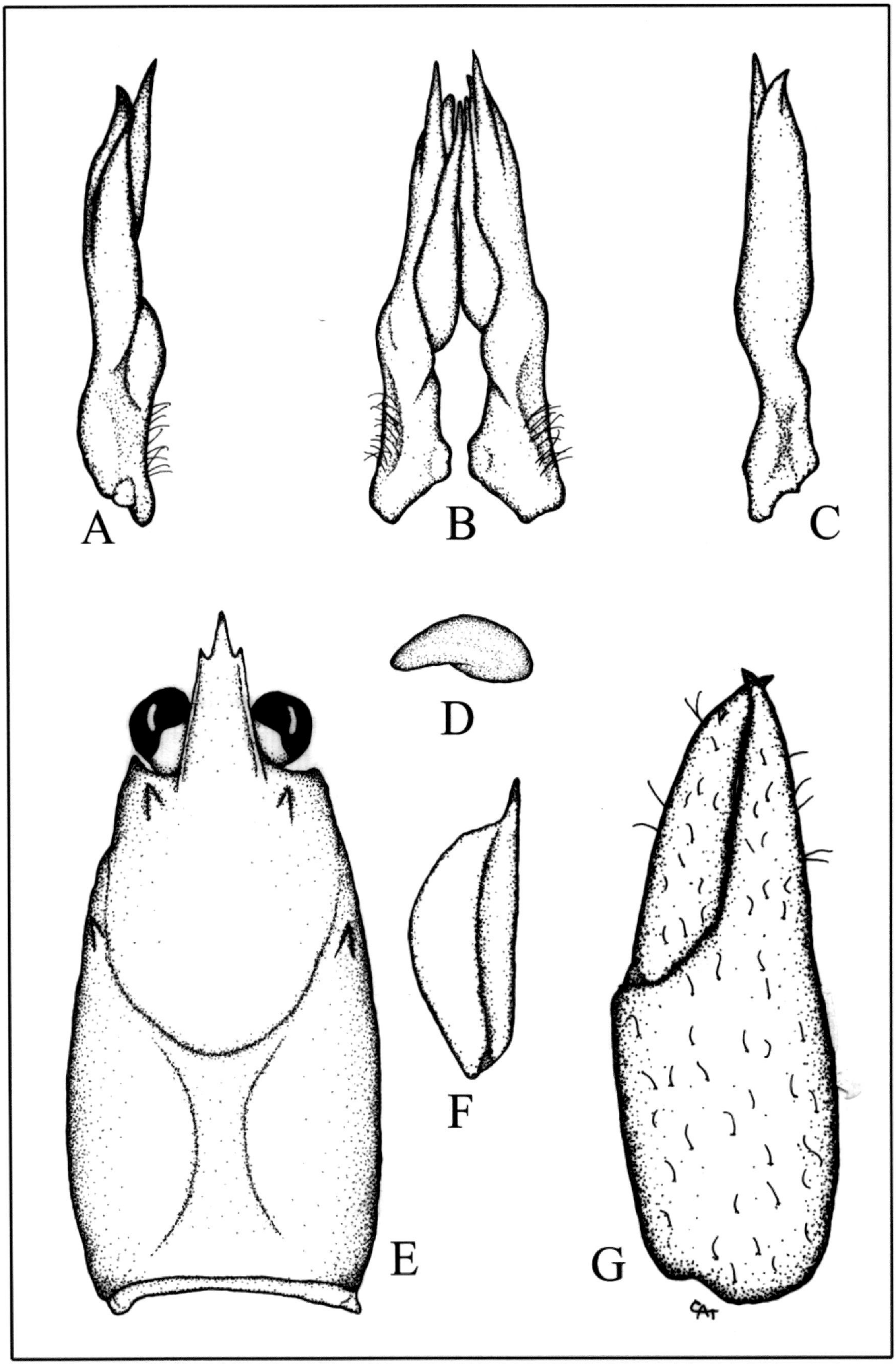

Fig. 27. *Cambarellus shufeldtii*: A, mesial view of form I gonopod; B, ventral view of form I gonopods; C, lateral view of form II gonopod; D, ventral view of annulus ventralis; E, dorsal view of carapace; F, dorsal view of right antennal scale; G, dorsal view of right chela.

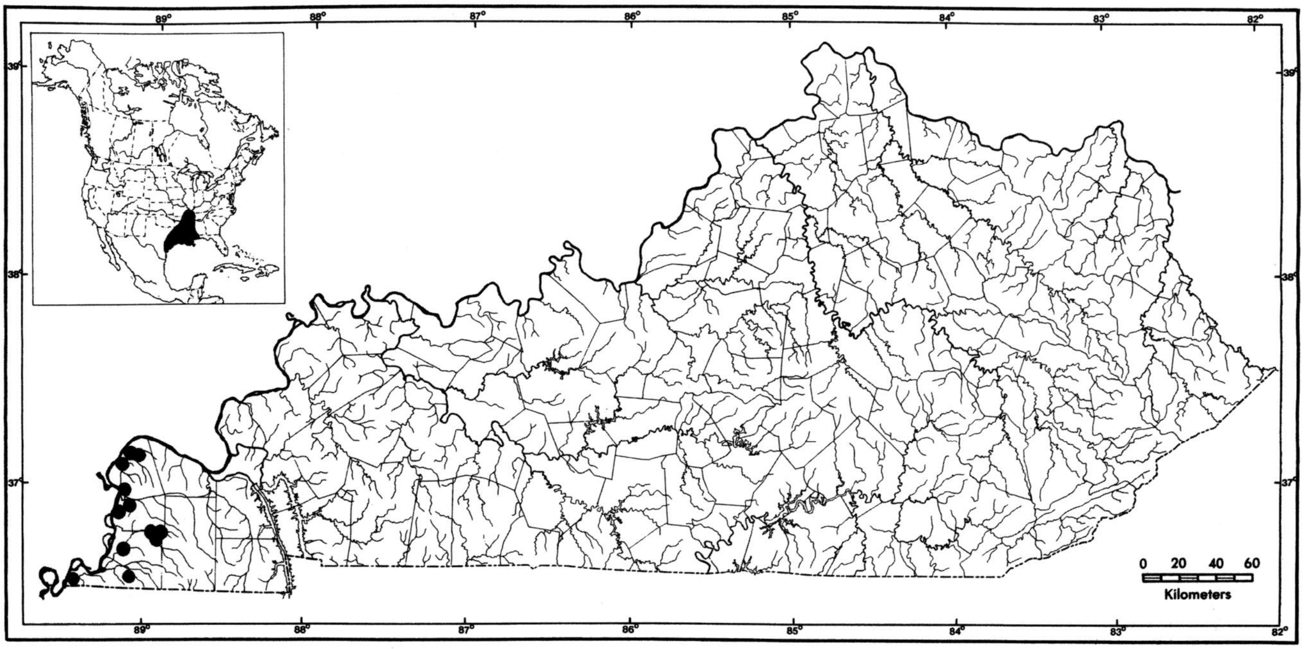

Fig. 28. Distribution of *Cambarellus shufeldtii* in Kentucky.

(Lowe 1961), thus suggesting earlier mating periods in more southern latitudes. Two ovigerous females collected in Kentucky in April measured 28.0 and 29.5 mm total length and carried 116 and 81 eggs, respectively. Egg diameter averaged 1.0 mm. A large amount of life history information has been collected for *Cambarellus shufeldtii* from Louisiana populations (Black 1966; Lowe 1961; Penn 1942, 1950). The following information is taken from those studies: Females mature at a length of approximately 18 mm at two months of age and may produce two clutches of eggs during their 1.5-year life spans. Clutch size averages 34 eggs and eggs stay attached for approximately three weeks. Males may breed twice by the time they are one year old, undergo 12 to 13 molts before reaching maturity, and may live to 18 months of age. While not a burrowing species, *C. shufeldtii* may build shallow "cells" in the mud to survive dry periods.

CRAYFISH ASSOCIATES – The following species have been collected from habitats containing *Cambarellus shufeldtii*: *Cambarus diogenes*, *C. ludovicianus*, *Orconectes lancifer*, *Procambarus acutus*, *P. clarkii*, and *P. viaeviridis*.

CONSERVATION STATUS – KSNPC Special Concern.

Barbicambarus Bouchard

The genus *Barbicambarus* contains a single member, *B. cornutus*. This species occurs in creeks and rivers in the Green River drainage of north-central Tennessee and west-central Kentucky. While very similar in overall chela, carapace, and gonopod shape to members of the genus *Cambarus, B. cornutus* is unique among all species in the family Cambaridae in that it has dense clumps of long setae along the entire length of its antennae.

Barbicambarus cornutus (Faxon)
Bottlebrush Crayfish

Fig. 29A

Fig. 29B

LIFE COLORS – (Figs. 29A, B) Two known color forms. One with large dark brown patches over entire body including the chelipeds; usually with five dark longitudinal bands going length of abdomen. This form known only from Nolin River drainage. The second form lacks splotches; bands on abdomen may be present, but usually not as evident. Both forms with overall base color olive green to brown; color highlights on rest of body very similar. Rostrum, postorbital ridges, large tubercles on the cheliped, and most joints of legs highlighted in bright orange to red. Venter cream to whitish in color.

CHARACTERS – (Fig. 30) Rostrum deeply excavated; wide proximally, margins converging, concave distally, ending with strong corneous spines. Acumen triangular with corneous spine at tip. Carapace dorsoventrally flattened; in dorsal view broadly oval. Areola very wide usually 4.2 to 4.8 times longer than wide at its narrowest point and sparsely punctated. Postorbital ridge well developed with strong corneous anterior spine. Antennae with dense fringe of long setae. Antennal scale widest at proximal to midpoint. Hepatic area covered with numerous small tubercles; cervical spines well developed; branchiostegal spines reduced or absent; suborbital angle obsolete, broadly rounded. Dactyl of chela longer than palm; propodus broadly rounded with serrated lateral tubercles, base strongly flattened laterally; dorsal surfaces of fingers with developed longitudinal ridges. Moderate gap when fingers closed. Mesial margin of palm with single row of 9–13 small tubercles; no subpalmar tubercles.

Central projection of gonopod of form I male corneous, recurved 90° to shaft of gonopod without subapical notch; mesial process elongated, narrow, and fingerlike. Annulus ventralis of female horizontally elongate with shallow trough separating right and left halves; fossa located on anterior face.

MAXIMUM SIZE – The maximum known size for the species in Kentucky is 134.5 mm total length. *Barbicambarus cornutus* is the largest species of crayfish likely to be encountered in Kentucky. While the maximum reported size for the *Procambarus clarkii* is 160.0 mm total length, specimens of *P. clarkii* exceeding 110 mm are rarely seen in the state.

MOST LIKE – This species is not readily confused with any other species living in Kentucky. The most distinguishing feature is the dense fringe of setae along the margins of the antennae. In very small individuals these hairs are not as noticeable, but are usually present.

DISTRIBUTION AND HABITAT – (Fig. 31) *Barbicambarus cornutus* is endemic to the upper Green River basin in Kentucky and Tennessee (Hobbs 1989). In Kentucky it is found most commonly in the Barren River drainage, the mainstem of the middle section of the Green River, and in the Nolin River above Nolin Lake.

The species is most commonly found under large, flat boulders along creek margins where there is a current. Occasionally, it can be found in shallow riffles if there are large boulders present.

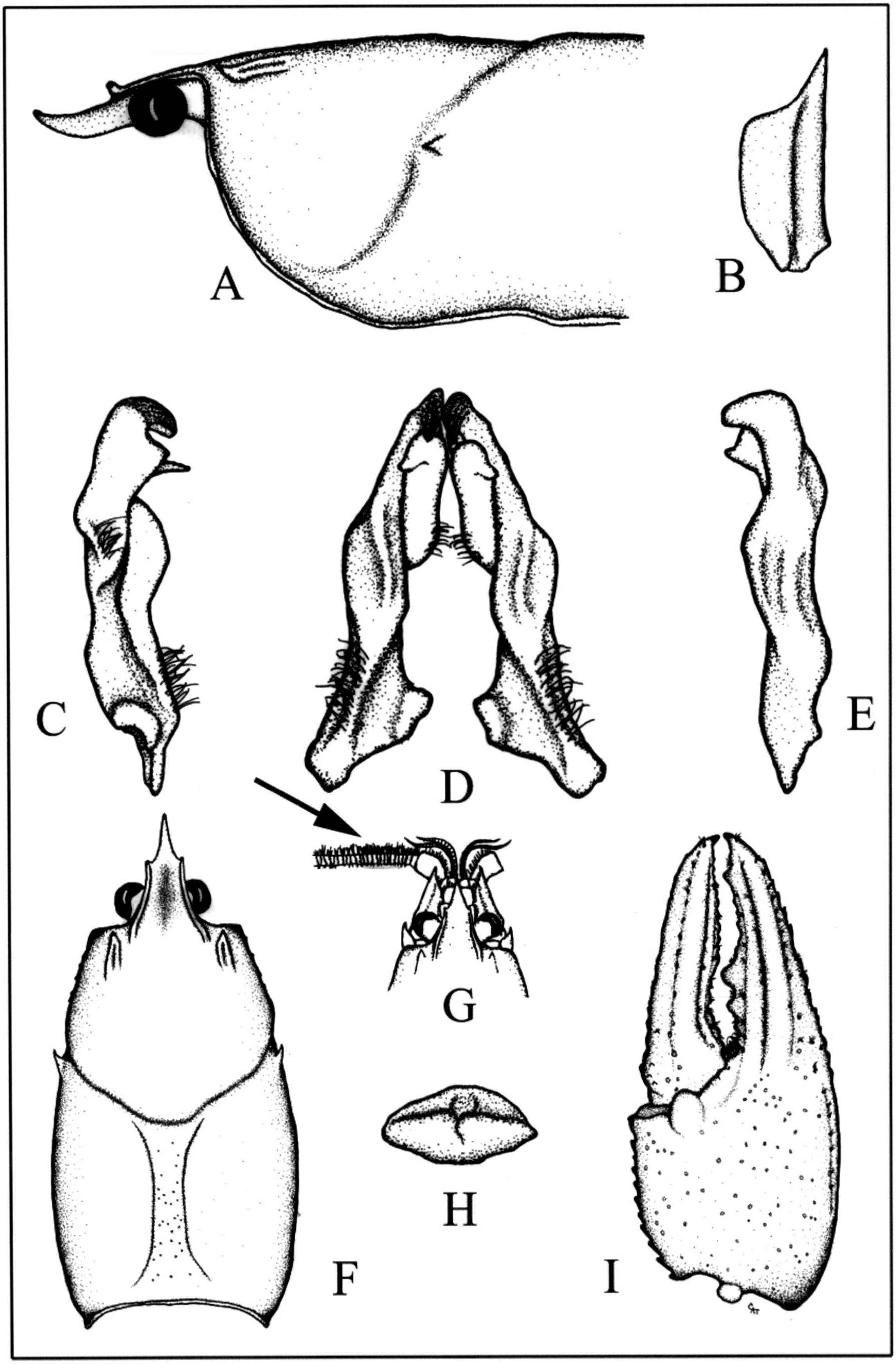

Fig. 30. *Barbicambarus cornutus*: A, lateral view of carapace; B, dorsal view of right antennal scale; C, mesial view of form I gonopod; D, ventral view of form I gonopods; E, lateral view of form II gonopod; F, dorsal view of carapace; G, dorsal view of head region with arrow denoting antennal setae; H, ventral view of annulus ventralis; I, dorsal view of right chela.

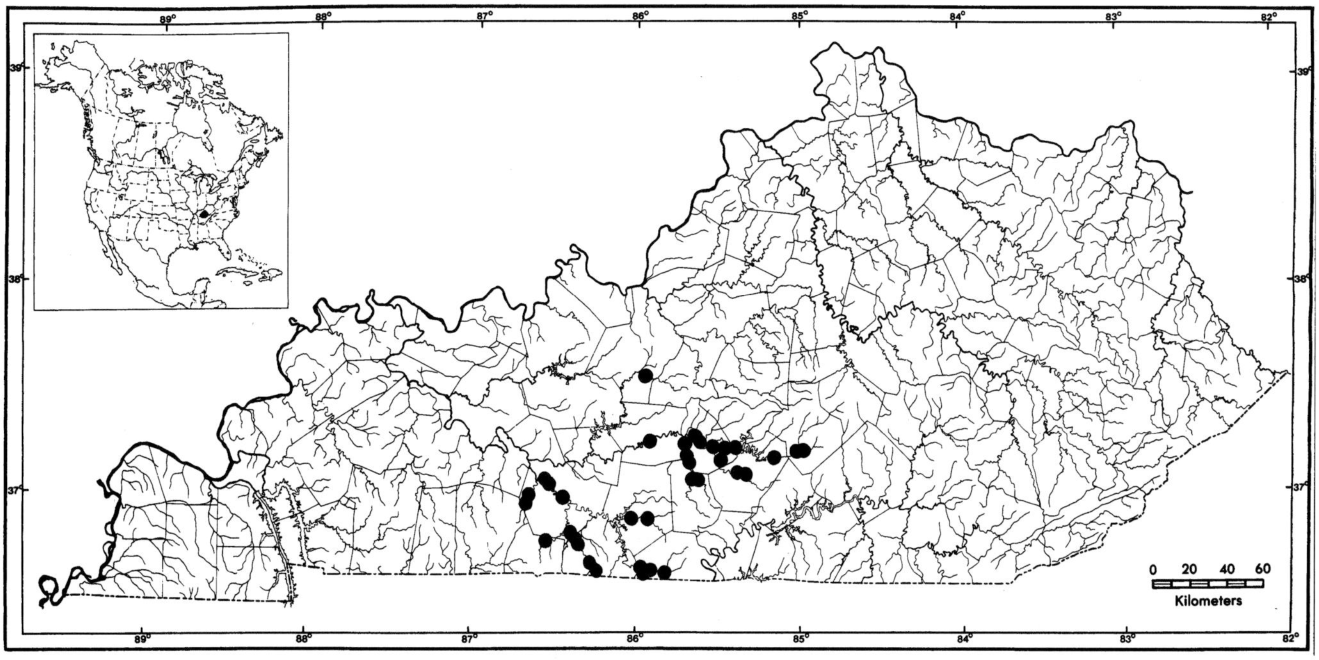

Fig. 31. Distribution of *Barbicambarus cornutus* in Kentucky.

LIFE HISTORY – In Kentucky form I males have been collected in April, March, July, September, and October. Two ovigerous females were collected on April 4, 2002: One with a total length of 92.4 mm had 114 eggs with a mean diameter of 2.8 mm; the other with a total length of 109.2 mm had 353 eggs with a mean diameter of 2.7 mm. A female collected on May 17, 2000, from the Green River had 98 newly hatched young attached to her pleopods. Upon examination of the young, few hairs could be seen on the antennae indicating that the fringe of hairs develops in subsequent molts. Little else is known about the life history of *Barbicambarus cornutus*.

CRAYFISH ASSOCIATES - We have collected *Barbicambarus cornutus* with *Cambarus cumberlandensis, C. graysoni, C. tenebrosus, C. rusticiformis, Orconectes barrenensis, O. putnami,* and *O. rusticus.*

CONSERVATION STATUS - KSNPC Special Concern. This status should probably be reevaluated given the discovery of new locations in the Green River drainage (Lawson 2003).

Cambarus graysoni **Photo by C. Lukhaup**

Cambarus Erichson

Cambarus is the second largest genus of crayfishes in the world. While most of its 91 species and subspecies occur in the southeastern United States, a handful of species do range north into southern Ontario and west to Colorado. *Cambarus* species occur in a wide range of habitat types from subterranean creeks to lowland ditches and swamps to upland creeks and rivers. Burrowing behavior is also variable within the genus; some members are primary burrowers and are rarely encountered above ground while the creek dwellers are tertiary burrowers. Male *Cambarus* have a gonopod that terminates in two recurved elements. Representatives of this genus are the most colorful crayfishes in Kentucky. Twenty-two described and one undescribed species of *Cambarus* occur in Kentucky. One of those, *C. batchi*, is endemic to the state.

Cambarus bartonii cavatus Hay
Appalachian Brook Crayfish

Fig. 32

LIFE COLORS – (Fig. 32) Individuals uniform light beige to dark brown in color. With exception of chelae, legs generally lighter in coloration, range from cream to light brown. Venter of body same color as legs.

CHARACTERS – (Fig. 33) Rostrum moderately to deeply excavated, wide at base; margins converging anteriorly at base of short acumen; no marginal spines. Acumen short. Carapace in dorsal view broadly oval; in lateral view posteriorly flattened, anterior somewhat vaulted. Areola wide, usually 4.2 to 5.6 times longer than wide at its narrowest point; sparsely punctated. Postorbital ridge somewhat reduced usually without anterior tubercle or tubercle reduced. Hepatic area with numerous rounded tubercles; cervical spines absent, but may have two to three small tubercles; branchiostegal spines present but small; suborbital angle acute, usually with small spine or tubercle distally. Antennal scale widest distal to midpoint. Chela with straight dactyl and propodus broadly rounded; fingers much longer than length of palm. Dorsal surfaces of fingers of chelae with poorly developed longitudinal ridges. Lateral margins of fingers smooth with three to six basal tubercles. Palm with one complete row of five to seven weak tubercles; may have second incomplete row of one to four tubercles; with one or two subpalmar tubercles.

Central projection of gonopod of form I male corneous, recurved approximately 90° to shaft with shallow to distinct subapical notch. Mesial process bulbous at base, tapering to a point. Annulus ventralis of female oval with deep, well-developed fossa and a short C-shaped sinus.
MAXIMUM SIZE – The maximum known size for the species is 110.7 mm total length.

MOST LIKE – By possessing a wide areola and weak tubercles along the mesial margin of the palm, *Cambarus bartonii cavatus* most closely resembles *C. sciotensis*. The ranges of the two overlap only in the Big Sandy drainage system. They are most easily distinguished by the greatly thickened rostral margins of *C. sciotensis*. In addition, *C. sciotensis* has a complete second row of tubercles on the palm, while *C. bartonii cavatus* may have an incomplete second row of one to four tubercles.

DISTRIBUTION AND HABITAT – (Fig. 34) *Cambarus bartonii cavatus* ranges from Tennessee and northwestern Georgia to southwestern Virginia (Hobbs 1989), West Virginia (Jezerinac et al. 1995), and west to Kentucky. The species is widespread but sporadic in the eastern half of Kentucky with records from all major river drainages except the Green.

This species is most commonly encountered in small upland creeks in eastern Kentucky where we have found it under larger rocks in areas with slower moving water. It is commonly found in small intermittent creeks. Under dry conditions it burrows into the substrate under large rocks. It will also occasionally occur in larger creeks. Hobbs (1981) indicated that in Georgia this species is associated with mountainous terrain where he found it commonly in seepage areas, fast-moving mountain creeks, and slower-moving valley creeks. He also indicated that it will invade littoral zones of reservoirs. Jezerinac et al. (1995) found *C. bartonii cavatus* in burrows in intermittent headwater creeks, roadside ditches and springs, as well as under large rocks along the banks of larger creeks.

LIFE HISTORY – In Kentucky form I males have been collected in March, April, May, August, and December. No ovigerous females have been collected in the state. Jezerinac et al. (1995) found form I males in West Virginia in April, June, July, September, and October. They suggested that form I males of this species could be collected throughout the year in West Virginia, as did Hobbs (1981) for Georgia. Jezerinac et al. (1995)

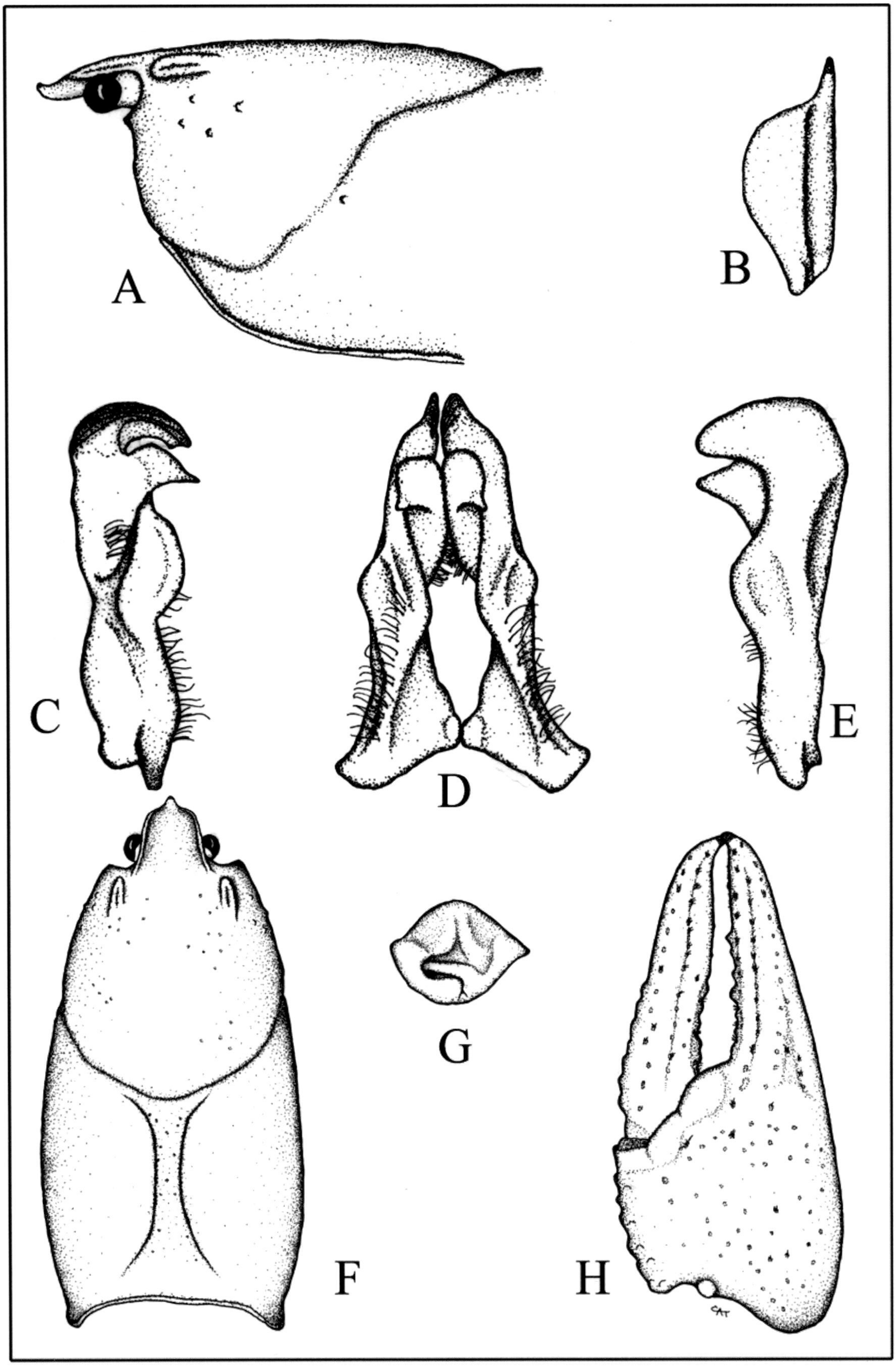

Fig. 33. *Cambarus bartonii cavatus*: A, lateral view of carapace; B, dorsal view of right antennal scale; C, mesial view of form I gonopod; D, ventral view of form I gonopods; E, lateral view of form II gonopod; F, dorsal view of carapace; G, ventral view of annulus ventralis; H, dorsal view of right chela.

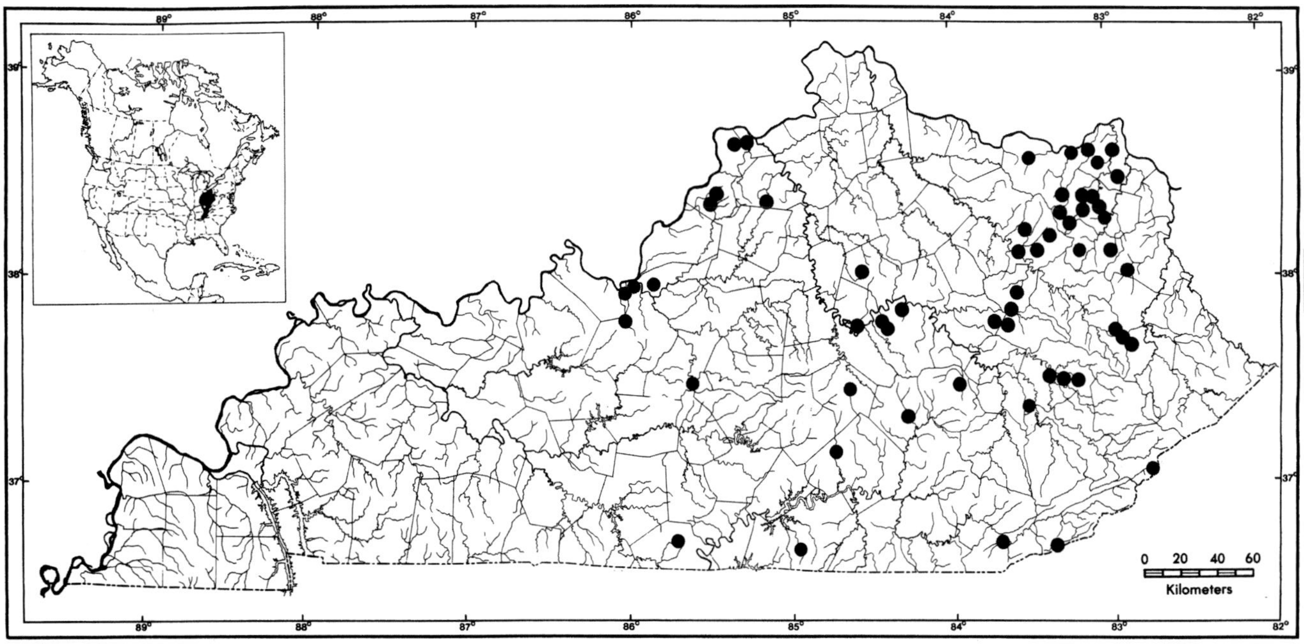

Fig. 34. Distribution of *Cambarus bartonii cavatus* in Kentucky.

collected ovigerous females in July, and reported one female with 304 eggs with diameters ranging from 2.4 to 2.5 mm. They also found a female in August with both eggs and hatched young. According to Jezerinac et al. (1995), in West Virginia this species copulates in June, eggs are laid and females are in berry during July, with the eggs hatching in August. In the populations they studied there was a 1:1 sex ratio. Given the proximity of West Virginia populations, we believe that timing of reproductive activities of those populations closely match those of eastern Kentucky populations. According to Hobbs (1981), in Georgia he found females with eggs in April and June, and he collected females with young in June and August. Hobbs (1981) reported that the smallest ovigerous female had 28 eggs while the largest had 117 eggs. The female with the largest number of eggs had 138. Hobbs further suggests that the hatchlings grow from the time they are born until the onset of winter, and during winter they cease growth until the following spring. He thought that this cohort does not reach sexual maturity until the second fall season with the females producing their first clutch of eggs in the spring when they are almost two years old. The longevity of *C. bartonii cavatus* in Georgia appears to be three years (Hobbs 1981). For additional information, Hamr and Berrill (1985) provided detailed life history data for the nominate subspecies *C. bartonii bartonii.*

CRAYFISH ASSOCIATES – We have collected *Cambarus bartonii cavatus* with *C. robustus, C. sciotensis, Orconectes cristavarius,* and *O. juvenilis.*

6/11/05 – Corn Cr., Trimble Co. Ky.

Cambarus batchi Schuster
Bluegrass Crayfish

Fig. 35

LIFE COLORS – (Fig. 35) Carapace bright red-brown, abdomen not as bright. Chelae bright red-brown, fingers have brightest color. All other legs much lighter in color from orange to straw color. Venter of body cream to pink. Intensity of coloration of individuals variable.

CHARACTERS – (Fig. 36) Rostrum deeply excavated, subtriangular; margins thickened and converging anteriorly; no marginal spines. Acumen short; margins slightly less thick than rostral margins. Carapace bullet-shaped from dorsal view; vaulted anteriorly from lateral view. Areola obliterated. Postorbital ridge very low with no terminal tubercle or spine. Hepatic area with very few small tubercles; cervical spines absent; branchiostegal spines absent or with small tubercle; suborbital angle absent. Antennal scale very small, spine reduced to rounded corneous tubercle; widest distal to midpoint. Chelae smooth with few setae; dactyl more than twice the length of palm; lateral margins of both fingers straight and smooth. Dorsal surfaces of fingers of chelae without longitudinal ridges. Mesial margin of palm with single row of four to seven tubercles, some may be fused to form rounded ridge; sometimes one or two tubercles lateral to main row; no subpalmar tubercles. Epistomal zygoma strongly arched. Spines on telson and uropod all minute; some absent. Abdomen much narrower than carapace.

Central projection of gonopod of form I male corneous, recurved approximately 90° to the shaft, with strong subapical notch. Mesial process bulbous constricting apically to narrow fingerlike process. Annulus ventralis of female with deep anterior fossa and S-shaped sinus.

MAXIMUM SIZE – The maximum known size for the species is 76.1 mm total length.

MOST LIKE – Of all Kentucky primary burrowers that have obliterated areolas and non-notched dactyls, *C. batchi* is most similar to *Cambarus deweesae.* The two species are readily separated by the shape and details of the chela. *Cambarus batchi* has a single row of tubercles on the palm, some of the basal tubercles may fuse to form a ridge. *Cambarus deweesae* has two well-developed rows of tubercles.

DISTRIBUTION AND HABITAT – (Fig. 37) *Cambarus batchi* is endemic to Kentucky and is known only from the outer Blue Grass Region of the state in the Kentucky River drainage. It is known from Estill, Garrard, Madison, and Rockcastle counties.

This species appears to be a primary burrower living in complex burrows that reach a depth of 1 to 2 meters (Schuster 1976). The burrows are constructed in hydric soils with a clay fragipan, usually in low-lying ditches or depressions. The water table is often within 0.5 m from the surface.

LIFE HISTORY – Form I males have been collected in February, March, April, May, September, October, and November. Schuster (1976) reported ovigerous females from May through June, and the first independent young were observed in May. He found young in the burrows of adult females throughout the summer until October. All males observed in burrows were by themselves.

Schuster (1976) reported two main molting periods: one in the spring and another in the fall. His observations of exuvia near the opening of burrows led him to conclude that these crayfishes molted outside of the burrows. This species seems to be inactive during much of the year. During the dry summer months and cold winter months, the burrow openings are plugged with mud and may remain plugged for several months. The species is most active at night during the spring and fall, especially during or just after periods of heavy rainfall, when indi-

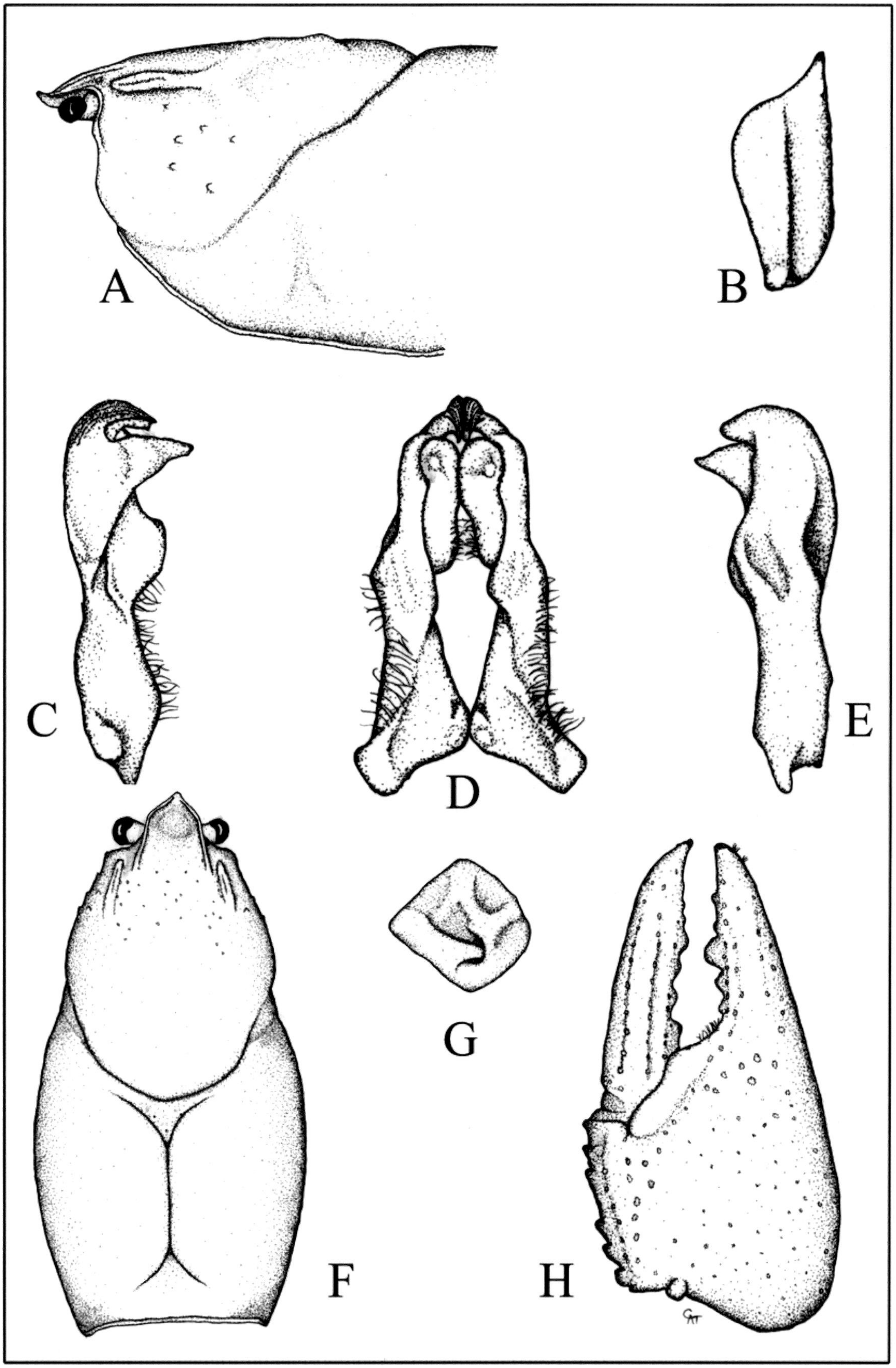

Fig. 36. *Cambarus batchi*: A, lateral view of carapace; B, dorsal view of right antennal scale; C, mesial view of form I gonopod; D, ventral view of form I gonopods; E, lateral view of form II gonopod; F, dorsal view of carapace; G, ventral view of annulus ventralis; H, dorsal view of right chela.

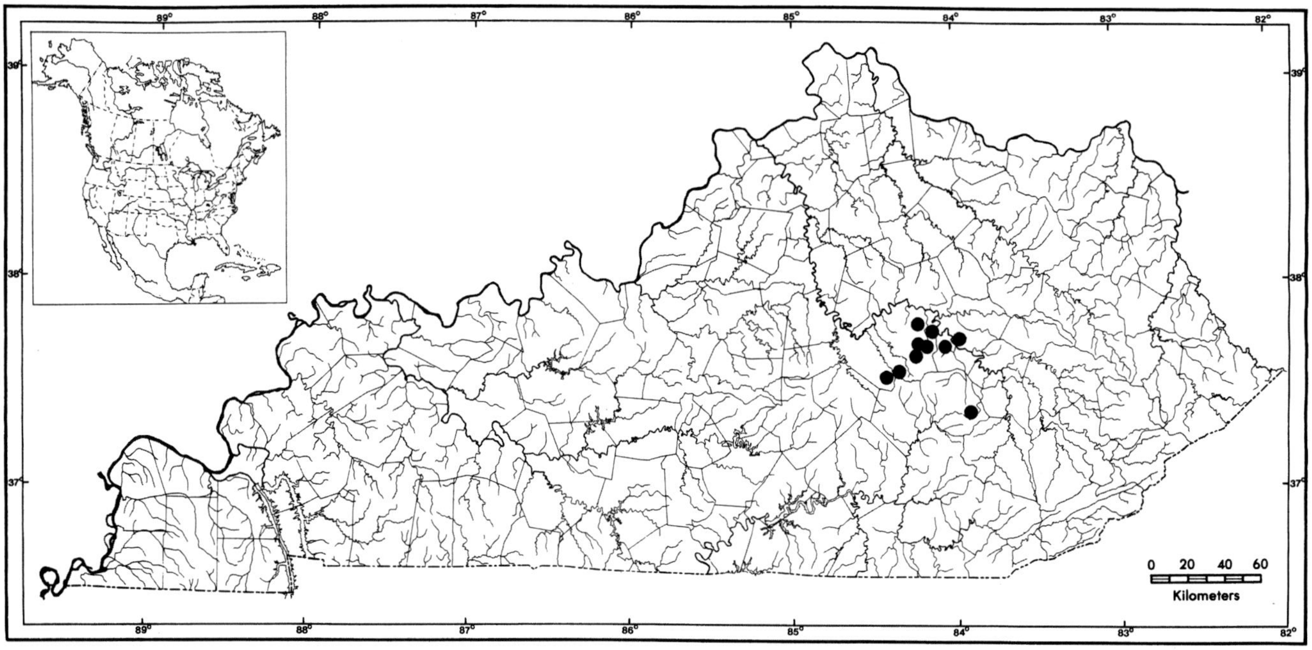

Fig. 37. Distribution of *Cambarus batchi* in Kentucky.

viduals are often seen wandering in the open fields where burrows are located. Schuster (1976) observed that most of the individuals walking around were adult form I males, presumably looking for mates.

Laboratory observations of burrowing activity in aquaria (Schuster, unpublished data) verified nocturnal activity patterns. These observations also indicate that individuals have little burrow fidelity. Individuals walking outside of the burrows, when disturbed, will quickly get into any available burrow. If the burrow is occupied the resident will usually evict the intruder. During the course of this study, however, individuals usually showed little aggression to others, even when in close proximity.

An interesting respiratory behavior was observed in all individuals. Periodically, an individual crayfish would cease burrowing activities and make its way to the top of the water table where it would prop itself with its legs pressed against the burrow wall on its side so that half the body would be in the water and the other half out of the water. When the crayfish did this, the observer could actually hear water draining from the exposed side of the carapace. The crayfish would then hold itself perfectly still for a period of one to two minutes. However, the observer could hear the bailers move and often see bubbles come from the anterior part of the gill chamber. It is hypothesized that this behavior increases the efficiency of respiration by exposing wet gill filaments to air at times of low dissolved oxygen levels in burrow water. Similar observations were made by Hobbs and Whiteman (1991) for *Fallicambarus devastator*, a primary burrower endemic to Texas. It is reasonable to conclude that this is a behavior common to all burrowers and may be to all crayfishes that are in respiratory distress.

CRAYFISH ASSOCIATES – We have collected *Cambarus batchi* in the same vicinity from which we have collected *C. ortmanni*.

CONSERVATION STATUS – AFS Special Concern.

Cambarus bouchardi Hobbs
Big South Fork Crayfish

Fig. 38

LIFE COLORS – (Fig. 38) Individuals more or less uniform yellowish-green to brown, with tips of chelae orange to red. Abdomen may be slightly lighter in coloration, often with darker reticulated pattern; may look banded. Most specimens covered with a thick black coating (Hobbs 1970), making color assessment difficult.

CHARACTERS – (Fig. 39) Rostrum elongate, almost straight and moderately excavated; margins slightly converging to acumen; margins thickened; some specimens with low rounded carina anteriorly. Acumen distinctly triangular with large corneous tubercle or small spine at apex. Carapace bullet-shaped, gradually narrowing anteriorly; from lateral view dorsal aspect straight, slightly vaulted anteriorly. Areola very wide, usually 2.8 to 4.4 times longer than wide at its narrowest point; room for 8 to 10 punctations, punctations usually very small and sparsely covering areola. Postorbital ridge moderately strong, ending with distinct tubercle with corneous spine at tip. Hepatic area with numerous small scattered tubercles; cervical spines moderately to very well developed; well-developed branchiostegal spines present; suborbital angle present, well developed. Antennal scale widest proximal to midpoint. Chelae elongated and rectangular in outline, covered with short hairs. Dorsal surfaces of fingers of chelae with well-developed longitudinal ridges on both fingers. Mesial margin of palm of chela usually with a single row of 7 to 10 large serrate tubercles; basal 2 or 3 fused to form a ridge; palm with additional 4 to 6 scattered tubercles lateral to mesial row; 1 to 3 subpalmar tubercles. Epistomal zygoma strongly arched.

Central projection of gonopod of form I male corneous, recurved 15° to 20° to shaft, flattened along entire length and slightly twisted, with a subapical notch. Mesial process bulbous at base, fingerlike apically; extends farther than central projection. Annulus ventralis of female subcircular in outline with deep and wide trough through cephalic half, deep transverse fossa and short, shallow sinus.

MAXIMUM SIZE – The maximum known size for the species is 80.6 mm total length.

MOST LIKE – In general appearance and color this species looks most like *C. distans*, and they are collected together where *bouchardi* occurs. The rostrum of *C. bouchardi* is narrower and more elongate in appearance than that of *C. distans*. *Cambarus bouchardi* can be easily separated from *C. distans* and all other Kentucky *Cambarus* species in possessing a central projection that is recurved 15° to 20° to shaft.

DISTRIBUTION AND HABITAT – (Fig. 40) O'Bara (1988) reported *C. bouchardi* to be endemic and restricted to the Roaring Paunch Creek drainage within the Big South Fork of the Cumberland drainage in Tennessee and Kentucky. In Kentucky this species is restricted to Roaring Paunch Creek proper in McCreary County.

The species occurs in upland mid-sized creeks littered with large boulders and interspersed with riffles. Specimens are most commonly found under larger boulders in swift current. Bouchard (1976a) reported that it preferred smaller substrates and that large rocks were limiting.

LIFE HISTORY – Very little is known about the life history of *Cambarus bouchardi*. Hobbs (1970) found first form males in April and September but no ovigerous females. In our collections form I males were collected in April and May. We collected three ovigerous females on April 19, 2002. The females were 60.2, 76.7, and 80.6 mm in total length and had 55, 223, and 195 eggs, respectively. The mean egg diameter for all three females was 2.2 mm.

CRAYFISH ASSOCIATES – We have collected *Cambarus bouchardi* only with *C. distans*.

CONSERVATION STATUS – AFS Endangered.

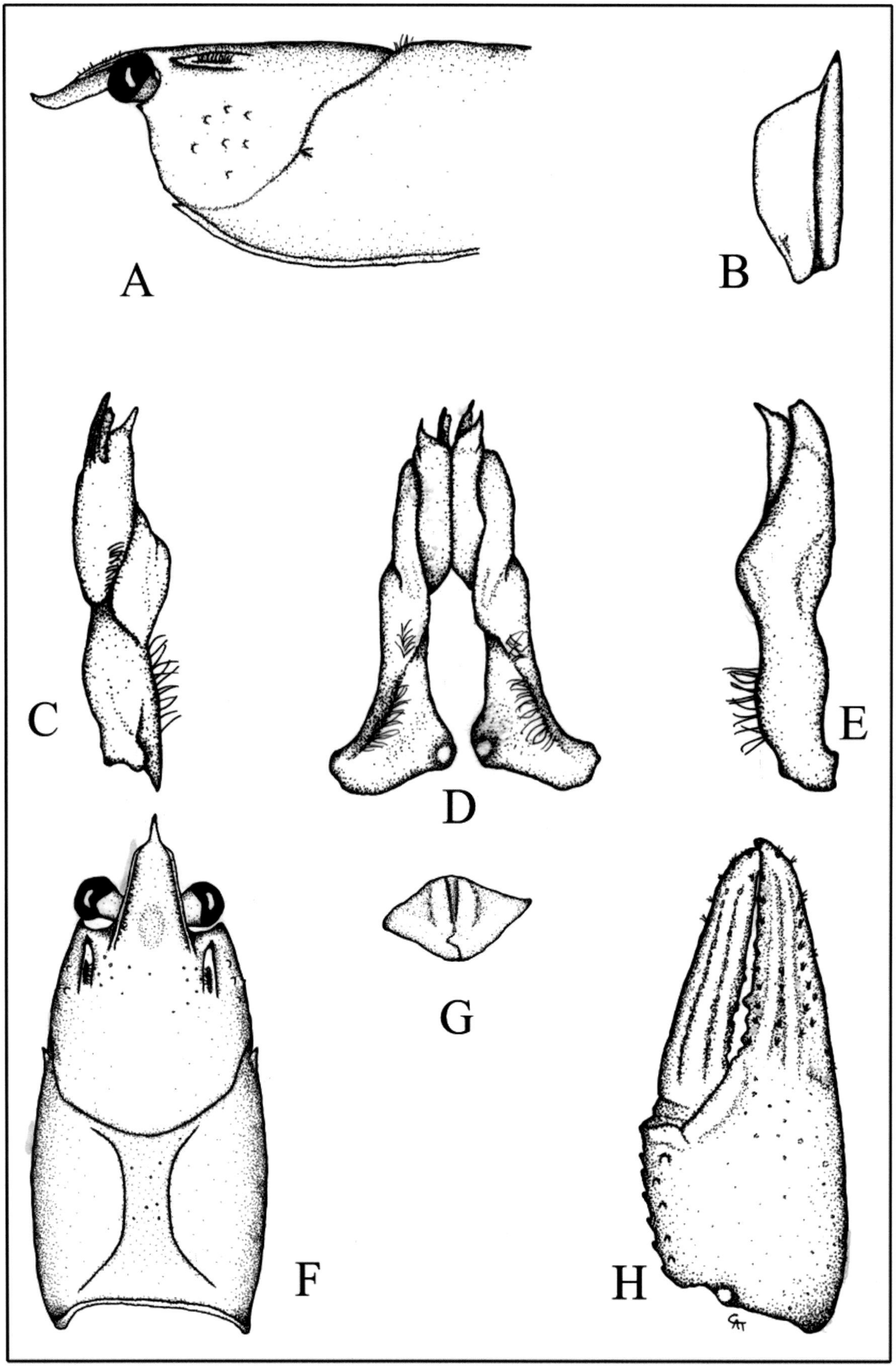

Fig. 39. *Cambarus bouchardi*: A, lateral view of carapace; B, dorsal view of right antennal scale; C, mesial view of form I gonopod; D, ventral view of form I gonopods; E, lateral view of form II gonopod; F, dorsal view of carapace; G, ventral view of annulus ventralis; H, dorsal view of right chela.

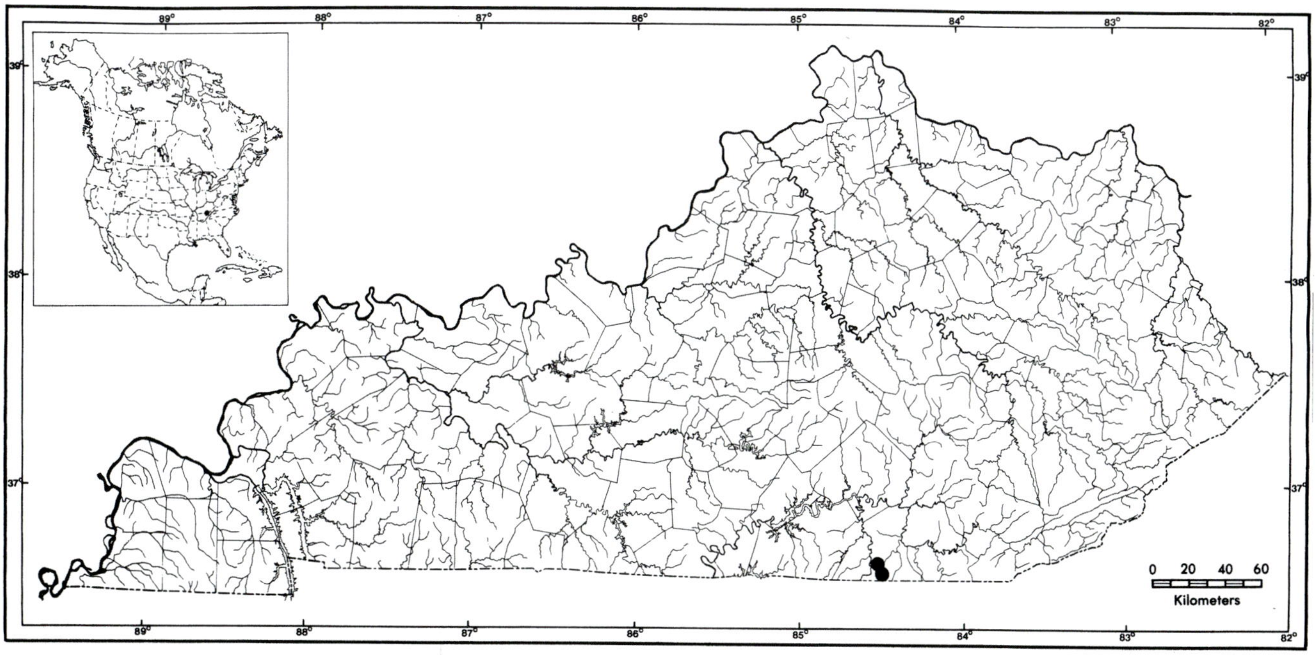

Fig. 40. Distribution of *Cambarus bouchardi* in Kentucky.

Cambarus buntingi Bouchard
Longclaw Crayfish

Fig. 41

LIFE COLORS – (Fig. 41) Color uniform light green to tan; darker in older individuals. Rostrum margins highlighted in cream to tan. Chelae same as base color; legs lighter in color, often a light cream. Abdomen with same base color, pleurites may be somewhat darker. Venter much lighter in color, cream to whitish.

CHARACTERS – (Fig. 42) Rostrum wide at base, moderately converging and moderately excavated; margins moderately thickened, distally rounded with small spines. Acumen triangular, with small apical tubercle. Carapace in dorsal view bullet shaped; in lateral view dorsoventrally flattened. Areola wide, usually 3.5 to 5.9 times longer than wide at its narrowest point, moderately punctate with room for five to seven punctations. Postorbital ridge moderately developed with small apical corneous spine. Hepatic area with many small to large tubercles; no cervical spines, but with four to five small tubercles; branchiostegal spines small but well developed; suborbital angle acute with corneous tip. Antennal scale greatest width distal to midpoint. Chelae with fingers at least twice as long as palm; dorsal surfaces of fingers with well-developed longitudinal ridges; lateral base of propodus with strong dorsoventral depression. Mesial margin of palm of chela with two rows of tubercles; mesialmost row with seven to nine tubercles. Three to four subpalmar tubercles. Epistomal zygoma strongly arched.

Central projection of gonopod of form I male corneous, recurved greater than 90° to 100° to shaft, with distinct subapical notch. Mesial process bulbous, gradually tapering to point. Annulus ventralis subrhomboidal with wide central fossa that leads to S-shaped sinus.

MAXIMUM SIZE – The maximum known size for the species is 100.7 mm total length.

MOST LIKE – By possessing a rostrum with marginal spines or tubercles and two rows of

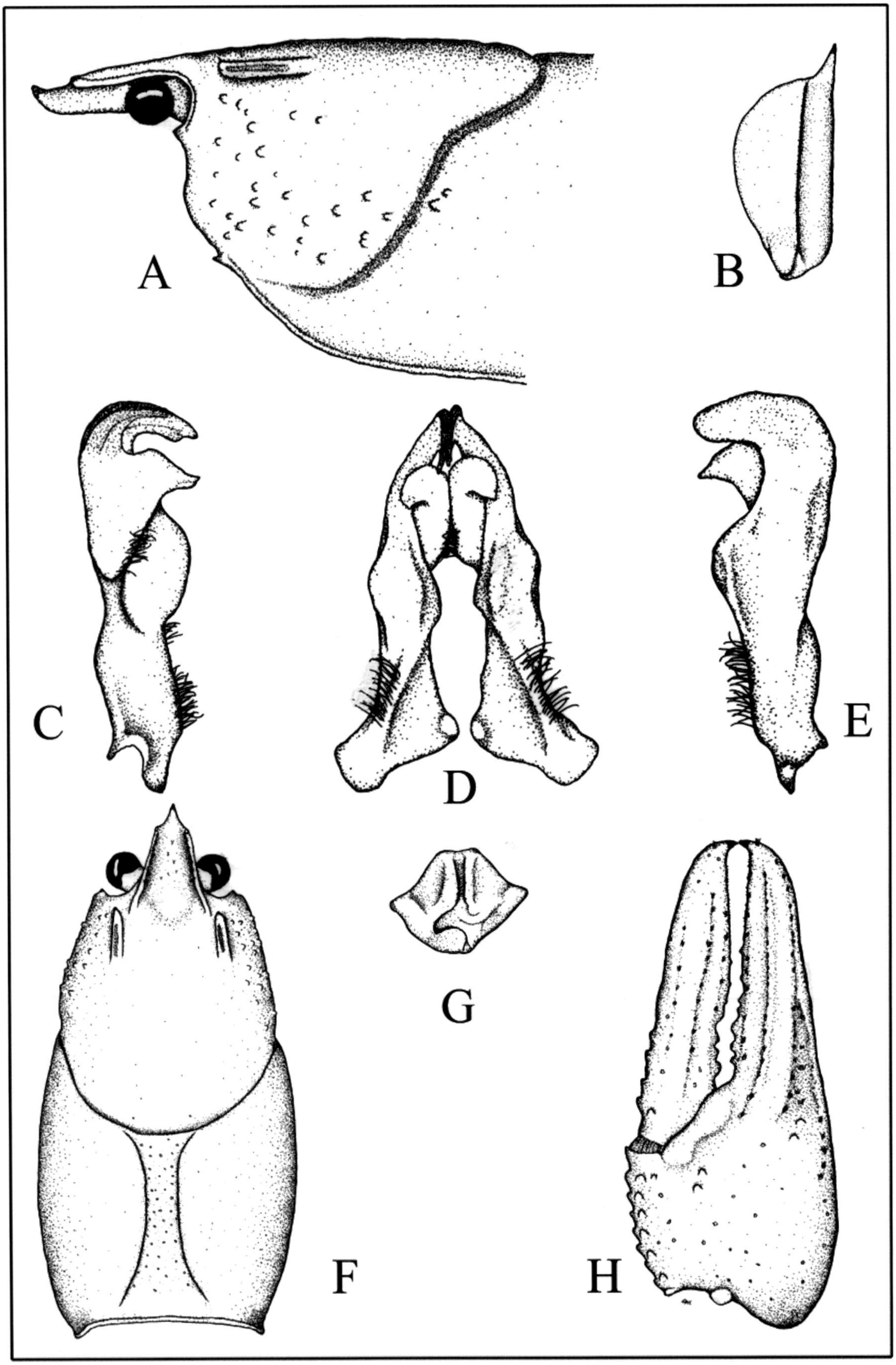

Fig. 42. *Cambarus buntingi*: A, lateral view of carapace; B, dorsal view of right antennal scale; C, mesial view of form I gonopod; D, ventral view of form I gonopods; E, lateral view of form II gonopod; F, dorsal view of carapace; G, ventral view of annulus ventralis; H, dorsal view of right chela.

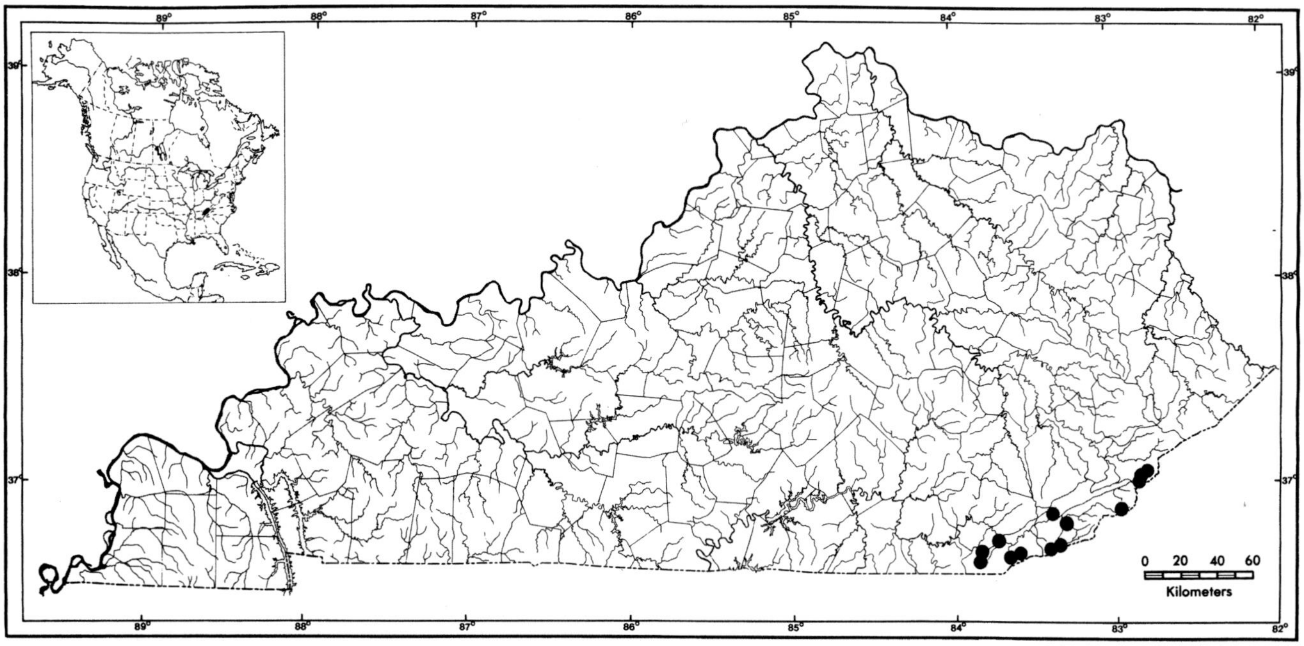

Fig. 43. Distribution of *Cambarus buntingi* in Kentucky.

tubercles along the mesial margin of the palm, this species is most like *Cambarus cumberlandensis*. The feature that most easily distinguishes the two species is the rostrum shape and the marginal spines or tubercles. In *C. buntingi* the rostral margins are rounded apically to form the acumen. At each rounded corner there is a single small tubercle. In *C. cumberlandensis* the rostral margins are thickened and more or less straight, apically ending in an abrupt angle with marginal spines. These spines are strongest in young individuals.

DISTRIBUTION AND HABITAT – (Fig. 43) *Cambarus buntingi* is restricted to the upper Tennessee and Cumberland River drainages in Tennessee and Kentucky (Bouchard 1973). In Kentucky it seems to be restricted to the Cumberland River drainage above Cumberland Falls.

Cambarus buntingi is most commonly collected from medium to large creeks and rivers with cobble substrates where it lives under large slab boulders. It also seems to do best in clean, more pristine creeks with little to no siltation.

LIFE HISTORY – We have collected form I males in April, August, and September. Bouchard (1973) also reported them for June, October, and November. No ovigerous females have been collected. Nothing else is known about its life history.

CRAYFISH ASSOCIATES – We have collected *Cambarus buntingi* with *C. distans*, *C. parvoculus*, and *Orconectes cristavarius*.

Cambarus cumberlandensis Hobbs and Bouchard
Cumberland Crayfish

Fig. 44

LIFE COLORS – (Fig. 44) Specimens usually uniform in color, ranging from blue-green to green to dark brown. Venter of body usually lighter, cream to whitish. Rostral margins highlighted in cream to gold. Carapace at mandibular muscle with round reticulation. Venter cream-colored.

CHARACTERS – (Fig. 45) Rostrum moderately to deeply excavated; margins straight, terminating in conspicuous corneous spines. Acumen elongate and triangular, greatly tapering, ending with large corneous spine. Carapace from dorsal view bullet shaped, from lateral view dorsoventrally flattened. Areola very wide, usually 3.5 to 4.9 times longer than wide at its narrowest point, moderately punctated with room for 10 to12 very small punctations at its narrowest part. Postorbital ridge flattened, anteriorly tipped with well-developed corneous spine. Hepatic area with numerous small tubercles; cervical spines well developed with several small tubercles around spine; branchiostegal spines well developed and corneous; suborbital angle acute with corneous spine. Antennal scale widest distal to midpoint. Chelae with long fingers, dactyl at least two times longer than palm. Dorsal surfaces of fingers with well-developed longitudinal ridges; lateral base of propodus with strong dorsoventral depression. Mesial margin of palm of chela with two rows of tubercles; mesialmost row with 8 to12 tubercles. One to three subpalmar tubercles. Epistomal zygoma strongly arched.

Central projection of gonopod of form I male corneous, recurved greater than 110° to shaft, with strong subapical notch. Mesial process bulbous and tapering distally. Annulus ventralis of female subrhomboidal with deep longitudinal trough through cephalic half, deep fossa and S-shaped sinus.

MAXIMUM SIZE – The maximum known size for the species is 112.5 mm total length.

MOST LIKE – *Cambarus cumberlandensis* is most similar to *C. buntingi* (see MOST LIKE for *C. buntingi*). The feature that most easily distinguishes the two species is the rostrum shape and the marginal spines or tubercles. In *C. cumberlandensis* the rostral margins are thickened and more or less straight, apically ending in an abrupt angle with marginal spines. These spines are strongest in young individuals. The acumen is clearly defined. In *C. buntingi* the rostral margins are rounded apically to form the acumen; in these rounded corners there is a single small tubercle.

DISTRIBUTION AND HABITAT – (Fig. 46) *Cambarus cumberlandensis* is widely distributed in the upper Cumberland River in Kentucky, and is found in that system downstream to the Roaring Fork River in Jackson County, Tennessee (Hobbs and Bouchard 1973). In Kentucky this species is most common below Cumberland Falls, but can be found sporadically just above the falls. *Cambarus cumberlandensis* is also known from several sites in the Little Barren River and Russell Creek of the Green River drainage.

Cambarus cumberlandensis lives in medium to large creeks and rivers, where it is most commonly found in swift-flowing water under very large slab boulders. Hobbs and Bouchard (1973) reported that this species lives under rocks and among debris in creeks from a few meters wide to large rivers. In larger creeks they observed that it lives in the same general habitats as *Orconectes* species.

LIFE HISTORY – In Kentucky form I males have been collected from April through November. An ovigerous female was collected in August and a female with young was collected in September. The ovigerous female measured 101.7 mm total length and carried 261 eggs. Egg diameter averaged 2.7 mm. Hobbs and Bouchard (1973) also reported form I males in November. They ob-

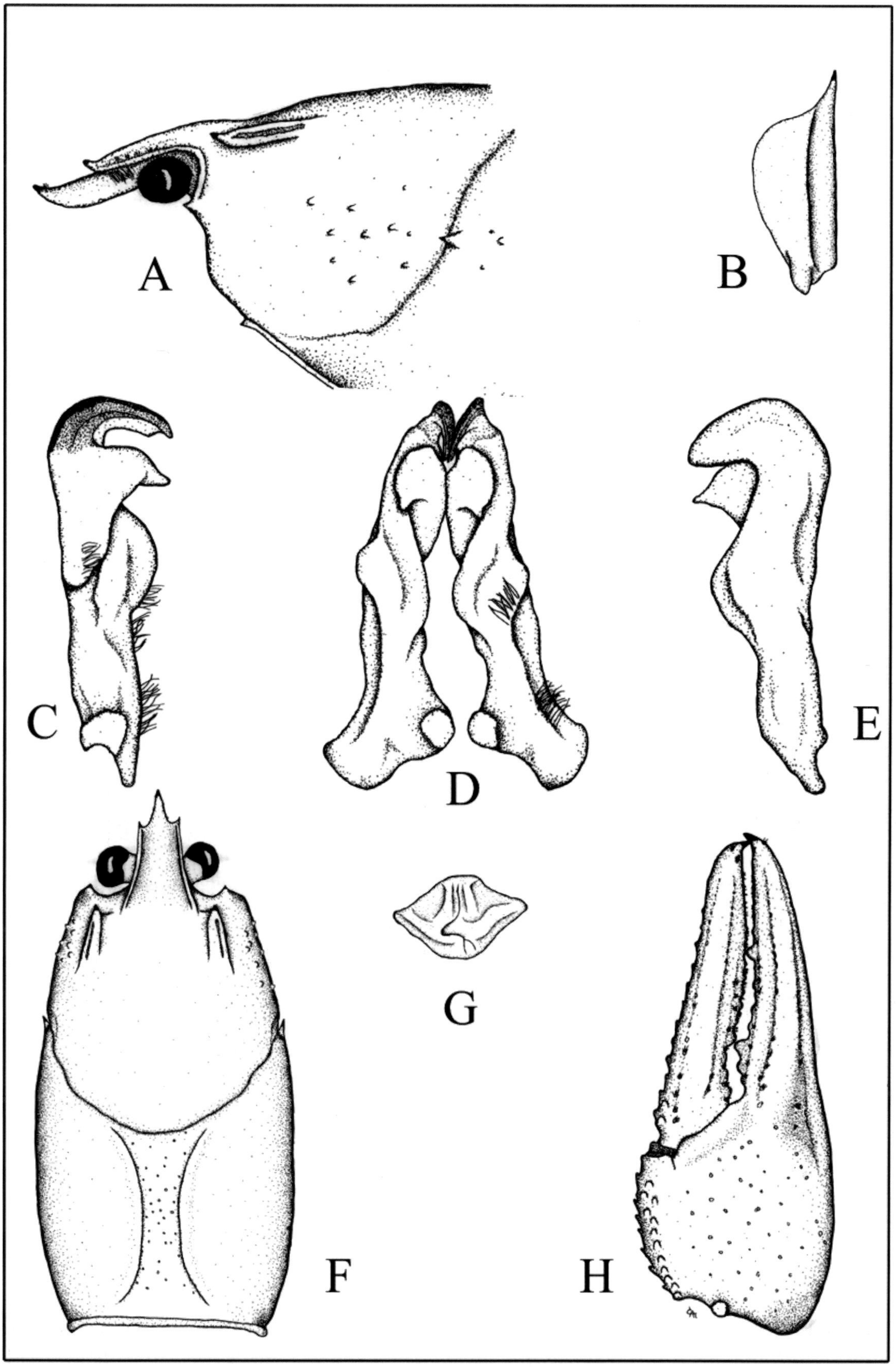

Fig. 45. *Cambarus cumberlandensis*: A, lateral view of carapace; B, dorsal view of right antennal scale; C, mesial view of form I gonopod; D, ventral view of form I gonopods; E, lateral view of form II gonopod; F, dorsal view of carapace; G, ventral view of annulus ventralis; H, dorsal view of right chela.

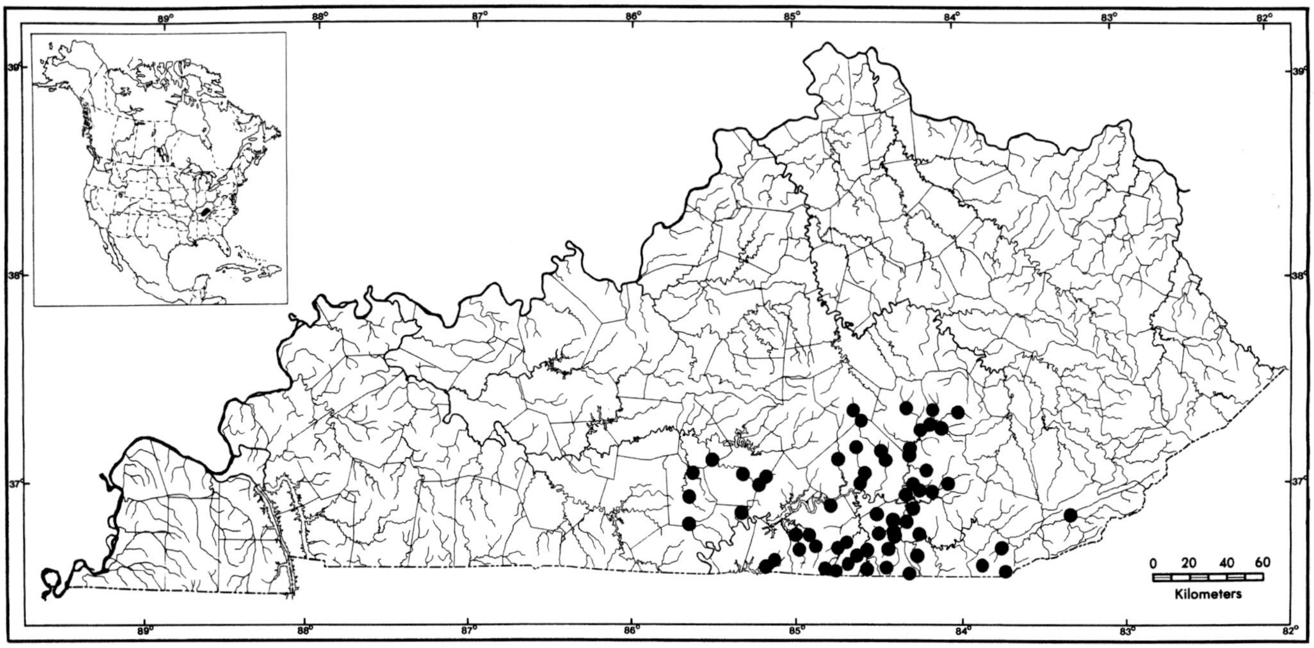

Fig. 46. Distribution of *Cambarus cumberlandensis* in Kentucky.

served ovigerous females during July, August, and September, and collected a female with young in August. Rhoades (1944) reported (as *C. extraneous*) a female with "eyed" eggs on August 15, 1939. Nothing else is known about the life history of this species.

CRAYFISH ASSOCIATES – We have collected *Cambarus cumberlandensis* with *C. distans, C. rusticiformis, C. sphenoides, C. graysoni, Orconectes placidus,* and *O. putnami*. Hobbs and Bouchard (1973) also reported it co-occurring with *C. striatus*.

Cambarus deweesae Bouchard and Etnier Valley Flame Crayfish

Fig. 47A: blue color form.

Fig. 47B: red/brown color form.

LIFE COLORS – (Figs. 47 A, B) Color extremely variable, with several distinct color patterns. Some populations concolorous. General base color from orange to red to a rich brown to a uniform deep blue. Some populations with coloration similar to some *Cambarus dubius* populations: carapace mostly dark blue; chelae blue with bright orange tips; rostrum and pereiopod joints highlighted in orange; legs cream color. Venter cream colored.

CHARACTERS – (Fig. 48) Rostrum deeply excavated; margins moderately thickened, straight, converging anteriorly. Acumen clearly defined, broadly triangular, gradually coming to a point. Carapace from dorsal view bullet shaped; in lateral view both anterior and posterior sections rounded dorsally. Areola generally obliterated, but occasionally open; if open then very narrow, width usually 38.5 to 100 times longer than wide at its narrowest point, no room for punctations. Postorbital ridge not well defined, only a thickening with no anterior tubercle or spine. Hepatic area with numerous small round tubercles; cervical spines absent, two or three small tubercles in the area where spine would be; branchiostegal spines small or absent; suborbital angle absent. Antennal scale widest distal to midpoint. Chelae broadly triangular in outline; longitudinal ridges on both fingers well developed. Mesial margin of palm of chela with two rows of tubercles; mesialmost row with five to seven tubercles, sometimes with basal two fused; second row with four to six tubercles, often incomplete with gaps. One to two subpalmar tubercles present. Epistomal zygoma strongly arched.

Central projection of gonopod of form I male corneous, recurved approximately 90° to shaft, with subapical notch. Mesial process bulbous at base, tapering apically. Annulus ventralis of female subrhomboidal with strong median ridge, deep C-shaped fossa and S-shaped sinus.

MAXIMUM SIZE – The maximum known size for the species is 79.6 mm total length.

MOST LIKE – Depending on individual coloration, specimens of *Cambarus deweesae* can look like several other *Cambarus* species with an obliterated areola. Those that are brown to red in color resemble *C. batchi* but the two can be separated based on the morphology of the chelae. In *C. batchi* the chela is subrectangular with a single row of tubercles on the palm, while in *C. deweesae* the chelae are distinctly triangular in outline with two rows of tubercles. The bluer forms of *C. deweesae* resemble *C. dubius*. However they may be separated by the areola. In *C. dubius* it can be quite narrow but is always open, while in *C. deweesae* the areola is obliterated. *Cambarus dubius* also lacks a second row of tubercles on its palm.

DISTRIBUTION AND HABITAT – (Fig. 49) *Cambarus deweesae* until recently was thought to have a very restricted range in the Clinch and Emory River drainages of northwestern Tennessee (Bouchard and Etnier 1979). However, in

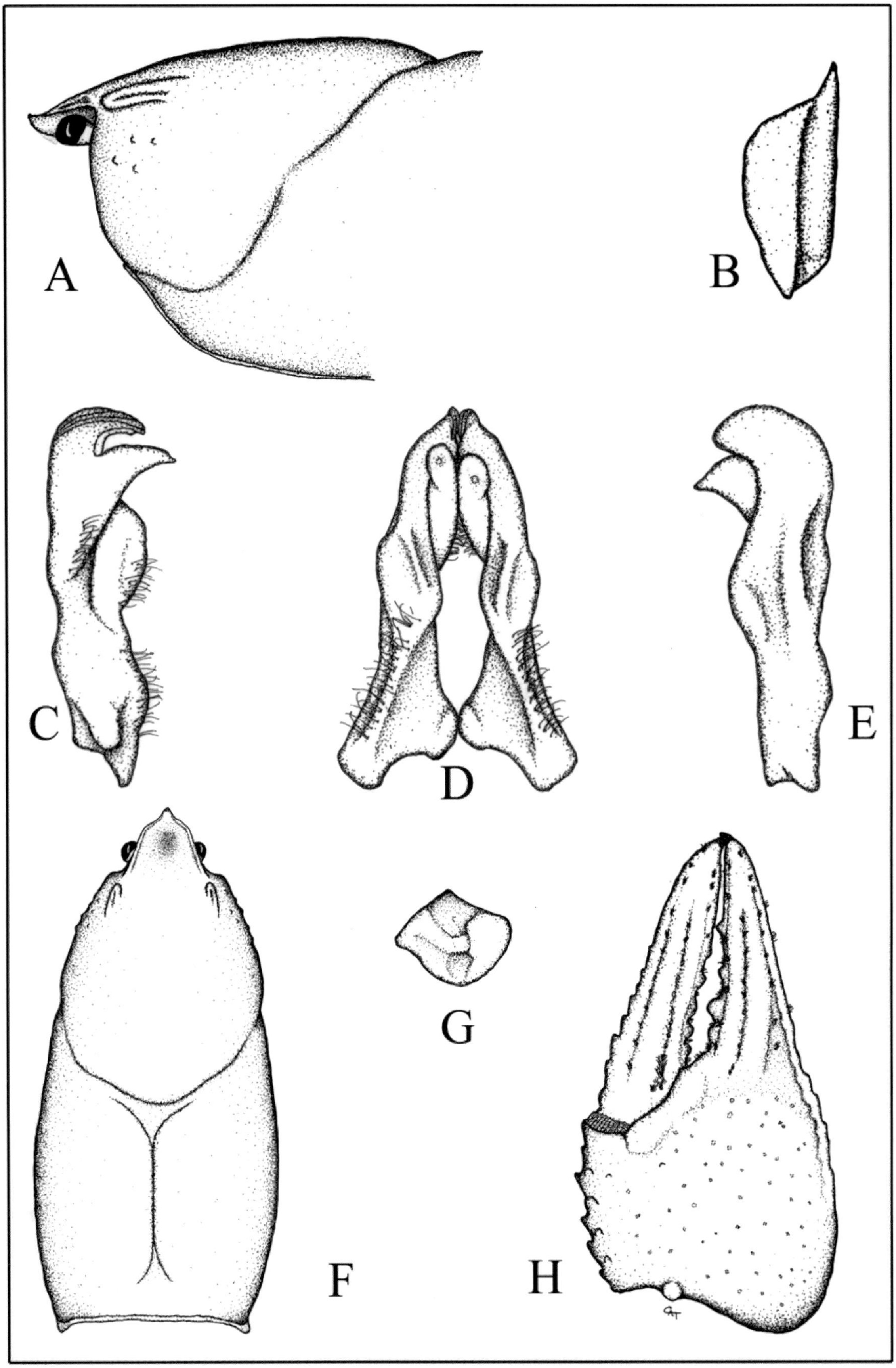

Fig. 48. *Cambarus deweesae*: A, lateral view of carapace; B, dorsal view of right antennal scale; C, mesial view of form I gonopod; D, ventral view of form I gonopods; E, lateral view of form II gonopod; F, dorsal view of carapace; G, ventral view of annulus ventralis; H, dorsal view of right chela.

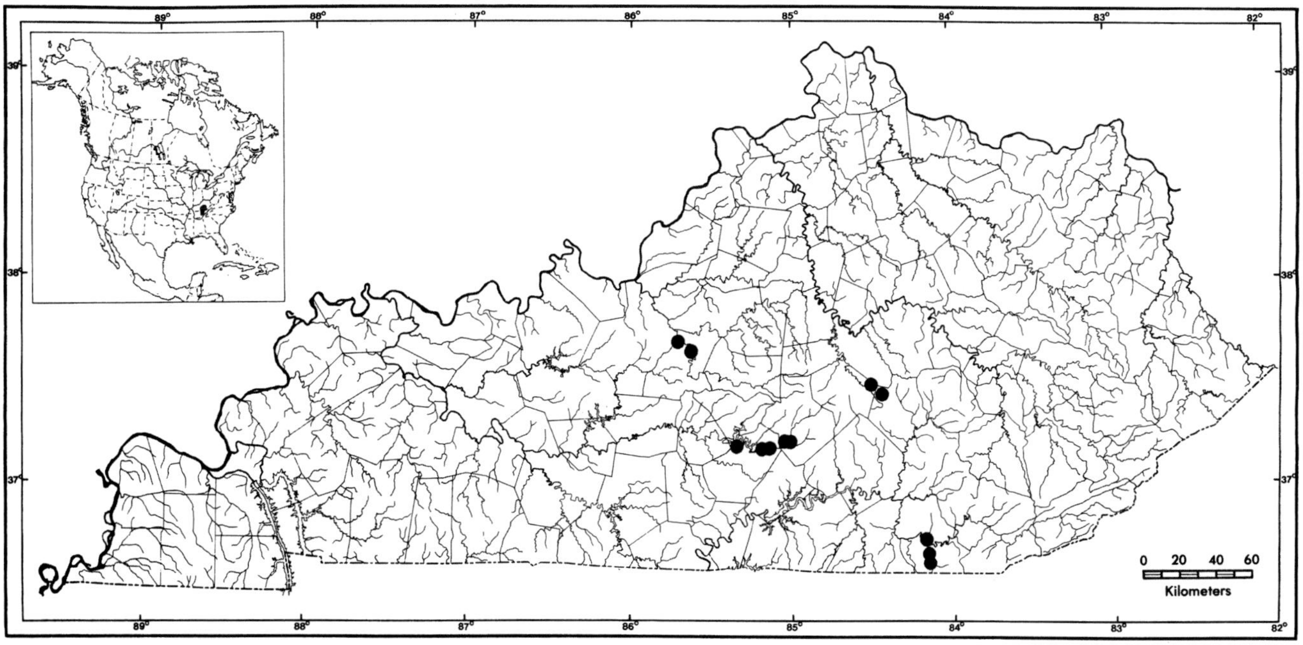

Fig. 49. Distribution of *Cambarus deweesae* in Kentucky.

recent years it has been widely but sporadically collected in central and southeastern Kentucky. It is known in Kentucky from low-lying fields within the Cumberland, Green, Salt, and Kentucky River drainages.

Cambarus deweesae is a primary burrower that most often has been found in fields composed of hydric soils where the water table is close to the surface. In Kentucky these fields usually lie next to creeks or wetland complexes. Bouchard and Etnier (1979) described the type locality in Tennessee as a poorly drained floodplain with clay soils interspersed with fragments of shale.

LIFE HISTORY – In Kentucky form I males of *C. deweesae* have been collected in April, May, October, November, and December. No ovigerous females have been collected. Bouchard and Etnier (1979) reported form I males from March. They also collected an ovigerous female in March. Nothing else is known about the life history of this species.

Bouchard and Etnier (1979) reported the burrows of *C. deweesae* to be complex and variable without well-defined chimneys. They indicated that the burrows typically have one large opening with a vertical shaft and one or more smaller openings that lead to horizontal cul-de-sacs. We have found the burrows to typically have chimneys and be 0.5 to 1.5 m in depth. Bouchard and Etnier (1979) reported that each burrow usually has only a single individual in it; we have made the same observation. We have never collected the species from open surface waters, although during spring and fall rains it has been collected at night, like *C. batchi* and *C. dubius*, walking on roads adjacent to the fields where burrows are found.

CRAYFISH ASSOCIATES – The only other species we have collected from burrows in the vicinity of *C. deweesae* burrows has been *C. ortmanni*.

CONSERVATION STATUS – AFS Endangered. The status should be reevaluated given the discovery of new locations across central and southeastern Kentucky.

Cambarus diogenes Girard
Devil Crayfish

Fig. 50

LIFE COLORS – (Fig. 50) Overall base color generally olive green to brown. Rostrum, tips of the chelae, telson, and uropods trimmed in deep burgundy. All leg joints and abdomen with burgundy trim. Juveniles with similar coloration; abdomen sometimes with three bright dorsal longitudinal stripes. Venter cream to whitish.

CHARACTERS – (Fig. 51) Rostrum deeply excavated, elongated and narrow; margins straight. Acumen very short, with small tubercle at tip. Carapace from dorsal view somewhat laterally compressed; head narrower than thorax. Areola obliterated. Postorbital ridge reduced and thickened with small rounded anterior tubercle. Hepatic area covered with several moderately large tubercles; cervical spines absent, but with five to eight small tubercles; branchiostegal spines small, but developed; suborbital angle usually well developed, sometimes rounded without spine. Antennal scale widest distal to midpoint. Chelae usually large and heavy; dactyl just longer than palm length. Dorsal surfaces of fingers of chelae with well-developed longitudinal ridges that have numerous tubercles, especially basally. Dactyl with broad notch at base, most prominent in form I males. Mesial margin of palm of chela with two complete rows of tubercles, a complete angled third row leaving a triangular area on the palm free of tubercles, also with scattered tubercles over main body of chela. The mesialmost row of tubercles with five to seven tubercles; second row with four to seven tubercles. Usually one to three subpalmar tubercles. Epistomal zygoma usually weakly rounded.

Central projection of gonopod of form I male corneous, recurved approximately 90° to shaft, without subapical notch. Mesial process bulbous basally, tapers to point apically. Annulus ventralis of female with cephalic margin broadly rounded and caudal margin triangular; with deep horizontal fossa, S-shaped sinus.

MAXIMUM SIZE – The maximum known size for the species is 126.3 mm total length.

MOST LIKE – By possessing an obliterated areola and a broad concavity at the base of the opposable margin of the dactyl, *Cambarus diogenes* is most like *C. ludovicianus, C. thomai,* and *C.* sp. A in general appearance. It may be separated from the latter two species by the presence of two complete rows of tubercles and a triangular area lateral to the second row that is devoid of tubercles or may have one tubercle. *Cambarus thomai* and *C.* sp. A have numerous tubercles that cover the entire surface of the palm without any areas devoid of tubercles. *Cambarus* sp. A is often found at the same location with *C. diogenes* but they are much more colorful than *C. diogenes. Cambarus diogenes* differs from *C. ludovicianus* in the length of the areola and shape of the abdomen. In *C. diogenes* the cephalic section of carapace is more than 1.4 times length of areola, while in *C. ludovicianus* the cephalic section of carapace is 1.4 times or less than the length of areola. The abdomen of *C. diogenes* is also more strongly developed than in *C. ludovicianus.* In a lateral view the abdomen of *C. diogenes* is much thicker than in *C. ludovicianus* and the lateral edge of the pleuron of the second abdominal segment is arched in *C. diogenes* while in *C. ludovicianus* it is straight.

DISTRIBUTION AND HABITAT – (Fig. 52) *Cambarus diogenes* is widely distributed east of the Rocky Mountains and south of the Great Lakes (Hobbs 1989). In Kentucky this species is generally distributed in the western half of the state.

Cambarus diogenes is most commonly encountered in burrows in or along creek banks but may frequently be collected in woody debris and submerged vegetation in creeks. *Cambarus diogenes* can be considered to be both a primary and secondary burrower (Penn and Marlow 1959).

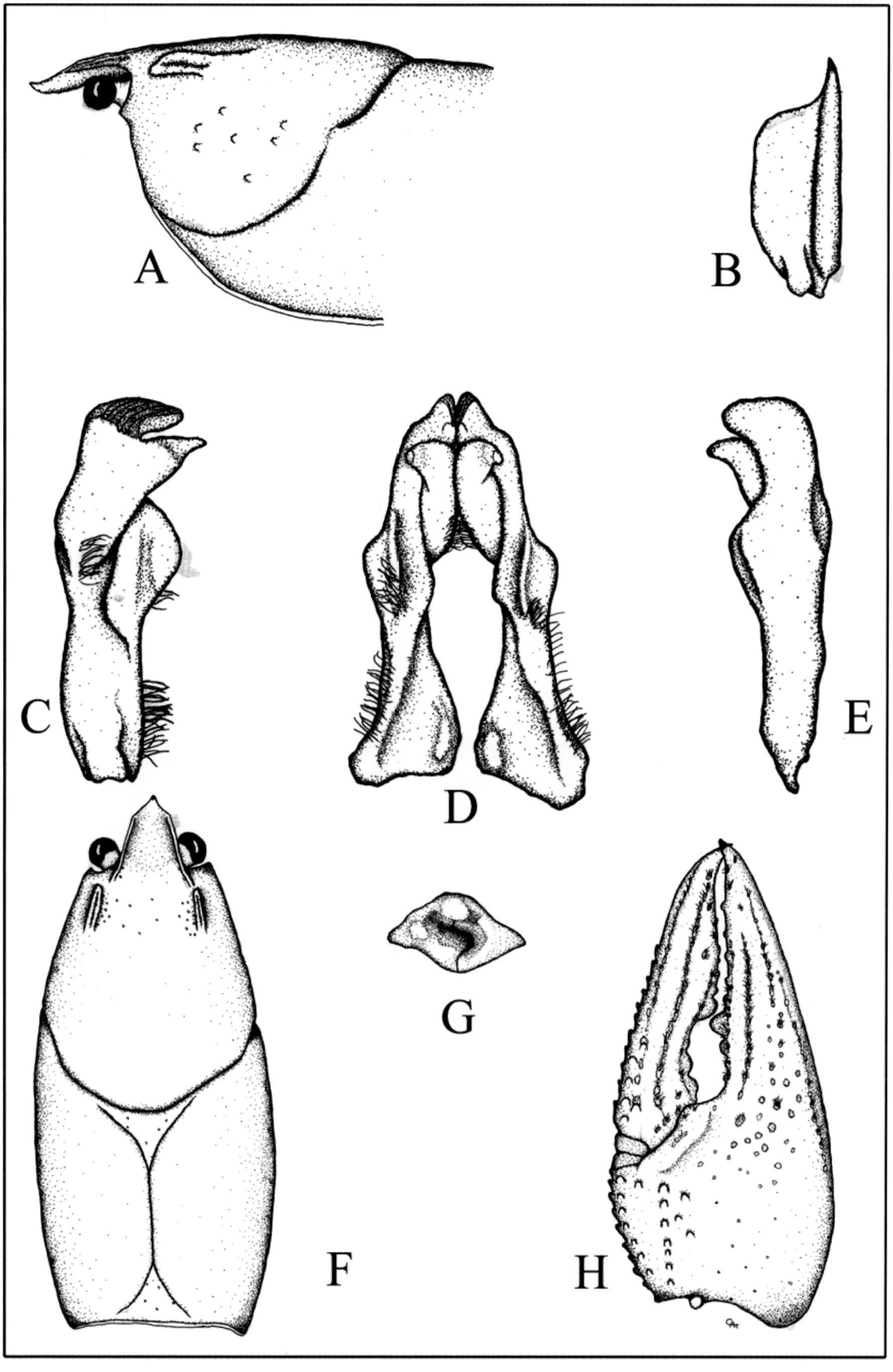

Fig. 51. *Cambarus diogenes*: A, lateral view of carapace; B, dorsal view of right antennal scale; C, mesial view of form I gonopod; D, ventral view of form I gonopods; E, lateral view of form II gonopod; F, dorsal view of carapace; G, lateral view of abdomen with arrow denoting lateral edge of pleuron; H, ventral view of annulus ventralis; I, dorsal view of right chela.

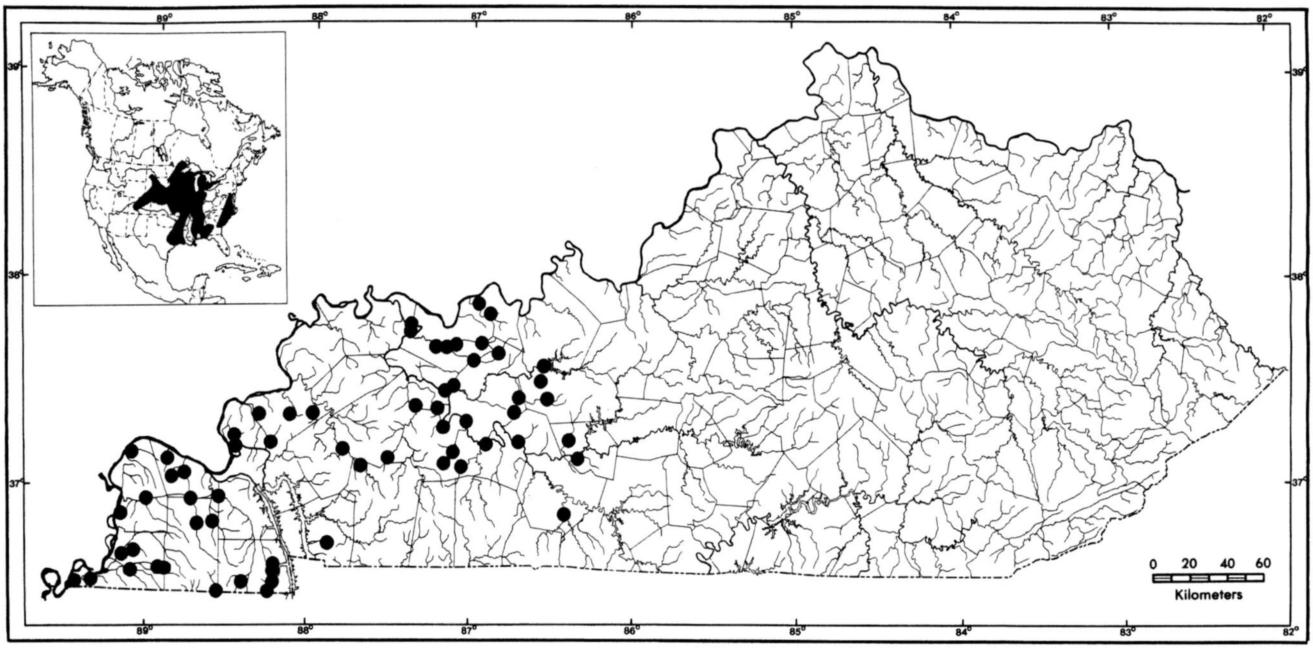

Fig. 52. Distribution of *Cambarus diogenes* in Kentucky.

LIFE HISTORY – In Kentucky form I males have been collected in March, April, and October. Hobbs (1942) reported first form males from Florida for the months of April, May, October, November, and December with a single female with young being found in May. Hobbs (1981) reported first form males from Georgia and Alabama occurring during the months of January, March, April, May, and October with a single ovigerous female with 40 eggs found in September. Penn and Marlow (1959) reported form I males from March through October. Hobbs and Marchand (1943) reported ovigerous females from Reelfoot Lake in western Tennessee during March and April and females with young in May. We collected two ovigerous females on April 8 and 9, 2001. The first female had a total length of 118.2 mm. It was carrying 218 eggs with a mean diameter of 2.7 mm. The second female had a total length of 114.8 mm, and was carrying 291 eggs with a mean diameter of 2.6 mm.

In summarizing the reproductive biology of *Cambarus diogenes* in Wisconsin, Hobbs III and Jass (1988) reported that mating occurs within burrows in late fall and winter and that females lay their eggs within their burrows in early spring. From March to May females will leave their burrows and move to open water to release their young. In West Virginia Newcombe (1929) reported that by mid-summer juveniles of the closely related *C. thomai* were beginning to construct their own burrows. The limited amount of data that we have for Kentucky specimens and our field observations suggest that Kentucky populations of *C. diogenes* have a reproductive cycle similar to that reported by Hobbs III and Jass (1988) and Newcombe (1929) for *C. thomai.*

Hobbs III and Jass (1988) reported that burrows occur singly or in colonies alongside water bodies. Burrows will commonly have two or occasionally three entrances that converge 0.33 m from the surface and then descend vertically, terminating in a flask-shaped chamber. They reported burrow depths of up to 1.5 m. In Kentucky we have observed similar burrow morphology.

Hobbs (1981) indicated that most of the ovigerous females he has collected of *C. diogenes* were found in open water; however, he reported that almost all other adults were collected from burrows. We have had a similar experience in our collection of this species in Kentucky. We have also commonly collected juveniles in open water (creeks and marshes), which was also reported by Hobbs (1981) and Penn and Marlow (1959). Most creeks, marshes, and swamps of western Kentucky have the banks lined with the burrows of the devil crayfish. Individual crayfish that burrow like this are considered to be secondary burrowers since they are often found in the adjacent water body, especially at night. It is not uncommon during the spring and fall reproductive seasons to see large females in debris dams in the water body even during the day. Large populations of *C. diogenes* are also common in fields adjacent to surface water.

CRAYFISH ASSOCIATES – The following species have been collected from sites containing *Cambarus diogenes*: *C.* sp. A, *C. graysoni*, *C. ludovicianus*, *C. striatus*, *C. tenebrosus*, *Fallicambarus fodiens*, *Orconectes barrenensis*, *O. compressus*, *O. immunis*, *O. lancifer*, *O. putnami*, *O. rafinesquei*, *O. ronaldi*, *O. rusticus*, *Procambarus acutus*, *P. clarkii*, and *P. viaeviridis*.

Cambarus distans Rhoades
Boxclaw Crayfish

Fig. 53

LIFE COLORS – (Fig. 53) Overall base color light to dark brown. Carapace uniformly colored, usually a rich medium brown. Chelae uniform golden color. Abdomen dark brown, often with darker markings to outline segments. Telson and uropods outlined in gold trim. Legs light cream in color. Venter cream to whitish.

CHARACTERS – (Fig. 54) Rostrum moderately excavated to flat; margins moderately thick, straight, rounded anteriorly to form acumen. Acumen clearly defined with corneous tubercle at tip. Carapace from dorsal view bullet shaped. Areola very wide, usually 3.4 to 4.2 times longer than wide at its narrowest point, with room for six to eight punctations; moderately punctate. Postorbital ridge flattened, anteriorly with or without small corneous tubercle. Hepatic area covered with small to moderately large tubercles; cervical spines usually absent, sometimes with one to three small tubercles in area; branchiostegal spines small but distinct; suborbital angle acute, sometimes with small corneous tubercle. Antennal scale widest distal to midpoint. Chelae stocky and rectangular in shape, covered with clumps of setae, and with rows of setae between ridges on fingers; apex of fingers with dense erect clumps of setae. Dorsal surfaces of both fingers of chelae with well-developed longitudinal ridges. Mesial margin of palm of chela with single complete row of tubercles, and second incomplete row with one to three very weak tubercles; six to nine large tear-shaped tubercles in mesialmost row, basal two to three fused to form ridge. Zero to four subpalmar tubercles on chela. Epistomal zygoma strongly arched.

Central projection of gonopod of form I male corneous, recurved to or slightly more than 90° to shaft, with strong subapical notch. Mesial process bulbous basally, tapering apically to a point. Annulus ventralis of female subrhomboidal, caudally rounded; with deep C-shaped fossa and distinct S-shaped sinus.

MAXIMUM SIZE – The maximum known size for the species is 83.2 mm total length.

MOST LIKE – With a very wide areola and stocky chelae, *Cambarus distans* most closely resemble *C. parvoculus*, and in first and second order creeks may be found together. *Cambarus parvoculus* tends to be a much smaller crayfish than *C. distans* as adults, often about half the size. The main morphological difference is in the curvature of the central projection of form I males. In *C. distans* the central projection is recurved approximately 90° to the shaft, while in *C. parvoculus* it is recurved much greater than 90°. In addition, the thickened rostral margins and square shape of the rostrum in *C. parvoculus* separates it from *C. distans*, where the margins of the rostrum are not as thickened and the rostrum is more elongated. Living specimens of *C. parvoculus* often have light blue-colored legs. This coloration has not been seen in *C. distans*.

DISTRIBUTION AND HABITAT – (Fig. 55) *Cambarus distans* has a distribution that ranges from eastern Kentucky south along the Appalachian uplands to northern Georgia and west to Sand Mountain in northeastern Alabama

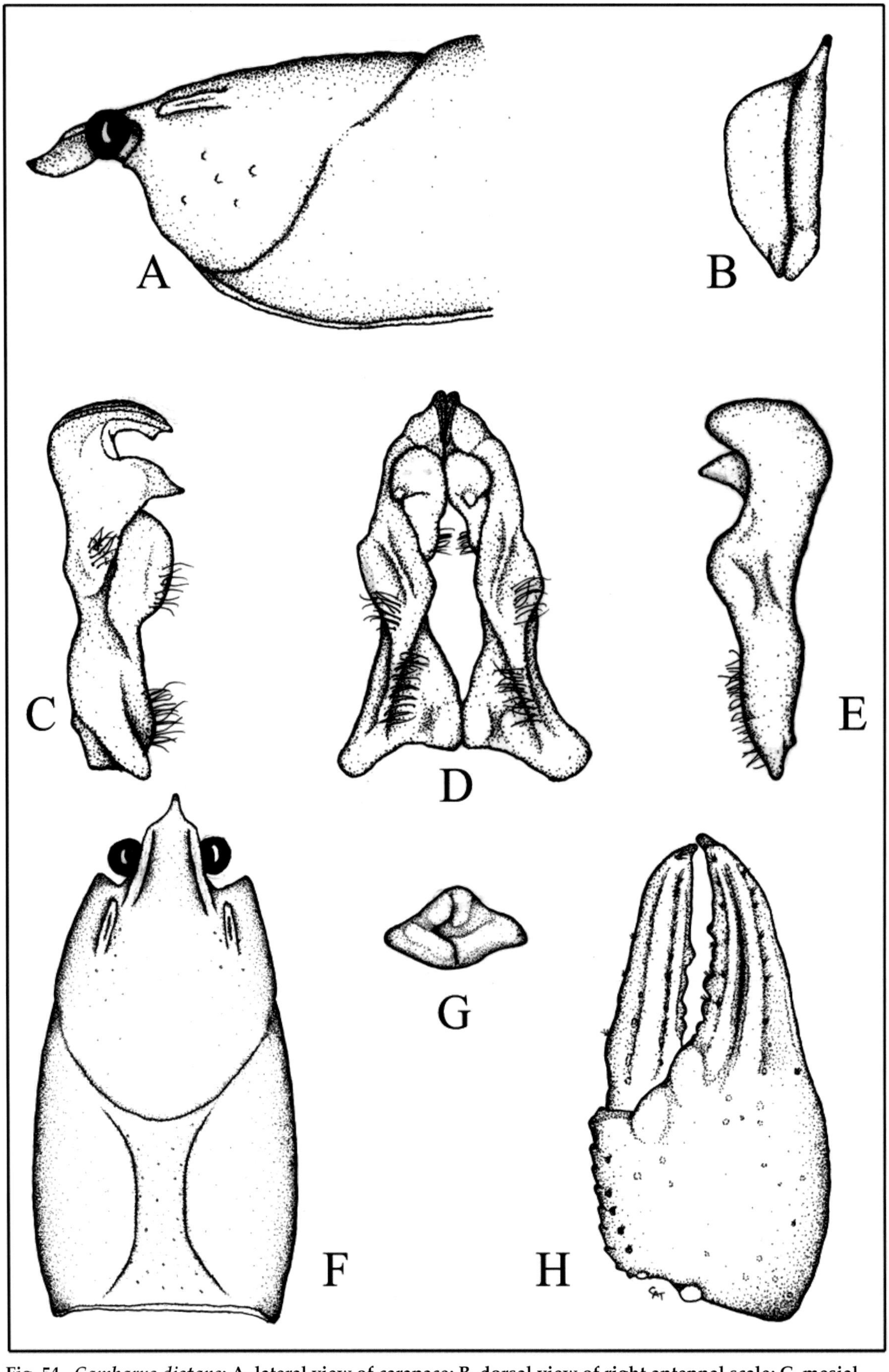

Fig. 54. *Cambarus distans*: A, lateral view of carapace; B, dorsal view of right antennal scale; C, mesial view of form I gonopod; D, ventral view of form I gonopods; E, lateral view of form II gonopod; F, dorsal view of carapace; G, ventral view of annulus ventralis; H, dorsal view of right chela.

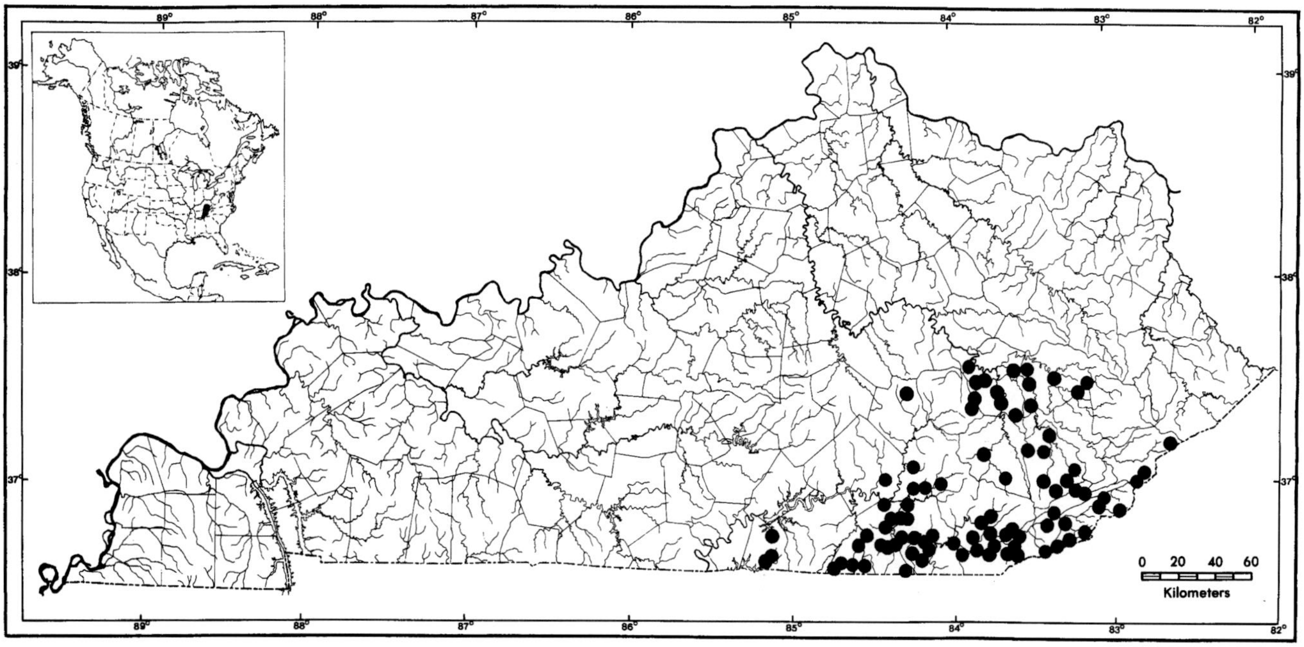

Fig. 55. Distribution of *Cambarus distans* in Kentucky.

(Bouchard 1976b, Hobbs 1989). In Kentucky it is found in the upper Cumberland and Kentucky River drainages.

Cambarus distans is common in small to medium montane creeks with clean rocky substrates and clear water. In these creeks it is found under cobble in riffles. Specimens can occasionally be found along the margins of creeks, but they seem to prefer the faster-flowing areas of the creek. If caught in receding pools, *C. distans* will make shallow single-chamber burrows under large rocks.

LIFE HISTORY – In Kentucky form I males have been collected monthly from March through December. Ovigerous females have not been collected from Kentucky or reported in the literature from other states. Nothing else is known about the life history of this species.

CRAYFISH ASSOCIATES – *Cambarus distans* has been collected with the following species: *C. bouchardi, C. cumberlandensis, C. parvoculus, C. rusticiformis,* and *Orconectes cristavarius.*

56A: red–orange to orange color form

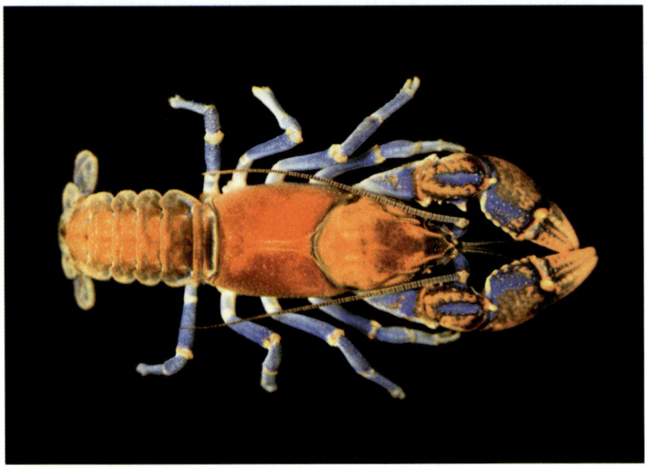

56B: red–brown with blue legs color form.

56C: blue color form.

56D: red claw with black body form.

Cambarus dubius Faxon
Upland Burrowing Crayfish

LIFE COLORS – (Figs. 56A, B, C, D) The color pattern of *Cambarus dubius* is extremely variable and has been a taxonomic issue for quite some time (Dewees 1972). Local populations of *C. dubius* usually are composed of similarly colored individuals, which was also noted by Dewees (1972); however, there is considerable interpopulation variation. In her detailed study of this species, Dewees (1972) described three distinct color patterns: 1) red-orange to orange; 2) blue with tips of chelae yellow to red-orange; and 3) red-brown with bright blue legs. Jezerinac et al. (1995) described three color forms from West Virginia: 1) orange or red body form; 2) blue body form; and 3) red claw, black body form. It is apparent from these two studies that these color variations represent four distinct color forms of *C. dubius*. We have seen all four of these color variations in Kentucky, plus a fifth color form described below.

The color pattern that is most common and widespread, and perhaps most associated with *C. dubius*, has a bright navy blue base color on the carapace, abdomen, and chelae, while the legs are lighter blue with the joints accented in red or orange (Fig. 56C). In this color form the chelae are highlighted in bright orange to red and the rostrum is trimmed in orange to red. Individuals like this have been widely collected from upland areas of southeastern and south-central Kentucky. A second color pattern which Dewees (1972) called the "red-brown with bright blue legs" form, has been found in McCreary County (Fig. 56B). It has an orange to red base color on the carapace, abdomen and chelae, while the legs are blue and the joints are highlighted in red. The chelae have patches of blue but are mostly red. A third color pattern, which Jezerinac et al. (1995) called the "red claw, black body color form," has a base color of orange to red on the carapace, abdomen, and chelae with the dorsal surface of the carapace and chelae royal blue to black (Fig. 56D). The legs are creamy in color with some red highlighting. This color pattern is found in the mountains of far eastern Kentucky. The fourth color pattern, which Dewees (1972) referred to as the "red-orange to orange" form, has a more or less uniform bright orange to brick red color, with the legs and venter light orange to cream (Fig. 56A). This form has been found in the Licking River drainage. Dewees (1972) also remarked that this is the color pattern that was found at the type locality. The fifth and previously undescribed color pattern is found in individuals from Lincoln County. This color morph has a uniform color of olive to dark green on the carapace, abdomen, and chelae. The chelae are tipped in yellow to orange. The legs are lighter green in color and the joints are highlighted in yellow.

In the field these color patterns can be problematic in identifying this species. It is possible that this is a complex of sibling species that are difficult to distinguish using morphological characters (Jezerinac et al. 1995), or according to Dewees (1972), these morphs may represent different subspecies of *C. dubius*. For the time being we are treating all of these color variations as those of a single taxon, *Cambarus dubius*.

CHARACTERS – (Fig. 57) Rostrum deeply excavated; margins thickened, straight, slightly converging, rounding anteriorly to form acumen. Acumen short, margins converging to corneous tubercle at tip. Carapace in dorsal view bullet shaped and relatively narrow; in lateral view both anterior and posterior sections dorsally rounded. Abdomen distinctly narrower than carapace. Areola narrow, usually 8.2 to 20.0 times longer than wide at the narrowest point, with room for one to two punctations, sparsely punctated. Postorbital ridge flattened with no anterior tubercle or spine. Cervical spines absent, with one or two small tubercles in area; branchiostegal spines very small, occasionally only very small tubercle present; suborbital angle obtuse with small to well-developed tubercle at angle. Antennal scale greatest width distal to midpoint. Chelae with both fingers straight, rectangular in outline; fingers of chelae with well-developed longitudinal ridges. Mesial margin of palm of chela with one complete row of six to eight small tear-shaped tubercles, basal two or three fused to form ridge; second row incomplete with one to three tubercles. No subpalmar tubercles. Epistomal zygoma strongly arched.

Central projection of gonopod of form I male corneous, recurved about 90° to shaft, with subapical notch. Mesial process bulbous at base, tapering to point. Annulus ventralis of female subrhomboidal with deep C-shaped fossa and S-shaped sinus.

MAXIMUM SIZE – The maximum known size for the species is 89.5 mm total length.

MOST LIKE – In Kentucky the species that most resembles *Cambarus dubius* is *C. deweesae*. Although these two species have considerable variation in color patterns, they also have patterns that resemble each other. The easiest way to distinguish these species from one another is in the form of the areola. In *C. dubius* the areola is always open. It may be narrow but there is always space for one to three punctations. In *C. deweesae* the areola is obliterated with no space for punctations. In addition, the chelae of the two species differ. The chelae of *C. dubius* are rectangular in shape when closed, while those of *C. deweesae* are triangular in shape when closed. *Cambarus deweesae* also has two complete rows of tubercles on the mesial margin of the palm chelae, while *C. dubius* has either one complete row or one compete row and a second incomplete row of three to four scattered tubercles.

DISTRIBUTION AND HABITAT – (Fig. 58) *Cambarus dubius* is known from western Pennsylvania, West Virginia, eastern Kentucky, southwestern Virginia, northwestern North Carolina, and eastern Tennessee (Dewees 1972; Hobbs 1989). It is generally distributed in southeastern Kentucky and is known from the Kentucky, Licking, Cumberland, and Big Sandy River drainages.

Cambarus dubius is a primary burrower that is most often associated with upland topography. Its burrows are commonly encountered in relatively flat fields, mountain-top bogs, in the drip zone under rock houses, and along the margins of first and second order creeks. This is one of the most difficult burrowers to collect. The burrows are often constructed in very rocky soils, and sometimes where there is only rock, they will

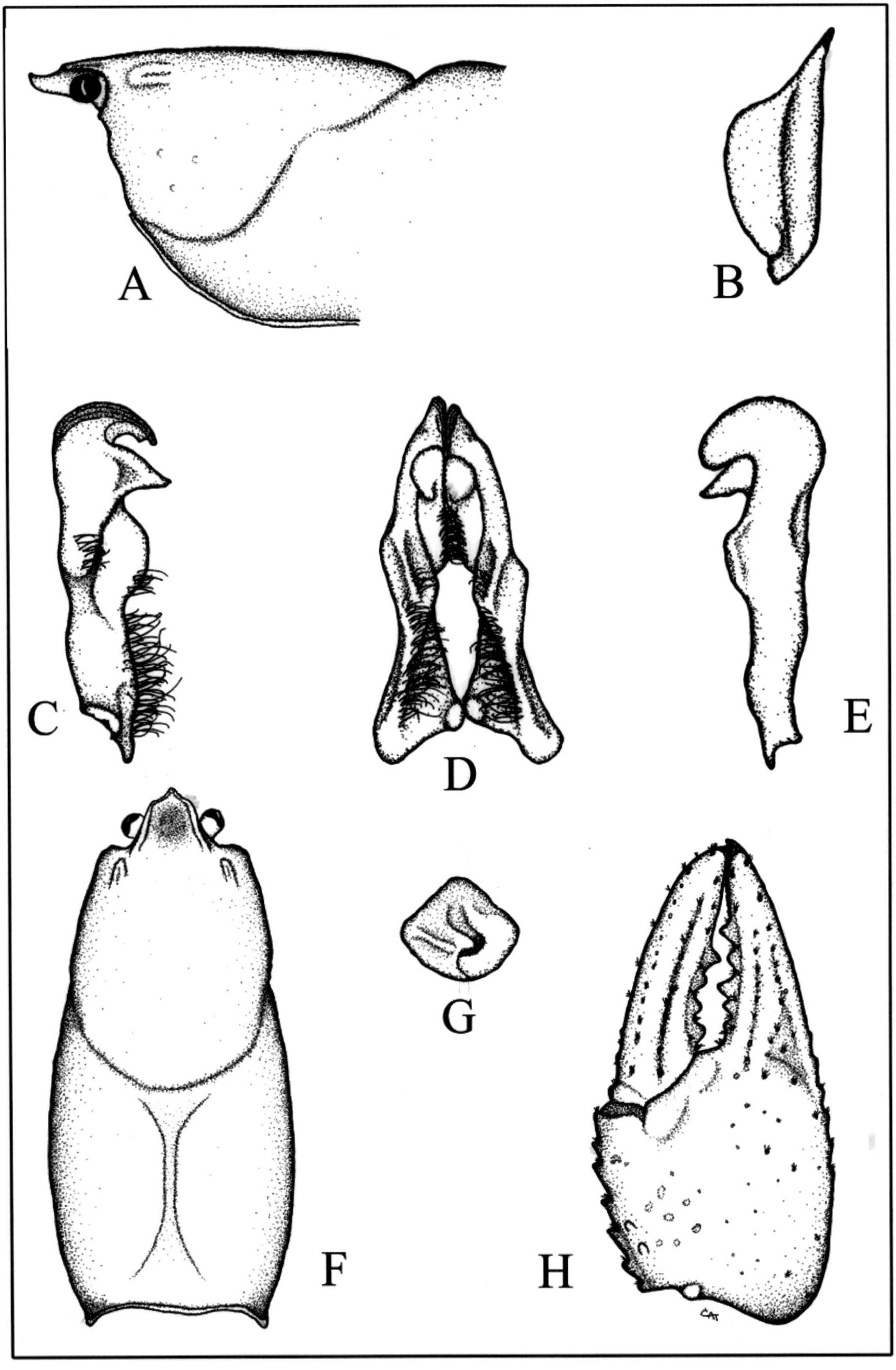

Fig. 57. *Cambarus dubius*: A, lateral view of carapace; B, dorsal view of right antennal scale; C, mesial view of form I gonopod; D, ventral view of form I gonopods; E, lateral view of form II gonopod; F, dorsal view of carapace; G, ventral view of annulus ventralis; H, dorsal view of right chela.

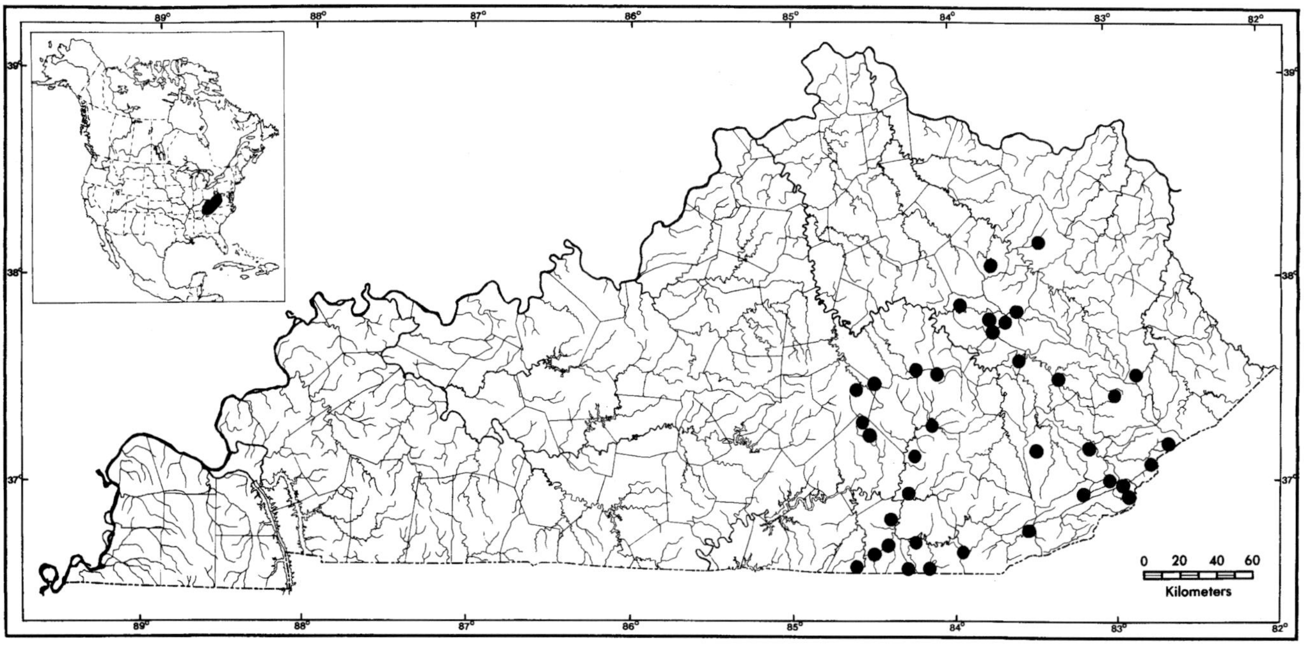

Fig. 58. Distribution of *Cambarus dubius* in Kentucky.

follow the fissures between the rocks. Under rock houses they are commonly found in very sandy soil that is rocky, or if made of pure sand, the burrows will collapse upon excavation.

LIFE HISTORY – In Kentucky form I males occur in February, April, May, July, August, September, October, and November. Dewees (1972) found form I males throughout the range of the species in all months except January and February, and surmised that they are present year-round. She also deduced that adult males of *C. dubius* molted twice annually, corresponding to their mating season. She collected ovigerous females ranging in size from 23.6 to 35.7 mm postorbital carapace length from April through July, described the eggs as salmon to amber in color and ranging from 2.0 to 2.6 mm in diameter. Jezerinac et al. found eggs in West Virginia *C. dubius* ranged from 1.4 to 2.1 mm in diameter. Dewees (1972) found females carried from 8 to 91 with a mean of 34.2 eggs per female (N=20), while Jezerinac et al. (1995) reported that the number of eggs ranged from 3 to 74 eggs per female. Dewees found that females generally molted shortly after the young became independent, usually within a month after hatching, and that the young remained attached to the mother for about a week after hatching. Independent young stayed close to the female for about 10 days after independence. She also found larger young in side tunnels off the main burrow and that in laboratory studies individuals of *C. dubius* did not reach sexual maturity until late into their second year of growth. Jezerinac et al. (1995) estimated longevity in the species to be three years or more. Dewees (1972) referred to *C. dubius* as an omnivorous scavenger, and Jezerinac et al. (1995) observed "stored leaves" in burrows implying that the crayfish consumed these leaves. Jezerinac et al. (1995) referred to this species as being "rather social" because they often found more than one adult per burrow. Dewees (1972) found them in complex burrow systems that were often interconnected and had several adult individuals, including ovigerous females, in the same complex. The burrows that we have excavated in Kentucky always had only one adult.

CRAYFISH ASSOCIATES – The rocky soils where *Cambarus dubius* occurs are generally unsuitable for other crayfish species. However, in lower-lying adjacent areas we have found *Cambarus thomai* and *C. ortmanni*. We have also found in adjacent surface waters the following crayfish species: *Cambarus parvoculus*, *C. distans*, *C. buntingi*, and *Orconectes cristavarius*.

Cambarus friaufi Hobbs
Hairy Crayfish

Fig. 59

LIFE COLORS – (Fig. 59) Overall base color golden-yellow to chestnut-brown. Lateral surfaces of carapace darker brown than rest of carapace. Chelae and all legs somewhat lighter in coloration. Abdomen green to brown with reticulated darker pattern. Venter cream to whitish.

CHARACTERS – (Fig. 60) Rostrum deeply excavated and triangular; margins greatly thickened, mesial row of long setae parallel to margins. Acumen short. Carapace in dorsal view bullet shaped; in lateral view slightly curved anteriorly. Areola very wide, width usually 3.6 to 4.5 times longer than wide at narrowest point, room for three to five punctations, sparsely punctated. Postorbital ridge flattened, not ending in tubercle or spine. Hepatic area with moderate number of small rounded tubercles; cervical spines absent, small tubercles present; branchiostegal spines small but distinct; suborbital angle acute ending in rounded tubercle. Antennal scale greatest width proximal to midpoint. Chelae subrectangular in outline; palm rounded and inflated; covered with long hairs dorsally and ventrally; fingers short and stubby, margins smooth with no tubercles. Dorsal surfaces of fingers of chelae with single well-developed longitudinal ridge. Mesial margin of palm of chela with single complete row of six to eight indistinct, low, and flattened tubercles, compressed mesially to form ridge; no subpalmar tubercles. Epistomal zygoma narrow and strongly arched.

Central projection of gonopod of form I male corneous, recurved approximately 90° to shaft, without subapical notch; well-defined corneous knob at lateral base of central projection. Mesial process fingerlike, hard basally, bulbous, and membranous apically. Annulus ventralis of female triangular cephalically, rounded caudally, with deep mesial trough and shallow fossa and sinus.

MAXIMUM SIZE – The maximum known size for the species is 42.3 mm total length.

MOST LIKE – The long setae on the chelae of *C. friaufi* are found on no other species of *Cambarus* occurring in Kentucky. Its small size, distinctive rostrum, corneous knob on the gonopod, and restricted distribution all aid in the identification of this species.

DISTRIBUTION AND HABITAT – (Fig. 61) *Cambarus friaufi* occurs in the Cumberland and Tennessee River drainages of the Highland Rim of western Tennessee and southern Kentucky (Bouchard and Bouchard 1995). In Kentucky the species is rare, occurring in the upper reaches of the Barren River drainage in Allen County and in the Little River, a tributary of the Cumberland River in Trigg County.

This species is found in small creeks and rivers with substrates of coarse, cherty gravel to small cobble, where individuals make shallow excavations into the substrate in shallow riffles. Members of the subgenus *Glarecola* are the only known species of *Cambarus* that utilize this type of habitat (Bouchard and Bouchard 1995). It is a habitat normally associated with *Orconectes* species such as *O. compressus*.

LIFE HISTORY – We have collected form I males in April and May. A single ovigerous female that measured 31.1 mm total length was collected in Christian County on May 3, 1988, with 29 eggs. The eggs had a mean diameter of 2.0 mm. Another ovigerous female with a total length of 32.0 mm from Hickman County, Tennessee that was collected on March 26, 1998, had 35 eggs with a mean diameter of 1.9 mm. This implies that the young emerge from late March through at least

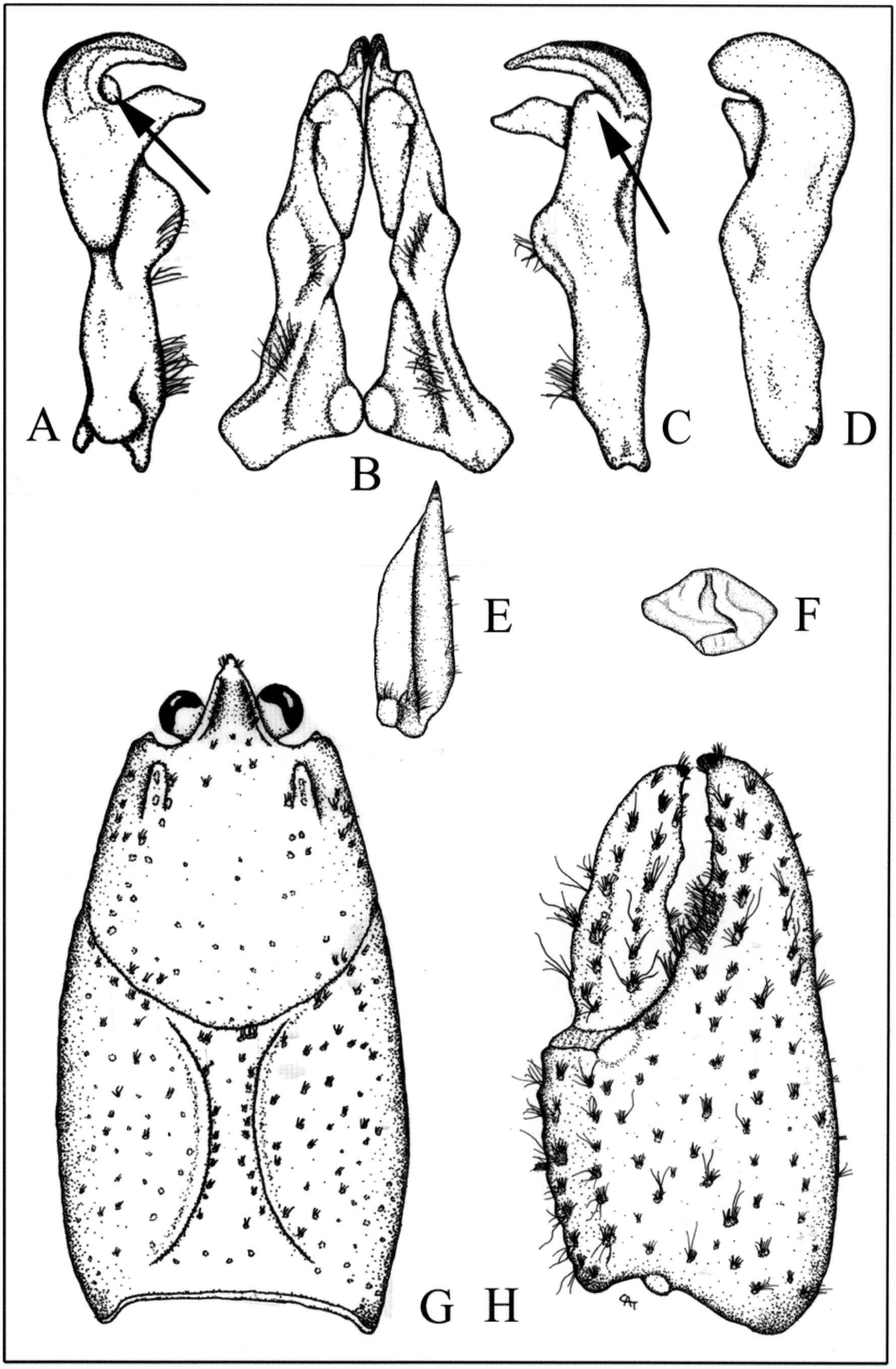

Fig. 60. *Cambarus friaufi*: A, mesial view of form I gonopod with arrow denoting corneous knob; B, ventral view of form I gonopods; C, lateral view of form I gonopod with arrow denoting cornous knob; D, lateral view of form II gonopod; E. ventral view of right antennal scale; F, ventral view of annulus ventralis; G, dorsal view of carapace; H, dorsal view of right chela.

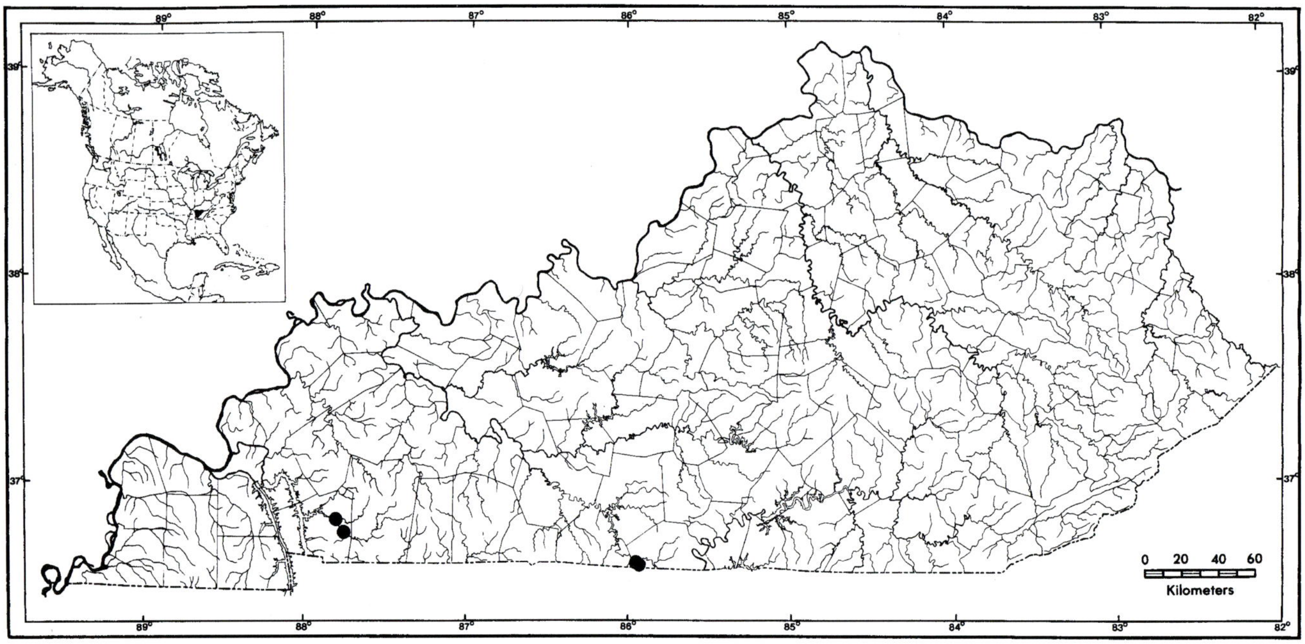

Fig. 61. Distribution of *Cambarus friaufi* in Kentucky.

May. Nothing else is known about the life history of this species.

CRAYFISH ASSOCIATES – We have collected the following species with *C. friaufi*: *C. rusticiformis, Orconectes compressus, O. durelli, O. placidus*, and *O. putnami*.

Cambarus graysoni Faxon
Twospot Crayfish

Fig. 62

LIFE COLORS – (Fig. 62) Overall base color brown to green-brown. Cephalothorax, abdomen, and chelae uniform brown color. Pleurites of first abdominal segment bright yellow or cream. Legs uniform light straw color. Venter cream to white.

CHARACTERS – (Fig. 63) Rostrum deeply excavated, subquadrate; margins straight and subparallel, rounded anteriorly into short acumen. Carapace in dorsal view bullet shaped; in lateral view anterior section rounded. Areola narrow, usually 8.1 to 16.4 times longer than wide at its narrowest point, with room for two to four punctations; sparsely punctated. Postorbital ridge flattened with no tubercle or spine. Hepatic area densely covered with small rounded tubercles; cervical spines absent; branchiostegal spines well developed with corneous tips; suborbital angle acute and well developed with spine. Antennal scale widest distal to midpoint. Chelae triangular with moderately long fingers. Dorsal surfaces of fingers of chelae with well-developed longitudinal ridges. Mesial margin of palm of chela with two complete rows of tubercles; mesialmost row with six to seven tubercles, with five to six tubercles that increase in size distally in most specimens. Two to three large well-developed subpalmar tubercles. Epistomal zygoma broad and only slightly arched.

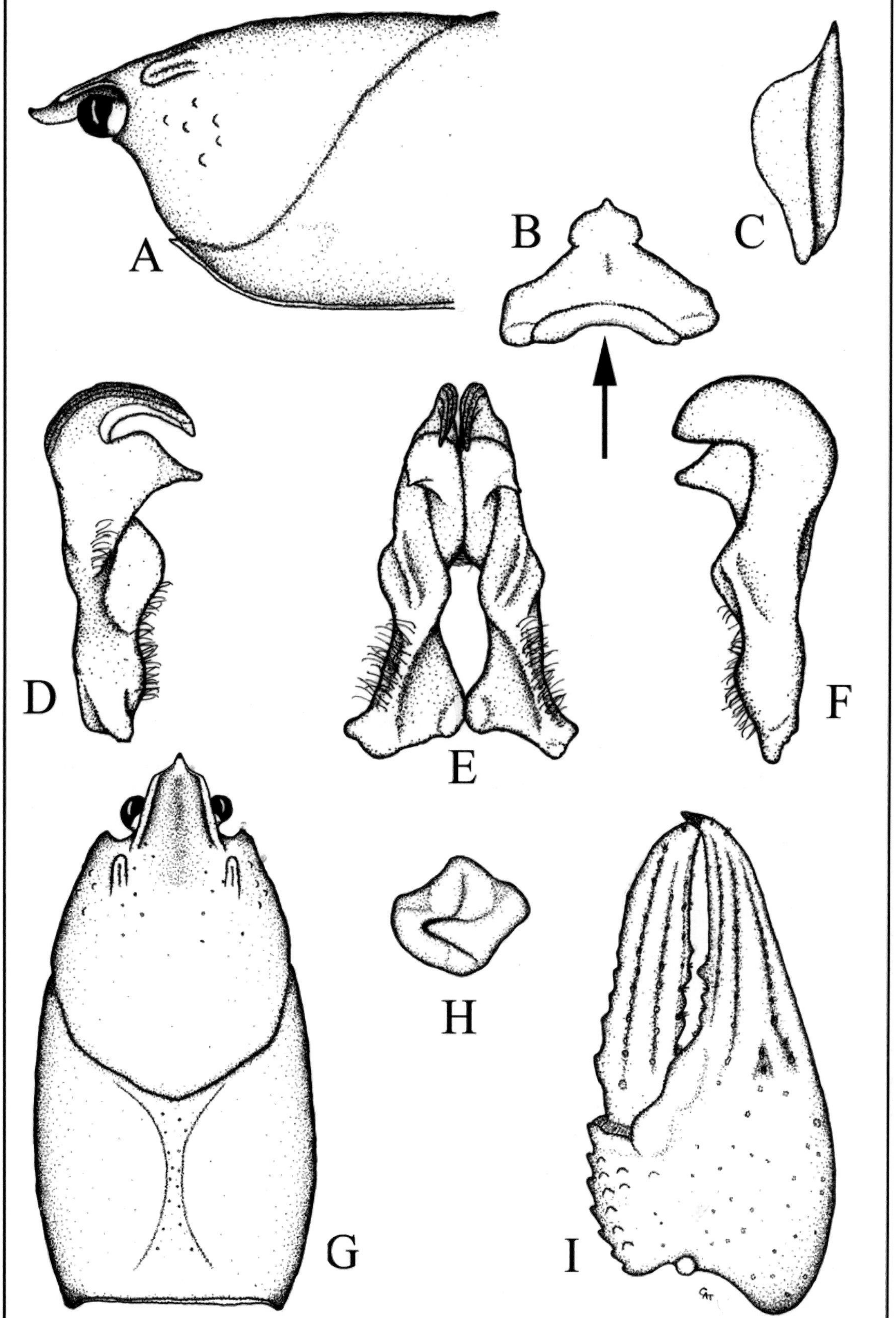

Fig. 63. *Cambarus graysoni*: A, lateral view of carapace; B, ventral view of epistomal zygoma; C, dorsal view of right antennal scale; D, mesial view of form I gonopod; E, ventral view of form I gonopods; F, lateral view of form II gonopod; G, dorsal view of carapace; H, ventral view of annulus ventralis; I, dorsal view of right chela.

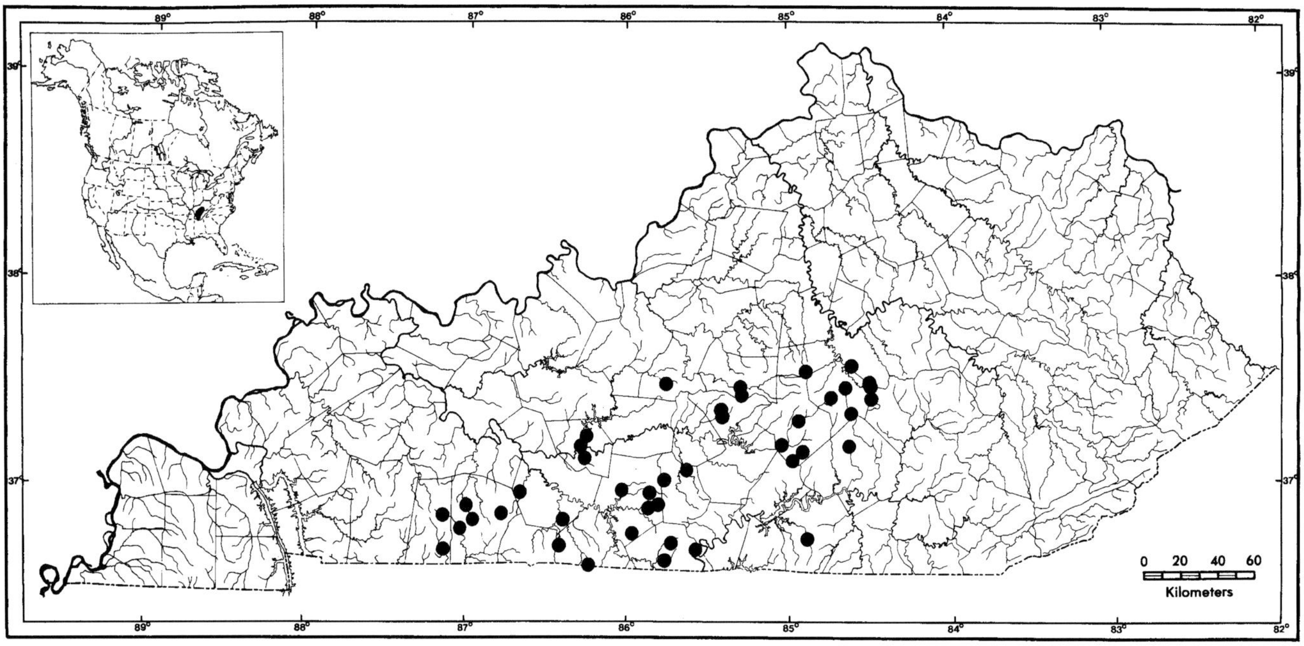

Fig. 64. Distribution of *Cambarus graysoni* in Kentucky.

Central projection of gonopod of form I male corneous, recurved 115° to 140° to shaft, no subapical notch. Mesial process bulbous at base, tapering to point. Annulus ventralis of female subrhomboidal and asymmetrical, with deep medial trough and very deep C-shaped fossa, short sinus.

MAXIMUM SIZE – The maximum known size for the species is 126.8 mm total length.

MOST LIKE – With two complete rows of tubercles along the mesial margin of the palm and narrow areola, *Cambarus graysoni* is most like *C. striatus*. The two can be distinguished in several ways. First, the areola of *C. striatus* is much narrower than that of *C. graysoni*. Second, the epistomal zygoma in *C. striatus* is strongly arched, while in *C. graysoni* it is very broad and almost straight. And third, in living or fresh dead specimens of *C. graysoni* the pleurites of the first abdominal segment are bright yellow or cream, giving the appearance of yellow spots; these spots are absent in *C. striatus*.

DISTRIBUTION AND HABITAT – (Fig. 64) The range of *Cambarus graysoni* includes the Nashville Basin, the Highland Rim, and the western edge of the Cumberland Plateau in Kentucky, Tennessee, and Alabama (Bouchard 1978; Hobbs 1989). In Kentucky, it is commonly encountered in the upper Green River drainage and sporadically in the Cumberland, Rolling Fork (Salt River), and Dix River (Kentucky River) drainages.

Cambarus graysoni is a secondary burrower that is commonly collected in the woody debris along the margins of medium to large creeks and rivers. It is sometimes also collected under cobble in riffle or swift-water areas of these creeks or in burrows dug into the creek banks.

LIFE HISTORY – In Kentucky first form I males have been collected in March, April, and July through December. One female with attached young was collected on July 11, 2002, with 99 young and 6 unhatched eggs. The mean total length of the young was 6.6 mm. Ovigerous females have not been collected from Kentucky, nor have they been reported from other states. Nothing else is known about the life history of this species.

CRAYFISH ASSOCIATES – *Cambarus graysoni* has been collected with *Barbicambarus cornutus*, *C. cumberlandensis*, *C.* sp. A, *C. ortmanni*, *C. rusticiformis*, *C. tenebrosus*, *Orconectes barrenensis*, *O. compressus*, *O. durelli*, *O. immunis*, *O. putnami*, *O. ronaldi*, and *O. rusticus*.

Cambarus ludovicianus Faxon
Painted Devil Crayfish

Fig. 65

LIFE COLORS – (Fig. 65) Base color on the carapace, chelae, and abdomen olive green to deep blue. Rostrum, cervical groove, and areola highlighted in bright red to burgundy. Posterior margin of each abdominal segment trimmed in red to burgundy or cream; with three evenly spaced cream to red longitudinal bands going the full length of abdomen to telson. Telson and uropods trimmed in red. Legs light red to cream in color. Venter cream to whitish. Penn and Marlow (1959) described it as being brilliant blue in Louisiana with three longitudinal red stripes, and with the base and margins of the telson and uropods trimmed in red.

CHARACTERS – (Fig. 66) Rostrum deeply excavated; margins thickened to base of short acumen; no rostral spines. Carapace in dorsal view bullet shaped and laterally compressed; in lateral view head region broadly rounded and vaulted. Areola obliterated. Postorbital ridge flattened without terminal tubercle or spine. Hepatic area with scattered small tubercles; branchiostegals covered with numerous small tubercles laterally; cervical spines absent; branchiostegal spines absent; suborbital angle acute. Antennal scale widest at midpoint. Chelae oval in outline. Dorsal surfaces of fingers of chelae with well-developed longitudinal ridges. Fingers long, about twice the length of palm. Mesial margin of palm of chela covered with tubercles, sometimes with two rows of tear-shaped, serrate tubercles; seven to eight tubercles in mesialmost row; four to six tubercles in second row; sometimes with triangular space lateral to second row without tubercles, rest of palm covered with small tubercles; two to four subpalmar tubercles, ventral surface of palm with additional scattered small tubercles. Epistomal zygoma strongly arched.

Gonopod of form I male with corneous central projection recurved approximately 90° to shaft, with subapical notch. Mesial process bulbous basally, tapers to point apically. Annulus ventralis of female subrhomboidal in outline; cephalic and caudal margins broadly rounded; with deep circular central area with oval fossa and S–shaped sinus.

MAXIMUM SIZE – The maximum known size for the species is 93.5 mm total length

MOST LIKE – *Cambarus lucovicianus* is most like *C. diogenes, C. thomai*, and *Cambarus* sp. A in general morphological appearance. It may be separated from the latter two species by the presence of one or more subpalmar tubercles in *C. ludovicianus. Cambarus thomai* and *C.* sp. A usually lack subpalmar tubercles or occasionally possess one. *Cambarus lucovicianus* also differs from *C. thomai* and *C.* sp. A in having longer fingers on the chelae. In *C. ludovicianus* the dactyl length is 1.8 times or greater than palm length. *Cambarus diogenes* differs from *C. ludovicianus* in the length of the areola and shape of the abdomen. In *C. diogenes* the cephalic section of carapace is more than 1.4 times length of areola, while in *C. ludovicianus* the cephalic section of carapace is 1.4 times or less than the length of areola. The abdomen of *C. ludovicianus* is also smaller, or less developed, than in *C. diogenes*. In a lateral view the abdomen is much thinner or flatter than those of *C. diogenes* and the lateral edge of the pleuron of the second abdominal segment is arched in *C. diogenes* while in *C. ludovicianus* it is straight. In live individuals the color pattern of *C. ludovicianus* is unique.

DISTRIBUTION AND HABITAT – (Fig. 67) Hobbs (1989) gives the range of *C. ludovicianus* as the lower Mississippi River drainage in Arkansas, Tennessee, Louisiana, Mississippi, and eastern Texas. In Kentucky we have found the species at a single location in the Mississippi River floodplain in Carlisle County. More collecting in the

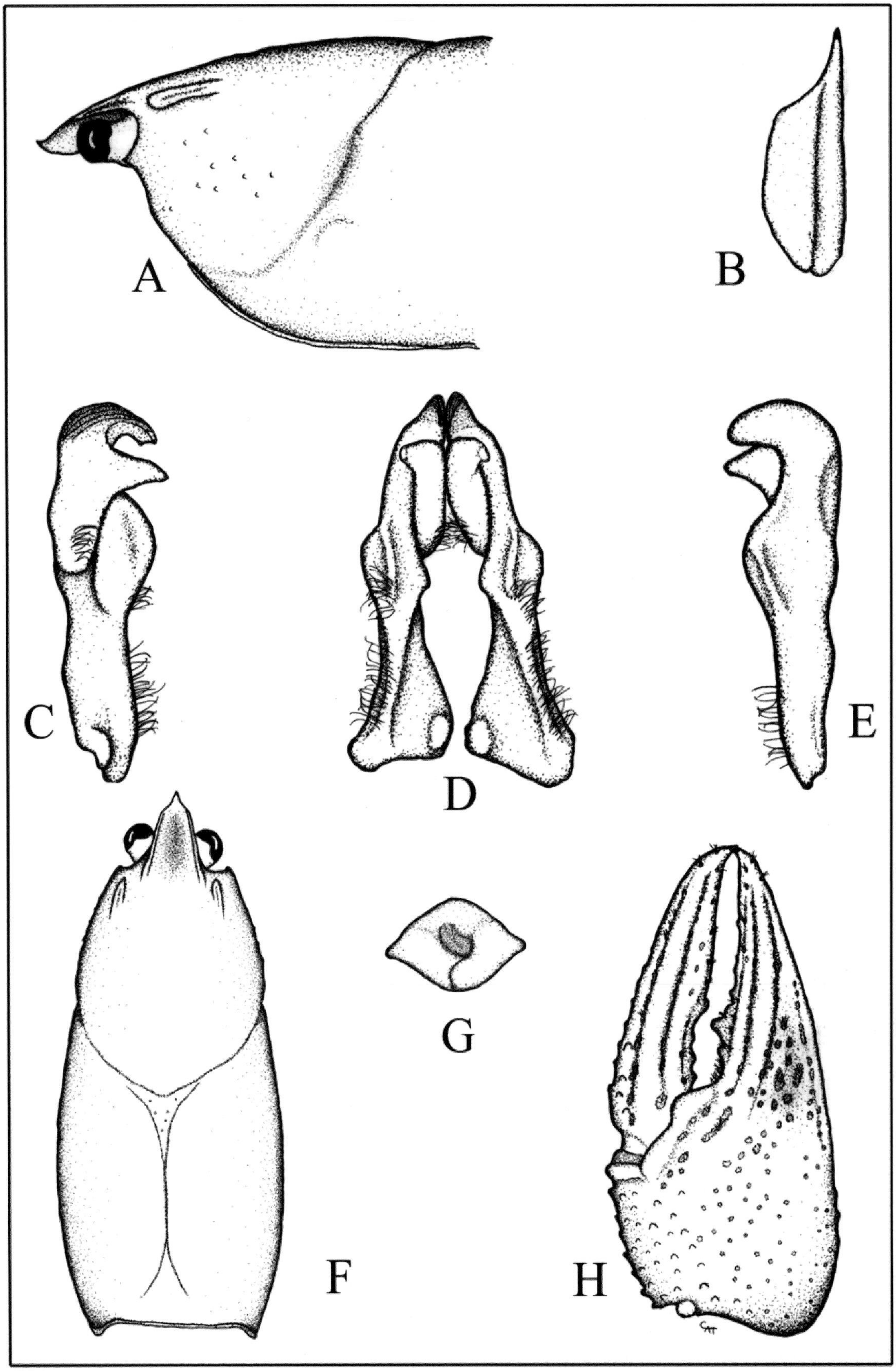

Fig. 66. *Cambarus ludovicianus*: A, lateral view of carapace; B, dorsal view of right antennal scale; C, mesial view of form I gonopod; D, ventral view of form I gonopods; E, lateral view of form II gonopod; F, dorsal view of carapace; G, lateral view of abdomen with arrow denoting lateral edge of pleuron; H, ventral view of annulus ventralis; I, dorsal view of right chela.

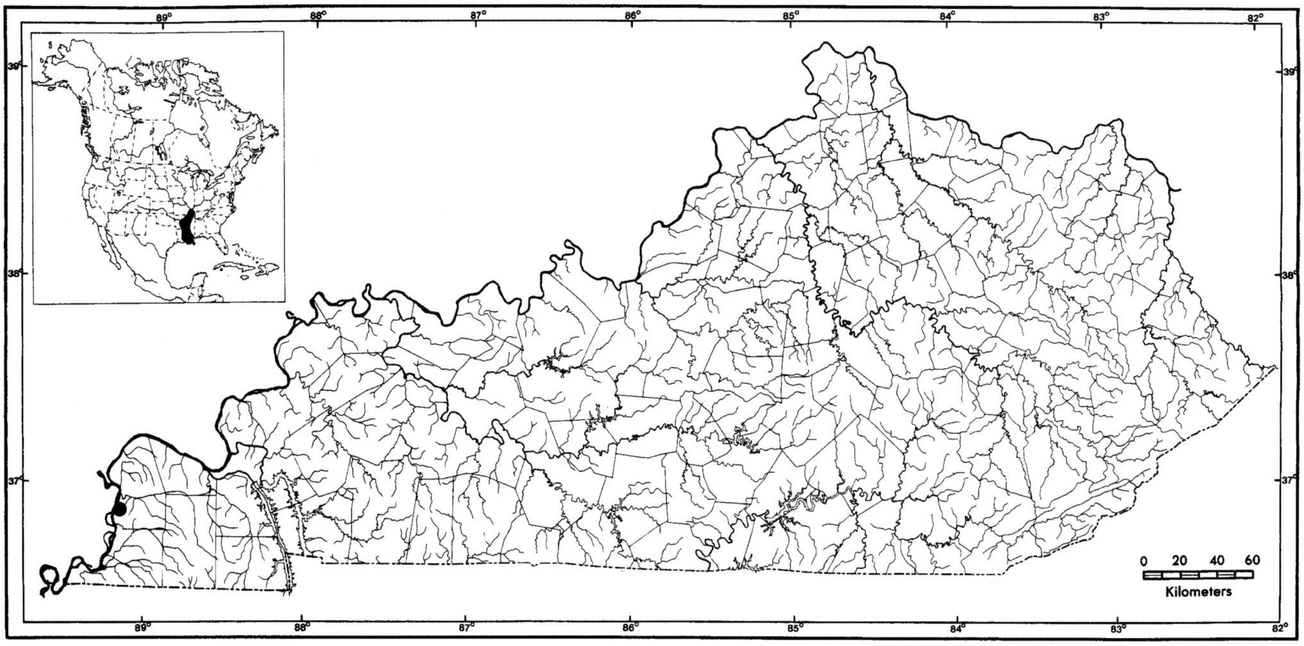

Fig. 67. Distribution of *Cambarus ludovicianus* in Kentucky.

Mississippi River floodplain of western Kentucky will most likely yield additional localities.

Cambarus ludovicianus is a primary burrower. Penn and Marlow (1959) described it occurring in Louisiana in the hardwood alluvial ridges and surrounding swamps, marshes, and lakes at or near "Gulf level." The single site from which it has been recorded in Kentucky is a cypress swamp near the Mississippi River. At that site simple burrows were found in the soft, wet, and exposed soil surrounding cypress trees approximately 1 m from standing water.

LIFE HISTORY – Little is known about the life history of *Cambarus ludovicianus.* Form I males or ovigerous females have not been recorded from Kentucky. Penn and Marlow (1959) collected form I males in Louisiana in February, May, July, August, and November. The smallest form I male was 32.8 mm total length. They found ovigerous females in December and January, and the largest number of young collected was taken in March. The smallest ovigerous female was 35.0 mm total length.

Penn and Marlow (1959) collected *C. lucovicianus* 14 times in their study, 4 times in open water, and 10 times from burrows. This suggests habits similar to those of *C. diogenes*, a closely related species. At the only known Kentucky location for the species, *C. ludovicianus* was collected from a burrow approximately 0.2 m from a burrow containing *C. diogenes*. They also reported that the burrows and chimneys do not differ from those of *C. diogenes*.

CRAYFISH ASSOCIATES – The single Kentucky location for *Cambarus ludovicianus* also contained *Cambarus diogenes*, *Cambarellus shufeldtii*, *Orconectes lancifer*, and *Procambarus clarkii*.

Cambarus ortmanni Williamson
Ortmann's Mudbug

Fig. 68

LIFE COLORS – (Fig. 68) Uniformly dark brown to almost olive-green on carapace, abdomen, and chelae. Abdomen sometimes with darker dorsal streaks or spots. Legs generally same color, but in newly molted individuals may be lighter. Freshly molted individuals of some populations may be slate blue–gray in color. Venter cream to whitish.

CHARACTERS – (Fig. 69) Rostrum moderately excavated; margins slightly thickened, straight, gently converging and rounded anteriorly to form short acumen. Carapace in dorsal view bullet shaped; in lateral view anterior section rounded and slightly vaulted. Areola very narrow, usually 12.8 to 41.7 times longer than wide at its narrowest point, space for one punctation. Postorbital ridge flattened with no tubercle or spine. Hepatic area covered with small rounded tubercles that extend over entire side of carapace; cervical spines absent, small tubercles present; branchiostegal spines represented by small triangular tubercle; suborbital angle obtuse, rounded, no spine. Antennal scale greatest width distal to midpoint. Chelae large and smooth. Dorsal surfaces of fingers of chelae with weakly developed longitudinal ridges, most evident on dactyl. Mesial margin of palm of chela with complete row of five to seven tubercles; second incomplete row also present with two to six tubercles with distinct gaps. One to two strong, large subpalmar tubercles. Epistomal zygoma wide, not strongly arched.

Central projection of gonopod of form I male corneous, recurved to approximately 90° to shaft, with very shallow subapical notch. Mesial process deep, bulbous at base, narrows to point. Annulus ventralis of female rhomboidal with deep C–shaped fossa, straight or slightly curved sinus.

MAXIMUM SIZE – The maximum known size for the species is 103.4 mm total length.

MOST LIKE – *Cambarus ortmanni* most closely resembles *C. bartonii cavatus* in general appearance and color. *Cambarus ortmanni* differs from that species by possessing a very narrow areola, which has room for only one punctation, while the areola of *C. bartonii cavatus* is wide. The suborbital angles also differ. In *C. ortmanni* it is rounded and obscure while in *C. bartonii cavatus* it is acute. The two species also occur in different habitats. *Cambarus ortmanni* is a primary or secondary burrower that gets into adjacent surface waters only in the fall or early spring. In contrast, *C. bartonii cavatus* is a creek-dwelling species.

DISTRIBUTION AND HABITAT – (Fig. 70) *Cambarus ortmanni* is known to occur in middle and lower Ohio River drainages of southern Indiana, southern Ohio, and Kentucky. In Kentucky the species occurs sporadically in the middle and lower reaches of most major watersheds draining into the Ohio River upstream of the Tradewater River. The species also inhabits the extreme upper reaches of the Green River drainage.

Cambarus ortmanni can be considered a primary and secondary burrower, depending on locality. Its burrows are commonly found in hydric soils along roadside ditches, creeks, ponds, and lowlying fields. In some of these habitats it is unlikely that individuals ever get into permanent bodies of water. At other locations the species can be encountered in medium-sized creeks in wet spring and fall months.

LIFE HISTORY – In Kentucky form I males have been collected in January, February, March, April, and September. No ovigerous females have been collected. Williamson (1907) reported form I males in April and May. Norrocky (1989) provided some life history notes for *C. ortmanni.* He

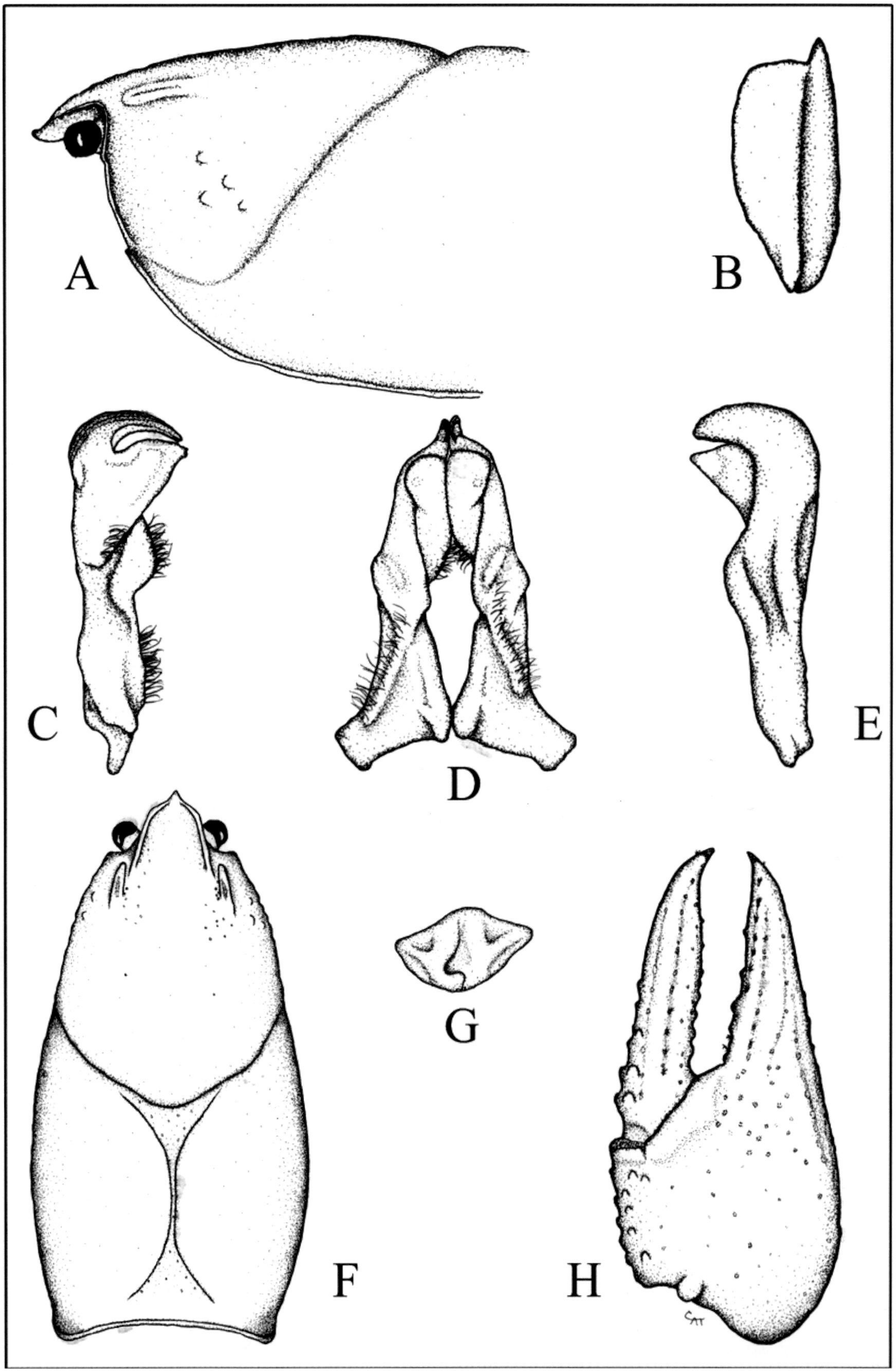

Fig. 69. *Cambarus ortmanni*: A, lateral view of carapace; B, dorsal view of right antennal scale; C, mesial view of form I gonopod; D, ventral view of form I gonopods; E, lateral view of form II gonopod; F, dorsal view of carapace; G, ventral view of annulus ventralis; H, dorsal view of right chela.

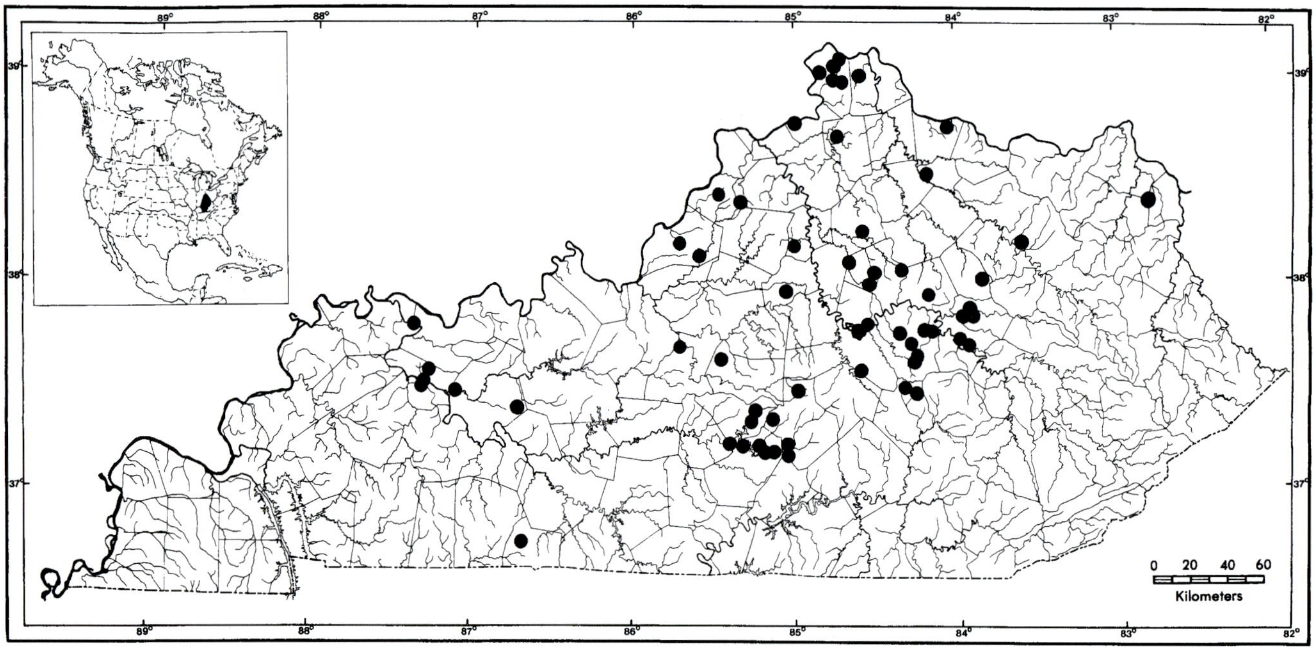

Fig. 70. Distribution of *Cambarus ortmanni* in Kentucky.

suggested that ovipositing takes place near mid–summer, and that juveniles grow to approximately 20 mm by fall and at the end of their second summer reach 30 to 35 mm when at least some molt to reproductive adults. He also suggested that this species lives three years or more. We have found that those individuals that burrow adjacent to temporary or permanent surface waters are often found in those waters during the fall and spring months. Young of the year *C. ortmanni* are commonly collected in ponds and creeks during the winter and spring.

CRAYFISH ASSOCIATES – *Cambarus ortmanni* has been collected in the same area as other burrowers such as *C. batchi*, *C.* sp. A,, and *C. thomai*. In flowing waters *C. ortmanni* has been collected with *C. tenebrosus*, *Orconectes immunis*, *O. jeffersoni*, and *O. rusticus*.

Cambarus parvoculus Hobbs and Shoup
Mountain Midget Crayfish

Fig. 71

LIFE COLORS – (Fig. 71) Overall base color golden to dark brown. Cephalothorax and chelae usually same color. Rostral margin highlighted in yellow to orange. Legs straw colored in some individuals to light gray or powder blue in other individuals. Abdomen may be same color or slightly darker than carapace. Venter cream to whitish.

CHARACTERS – (Fig. 72) Rostrum moderately excavated, subquadrate; margins greatly thickened to base of acumen, margins of acumen not thickened. Acumen triangular, very short, ending in corneous tubercle. Carapace in dorsal view bullet shaped, rounded laterally, gradually narrowing anteriorly. Areola very wide, usually 3.0 to 5.0 times longer than wide at its narrowest point; room for five to six punctations, sparsely

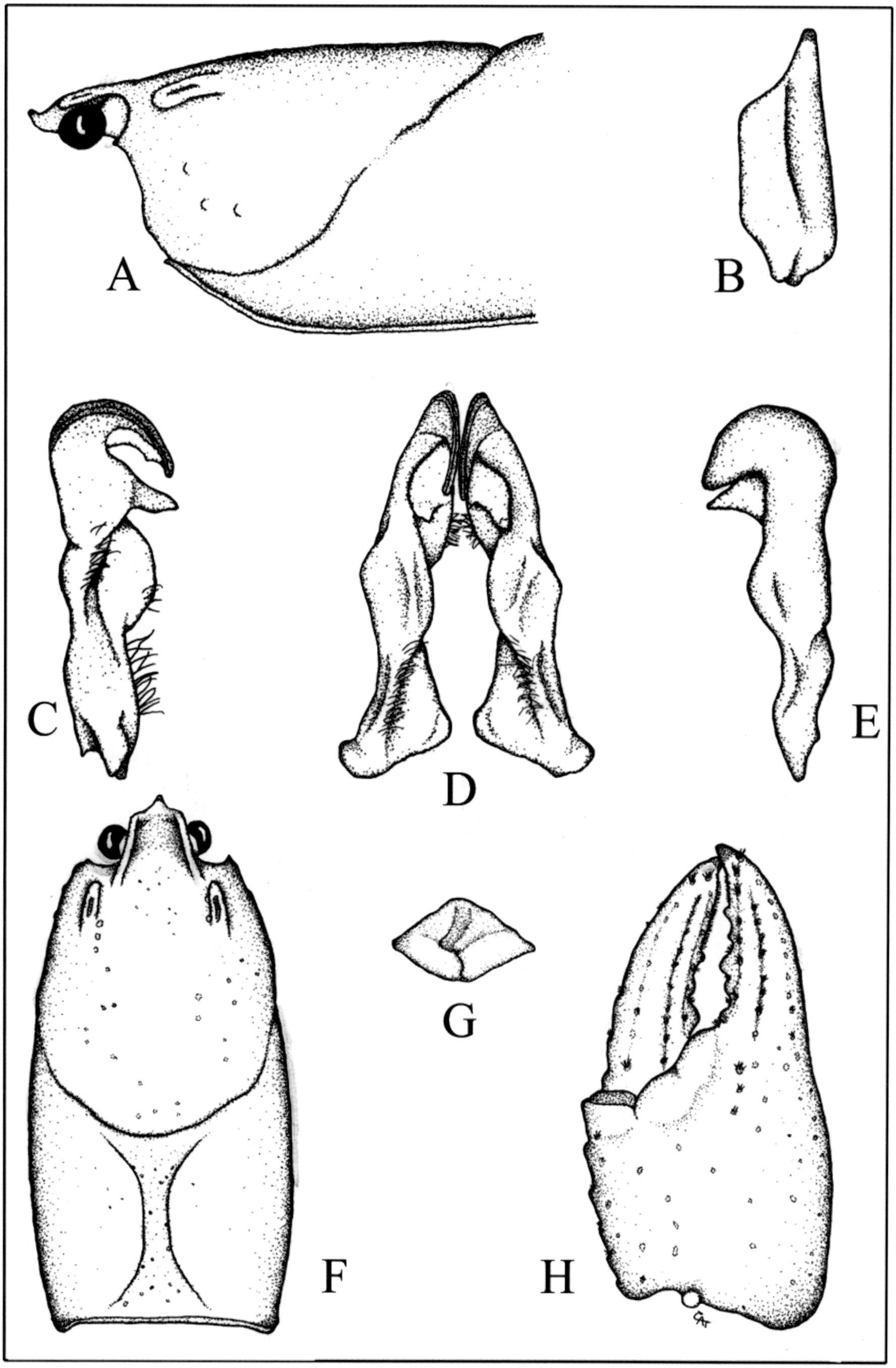

Fig. 72. *Cambarus parvoculus*: A, lateral view of carapace; B, dorsal view of right antennal scale; C, mesial view of form I gonopod; D, ventral view of form I gonopods; E, lateral view of form II gonopod; F, dorsal view of carapace; G, ventral view of annulus ventralis; H, dorsal view of right chela.

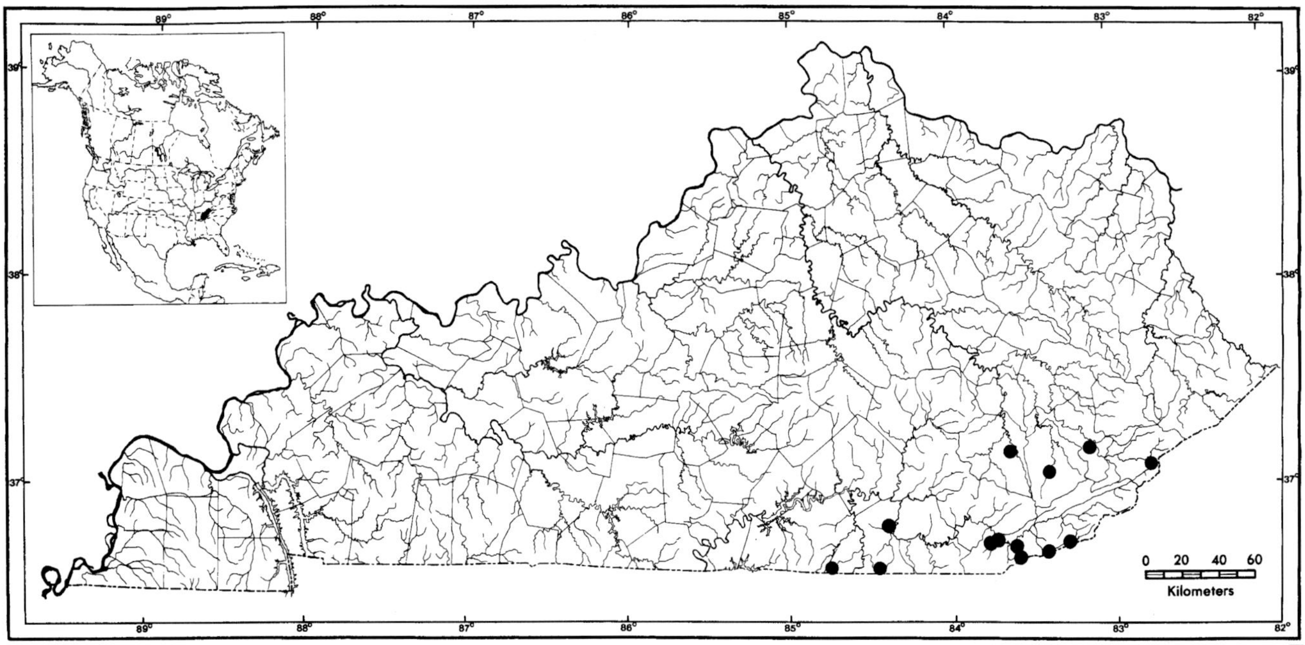

Fig. 73. Distribution of *Cambarus parvoculus* in Kentucky.

punctated. Postorbital ridge flattened, apically with small rounded tubercle. Hepatic area covered with relatively large tubercles; cervical spines absent, but may have small tubercle in area; branchiostegal spines absent, but with small round tubercles; suborbital angle well developed, tip with rounded tubercle. Antennal scale with greatest width proximal to midpoint. Chelae rectangular in outline. Dorsal surface of dactyl of chelae with poorly developed longitudinal ridges. Mesial margin of palm of chela with a single row of six to seven large round tubercles, the basal two or three fused to form ridge. No subpalmar tubercles present. Epistomal zygoma strongly arched.

Central projection of gonopod of form I male corneous, recurved much greater than 90° to shaft, with strong subapical notch. Mesial process bulbous at base, tapers apically. Annulus ventralis of female subrhomboidal with deep C–shaped fossa, S–shaped sinus.

MAXIMUM SIZE – The maximum known size for the species is 60.7 mm total length, although most adults collected in Kentucky are much smaller than this.

MOST LIKE – This species most closely resembles *Cambarus distans* in overall appearance and coloration. They are often collected together. The most strikingly obvious difference between the two is the square shape of the rostrum with noticeably thickened margins in *C. parvoculus*, while in *C. distans* it is usually much more elongated, and the margins are generally not as thickened. In form I males the greatest distinction between the two species is in the arch of the central projection. In *C. distans* it is usually recurved about 90° to the main axis, while in *C. parvoculus* it is more broadly recurved, and much greater than 90°.

DISTRIBUTION AND HABITAT – (Fig. 73) Hobbs (1989) listed this species as being largely restricted to the Cumberland Plateau where it occurred in the Cumberland River basin in eastern Tennessee, southeastern Kentucky, and northwestern Georgia. In Kentucky, *Cambarus parvoculus* is currently known sporadically from the Big South Fork of the Cumberland drainage, the upper Cumberland River drainage above Cumberland Falls, and upper Kentucky River drainage.

It is most commonly found in small, cool, headwater creeks, springs, and seepages. It is typically found under larger rocks in riffle areas and rarely in the open.

LIFE HISTORY – Form I males were found in Kentucky in March, April, July, August, and September. In Georgia, Hobbs (1981) indicated that form I males were known from April, September, and November, and that ovigerous females have been collected in April and May. Hobbs, however, does not provide data indicating the number of eggs found on these females. We collected no ovigerous specimens. Nothing else is known about the life history of this species.

CRAYFISH ASSOCIATES – *Cambarus parvoculus* has been collected with *C. buntingi*, *C. distans*, and *Orconectes cristavarius*.

TAXONOMIC NOTE – Recently a new species, *Cambarus jezerinaci*, was described by Thoma (2000) from spring-fed creeks in the Powell River drainage in eastern Tennessee and western Virginia. The type locality of the species (tributary to Dry Creek, Lee County) in western Virginia is south of Cumberland Mountain and less than 2.0 mi from the Kentucky border. Thoma described *C. jezerinaci* as the sister species to *C. parvoculus*, and indicated that it could be have been derived from populations that crossed the drainage divide from the Cumberland River in Kentucky. According to Thoma, *C. jezerinaci* is generally smaller than *C. parvoculus*, and it can be distinguished from *C. parvoculus* by the lack of setae on the chelae, an upturned rostrum, and four enlarged tubercles on the propodus. We have examined the type specimens of both species and it appears that there is overlap in these and other characters. Genetic studies need to be done in order to determine the exact relationship between these two forms.

Cambarus robustus Girard
Big Water Crayfish

Fig. 74A: juvenile color form.

Fig. 74B: adult color form.

LIFE COLORS – (Figs. 74 A, B) Overall base color olive green to dark brown. Carapace uniform dark green to brown; chelae same color or slightly lighter than carapace; legs lighter in color and generally cream to straw color; abdomen may be slightly darker than carapace. Younger individuals may have a reticulated green color pattern over the entire body, abdomen often with a pair of very distinct black longitudinal stripes that extend to base of telson. Venter cream to whitish.

CHARACTERS – (Fig. 75) Rostrum moderately excavated; margins moderately thickened, straight to slightly converging, rounded anteriorly to form acumen. Acumen triangular with corneous tubercle at tip, margins not thickened. Carapace in dorsal view with branchiostegal area very wide and rounded, head region narrower and also rounded; in lateral view posterior section straight dorsally and somewhat vaulted and rounded anteriorly. Areola moderately wide, usually 5.6 to 8.8 times longer than wide at its narrowest point, room for seven to nine punctations. Postorbital ridge not well developed, flattened, apical tubercle small or absent. Hepatic area covered with numerous large tubercles, branchiostegals densely covered with many small tubercles; cervical spines very small to absent, or present as tubercles; branchiostegal spines absent but with small tubercle; suborbital angle acute. Antennal scale widest distal to midpoint. Chelae large and with long fingers, form elongated

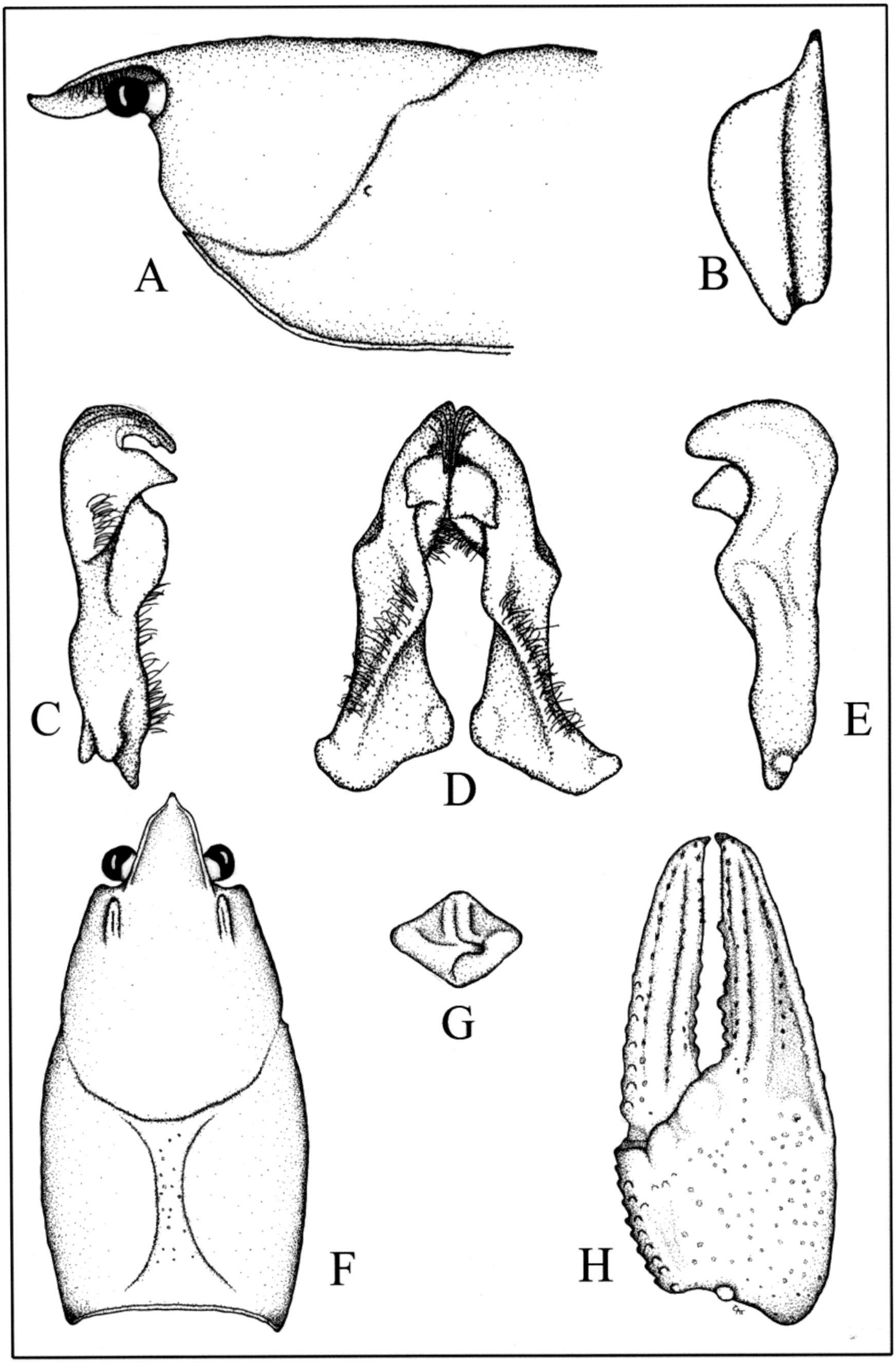

Fig. 75. *Cambarus robustus*: A, lateral view of carapace; B, dorsal view of right antennal scale; C, mesial view of form I gonopod; D, ventral view of form I gonopods; E, lateral view of form II gonopod; F, dorsal view of carapace; G, ventral view of annulus ventralis; H, dorsal view of right chela.

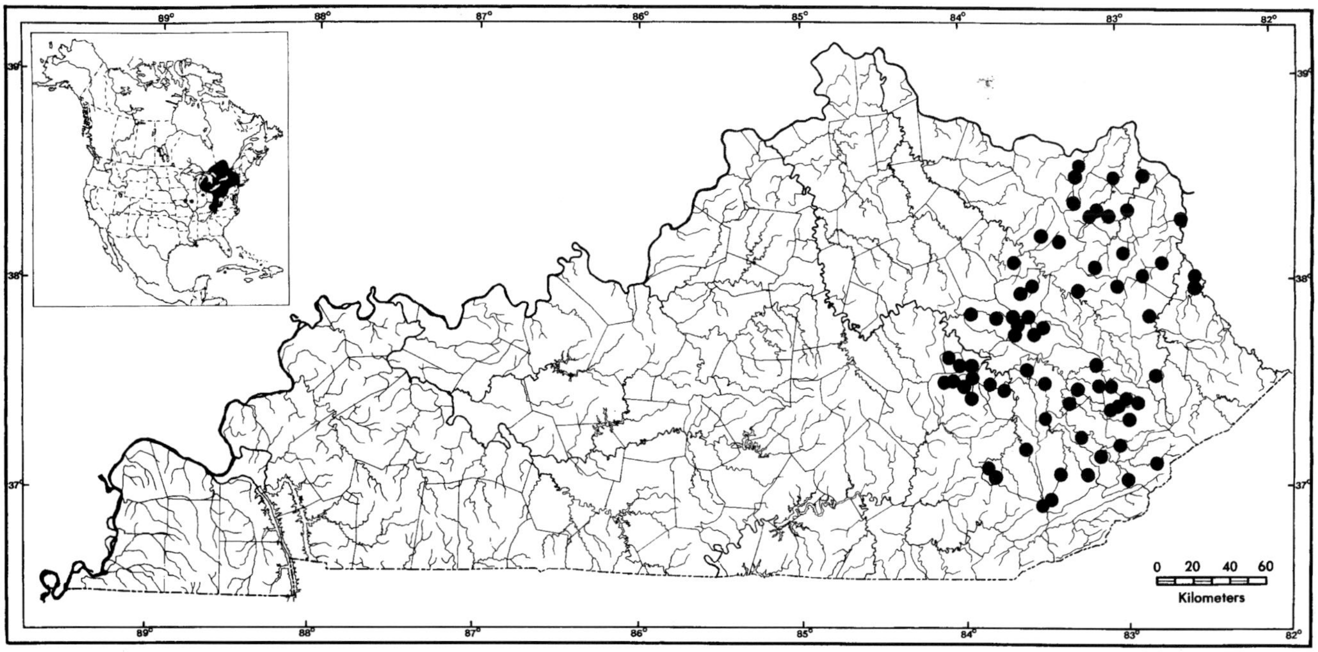

Fig. 76. Distribution of *Cambarus robustus* in Kentucky.

triangle in outline; base of propodus with very distinct dorsoventral depression. Dorsal surfaces of fingers of chelae with moderately deep setal furrows producing moderately developed longitudinal ridges. Mesial margin of palm of chela with two complete rows of tubercles, sometimes with additional tubercles on palm; mesialmost row with eight large serrate tubercles, second row with five to seven smaller tubercles. Occasional specimens with fewer than eight tubercles in mesialmost row. Two to three subpalmar tubercles present. Epistomal zygoma strongly arched.

Central projection of gonopod of form I male corneous, recurved more than 90° to shaft, with a subapical notch. Mesial process bulbous basally, tapers apically into short tip. Annulus ventralis of female subrhomboidal, with wide medial longitudinal trough that expands caudally, deep C–shaped fossa and short S–shaped sinus.

MAXIMUM SIZE – The maximum known size for the species is 103.0 mm total length.

MOST LIKE – *Cambarus robustus* is most like *C. buntingi* and *C. cumberlandensis* in appearance. The most striking common feature of all three species is the elongated chelae with a distinct dorsoventral depression at the base of the propodus. The main distinguishing character that separates *C. robustus* from the other two is the lack of any lateral spines or tubercles on the rostrum. The other two species have either well-developed spines (*C. cumberlandensis*) or tubercles (*C. buntingi*). Another species that *Cambarus robustus* can be confused with is *Cambarus bartonii cavatus*. These two species are sometimes collected together, and larger individuals are very similar. In *C. robustus* the rostrum is more elongated and moderately excavated, while in *C. bartonii cavatus* the rostrum is shorter and wider and is usually deeply excavated and spoonlike. *Cambarus robustus* has two complete rows of tubercles on the mesial margin of the palm of the chelae, while *C. bartonii cavatus* has a single complete row of weak tubercles with an incomplete second row of several scattered tubercles. Lastly, *C. bartonii cavatus* usually does not have the deep dorsoventral depressions at the base of the propodus of the chelae except in large form I males.

DISTRIBUTION AND HABITAT – (Fig. 76) A widespread species, *Cambarus robustus* occurs from southern Ontario east to New York, west to Illinois, and south to North Carolina and Virginia (Hobbs 1989). *Cambarus robustus* is widespread in eastern Kentucky from the Kentucky to Big Sandy River drainages. Jezerinac et al. (1995) reported it being uncommon in the Big Sandy River in West Virginia, and attributed this to competition with *C. sciotensis.* We, however, have found it from at least 11 locations within that drainage in Kentucky. In the Licking and Kentucky River drainages it has been found only in the upper half of each system.

Cambarus robustus is most commonly found in medium- to large-sized creeks and rivers. In those creeks it is most commonly found under very large, flat boulders with generally one crayfish per boulder. It seems to prefer a completely different microhabitat from *Orconectes cristavarius*, a species that is often collected with it and that has a similar distribution. *Orconectes cristavarius* is most commonly found under smaller rocks in the main parts of riffles, while *C. robustus* is found under boulders along the margins, generally outside of strong currents.

LIFE HISTORY – Form I males of *Cambarus robustus* have been collected from Kentucky in January, March, April, May, June, July, September, October, and December. No ovigerous females have been reported from the state. Jezerinac et al. (1995) reported form I males for West Virginia in April, May, June, July, August, September, and October, and ovigerous females in May, July, and September. Egg counts ranged from approximately 50 eggs to 290 eggs per female with the egg diameters ranging from 2.2 to 2.9 mm. Life history data have been reported for several more northern populations of *C. robustus*, two in Ontario (Hamr and Berrill 1985; Corey 1990) and several in New York (Crocker 1957). In all three studies ovigerous females were encountered in June and July. Copulation was observed in New York populations in October and in Ontario populations in June and July. Noteworthy differences were observed between the Ontario populations. Hamr and Berrill reported that *C. robustus* first reproduced at four years of age at sizes ranging from 34 to 45 mm carapace length, while Corey reported reproduction occurring at three years of age when individuals ranged in size from 24 to 34 mm carapace length. The observations of body size at sexual maturity of Corey agree with those made by Crocker in New York. Hamr and Berrill reported a five-year life span for the species while Corey estimated it to be four years. Hamr and Berrill also suggested that summer and early fall were major molting periods and indicated that the young stayed with the female for about a month after hatching. Jezerinac et al. (1995) reported a 1:1 sex ratio in the populations they studied in West Virginia.

CRAYFISH ASSOCIATES – We have collected the following species with *Cambarus robustus*: *C. bartonii cavatus*, *C. sciotensis*, *C. veteranus*, *Orconectes cristavarius*, and *O. rusticus*.

Cambarus rusticiformis Rhoades
Depression Crayfish

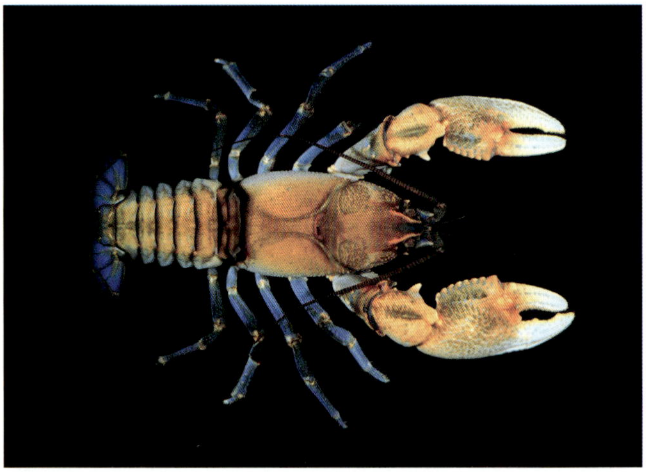

Fig. 77A: mature, recently molted adult.

Fig. 77B: immature adult.

LIFE COLORS – (Figs. 77A, B) Overall base color golden brown to dark brown. Carapace uniform in color, except area of mandibular muscle where usually vermiculated and contrasting. Dark brown saddle at caudal margin of carapace present in more eastern populations. Rostral margins yellow to orange; chelae generally same color as carapace; larger individuals with cream to whitish fingers; chelae tips cream to orange. Abdomen usually lighter in color; legs generally cream colored, sometimes powder to cobalt blue in large newly molted individuals. Venter cream to whitish.

CHARACTERS – (Fig. 78) Rostrum moderately excavated; margins thickened, wide at the base and greatly narrowing apically, making margins concave at mid-length; margins ending in corneous spinelike tubercles. Acumen elongated and narrow with corneous tip. Carapace in dorsal view bullet shaped, gradually narrowing anteriorly; in lateral view dorsoventrally flattened posteriorly and slightly vaulted anteriorly. Areola wide, usually 4.3 to 6.9 times longer than wide at its narrowest point; room for seven to nine punctations. Postorbital ridge not well developed, flattened, apical large rounded or pointed tubercle present. Hepatic area covered with large tubercles, same tubercles present on branchiostegals; cervical spines present and well developed; branchiostegal spines represented as small tubercles; suborbital angle broadly rounded. Antennal scale with greatest width distal to midpoint. Chelae large, elongate and rectangular in outline. Dorsal surface of dactyl with a single well-defined longitudinal ridge; propodus with moderately developed ridge or no ridge. Palm of chelae long. Mesial margin of palm of chela with a single row of 8 to 10 small, rounded tubercles; tubercles in some specimens minute and difficult to count. No subpalmar tubercles present. Epistomal zygoma moderately arched.

Central projection of gonopod of form I male corneous, recurved approximately 90° or slightly more than 90° to shaft; with subapical notch. Mesial process bulbous at base, apically narrows at tip. Annulus ventralis of female subrhomboidal, with deep medial longitudinal trough expanding caudally, fossa deep and C–shaped, sulcus S–shaped.

MAXIMUM SIZE – The maximum known size for the species is 84.5 mm total length.

MOST LIKE – With a long palm region and poorly developed longitudinal ridges on the dactyl, *Cambarus rusticiformis* in general appearance most resembles *C. tenebrosus*. Chelae and rostral characters can be used to separate the species. In *C. tenebrosus* the rostral margins are almost straight and gradually converging into the acumen, while in *C. rusticiformis* they are concave, greatly narrow distally, and terminate in corneous tubercles. *Cambarus rusticiformis* also has only a single row of tubercles along the mesial margin of the palm of the chelae, while *C. tenebrosus* has two rows.

DISTRIBUTION AND HABITAT – (Fig. 79) *Cambarus rusticiformis* is known from the Cumberland and Green River drainages in Kentucky and Tennessee (Bouchard 1976b; Hobbs 1989) and from the Ohio River proper bordering southern Illinois. In Kentucky it is known from the Barren, upper Green, Nolin, and Rough River drainages. It is also widespread in the Cumberland River system from eastern Kentucky to creeks draining into Lake Barkley in western Kentucky.

Cambarus rusticiformis is generally found in medium-sized creeks and rivers that have large, flat boulders strewn along the bottom. It is most often found under these boulders where it excavates bowl-shaped depressions. Rhoades (1944) made a similar observation. Specimens of this species have often been seen in these depressions, after the rocks have been removed, sitting with all of their legs drawn close to the body and the abdomen tucked underneath them. Unlike most other crayfishes, which quickly try to escape when exposed, these crayfishes do not attempt to flee; instead, they remain still in their depressions.

LIFE HISTORY – In Kentucky Form I males are known from January, March, April, May, June, August, September, October, and November. A single ovigerous female with a total length of 81.4 mm was collected on April 22, 1972. She had 36 eggs attached to her pleopods with a mean diameter of 2.8 mm. Nothing is else is known about the life history of this species.

CRAYFISH ASSOCIATES – Rhoades (1944) indicated that *Cambarus rusticiformis* was collected with *C. striatus*, *Orconectes compressus*, and *O. placidus* in the western part of the state. We have collected it in association with *Barbicambarus cornutus*, *C. cumberlandensis*, *C. distans*, *C. tenebrosus*, *C. graysoni*, *Orconectes putnami*, *O. rafinesquei*, *O. rusticus*, and *O. tricuspis*.

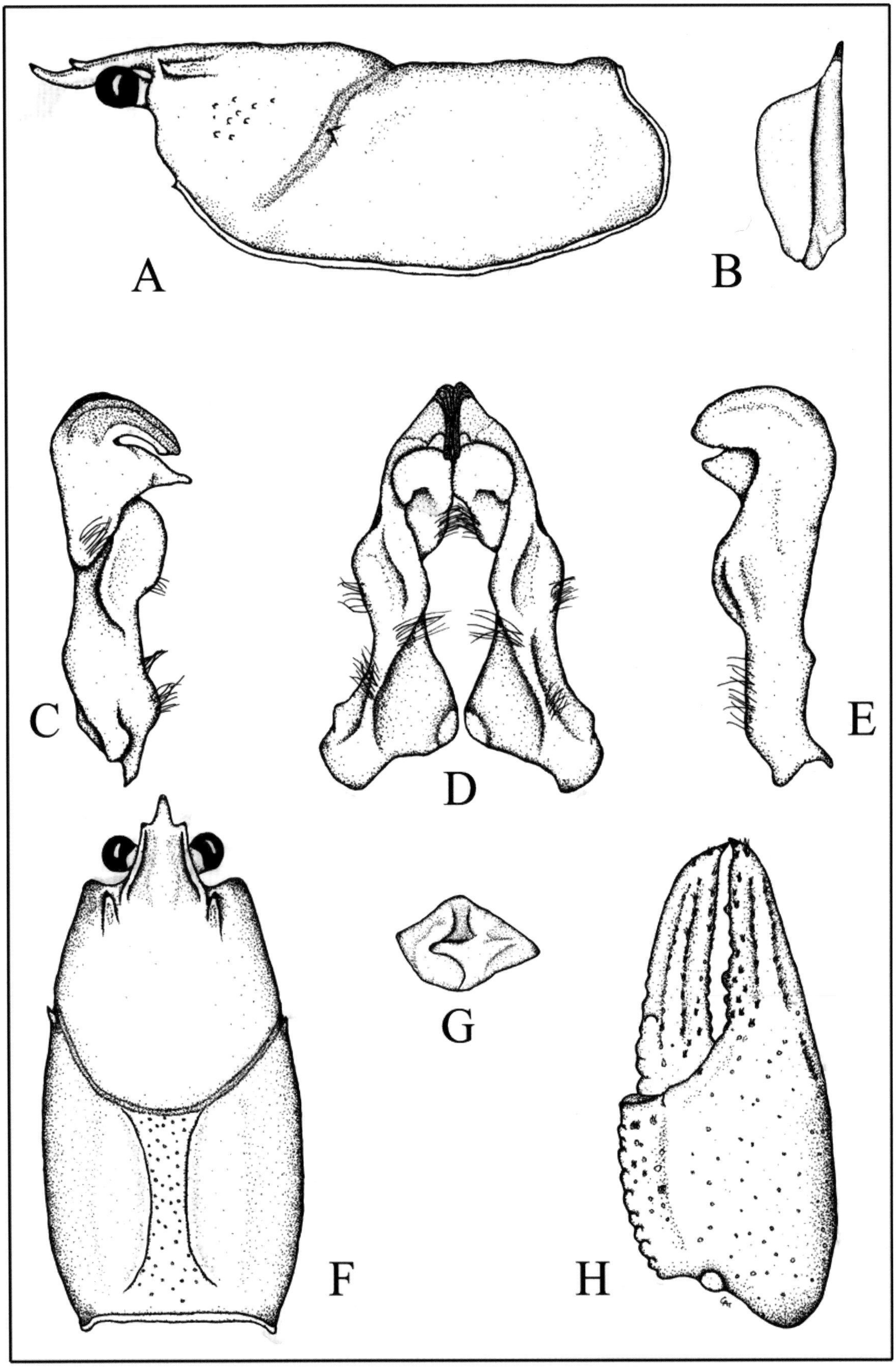

Fig. 78. *Cambarus rusticiformis*: A, lateral view of carapace; B, dorsal view of right antennal scale; C, mesial view of form I gonopod; D, ventral view of form I gonopods; E, lateral view of form II gonopod; F, dorsal view of carapace; G, ventral view of annulus ventralis; H, dorsal view of right chela.

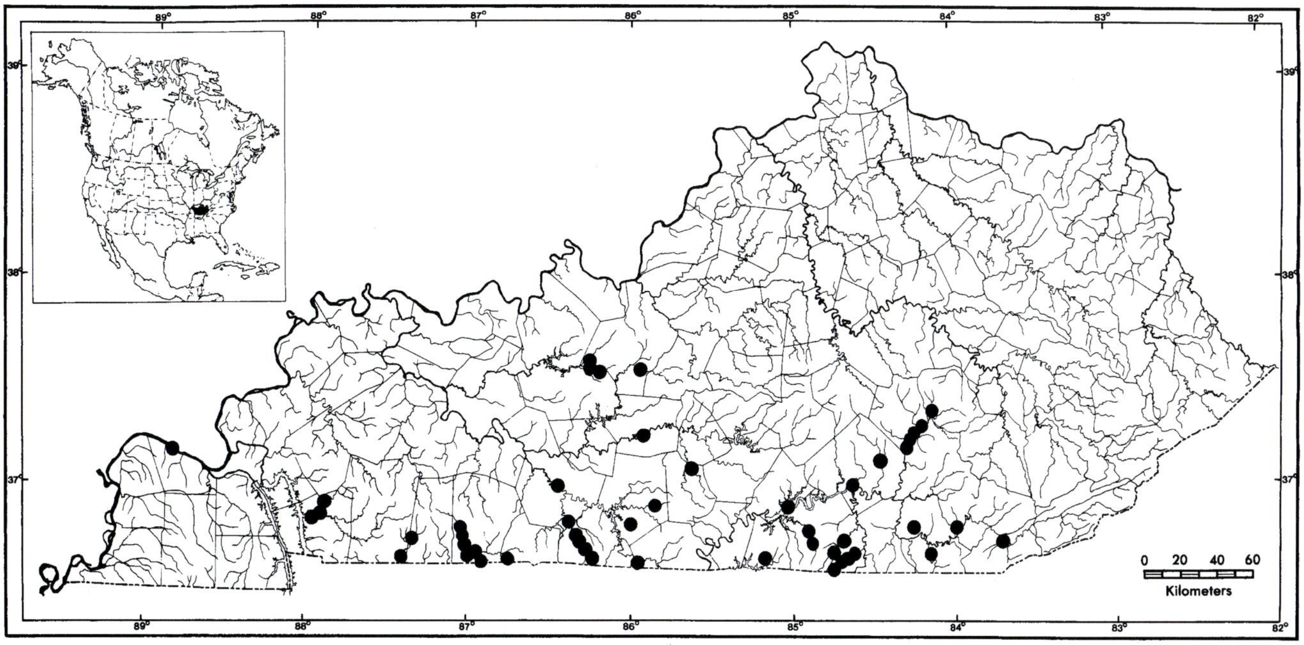

Fig. 79. Distribution of *Cambarus rusticiformis* in Kentucky.

Cambarus sciotensis Rhoades
Teays River Crayfish

Fig. 80

LIFE COLORS – (Fig. 80) Overall base color olive to dark brown. Carapace uniform color except conspicuous darker vermiculations of mandibular muscle; rostral margins usually lighter than carapace, cream in color; chelae uniform color, somewhat lighter than carapace; abdomen uniform in color, slightly darker than carapace; legs cream to whitish. Venter cream to whitish.

CHARACTERS – (Fig. 81) Rostrum moderately excavated, subquadrate in shape; margins distinctly thickened, no rostral spines. Acumen triangular in shape, margins not thickened, corneous tubercle at tip. Carapace in dorsal view very wide at branchiostegals, narrows somewhat in hepatic region, dorsoventrally flattened in lateral view. Areola very wide, usually 3.5 to 6.2 times longer than wide at its narrowest point, with room for 8 to 10 punctations; densely punctate. Postorbital ridge flattened with small pointed tubercle at tip; tubercle and spine better developed in smaller individuals. Hepatic area covered with large rounded tubercles; branchiostegals completely covered with smaller rounded tubercles. Cervical spines weakly developed or absent; branchiostegal spines represented by small pointed tubercle; suborbital angle acute, with tubercle or spine. Antennal scale greatest width distal to midpoint. Chelae elongated and triangular in outline, fingers very long, especially in form I males. Dorsal surfaces of fingers of chelae with moderately developed longitudinal ridges. Mesial margin of palm of chela with two complete rows of tubercles; 8 to 10 tubercles in mesial–most row, second row with six to eight tubercles, additional scattered tubercles on palm. Zero to three well-developed subpalmar tubercles. Epistomal zygoma strongly arched.

Central projection of gonopod of form I male corneous, recurved about 90° or slightly more to shaft, with well-developed subapical notch.

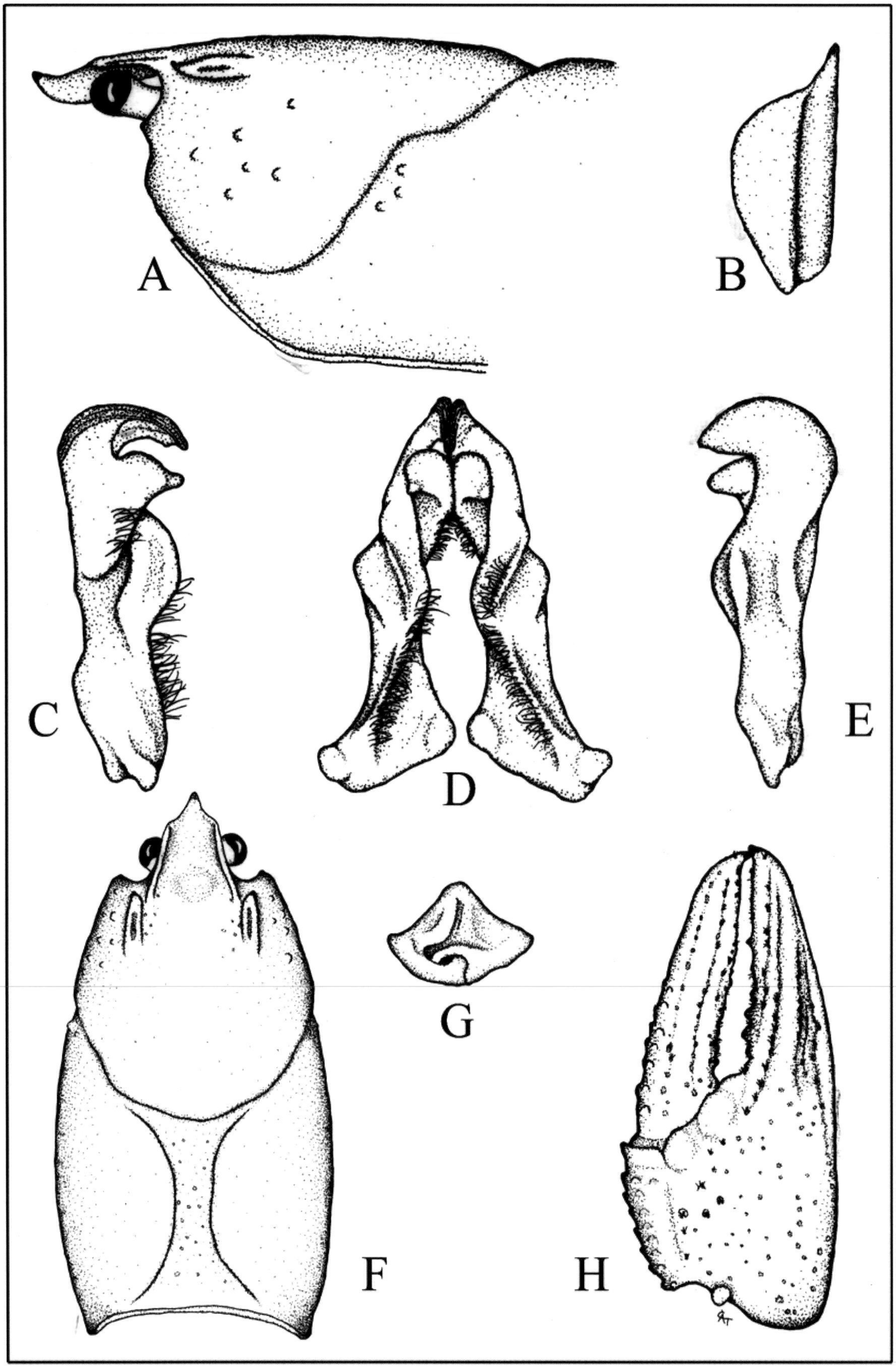

Fig. 81. *Cambarus sciotensis*: A, lateral view of carapace; B, dorsal view of right antennal scale; C, mesial view of form I gonopod; D, ventral view of form I gonopods; E, lateral view of form II gonopod; F, dorsal view of carapace; G, ventral view of annulus ventralis; H, dorsal view of right chela.

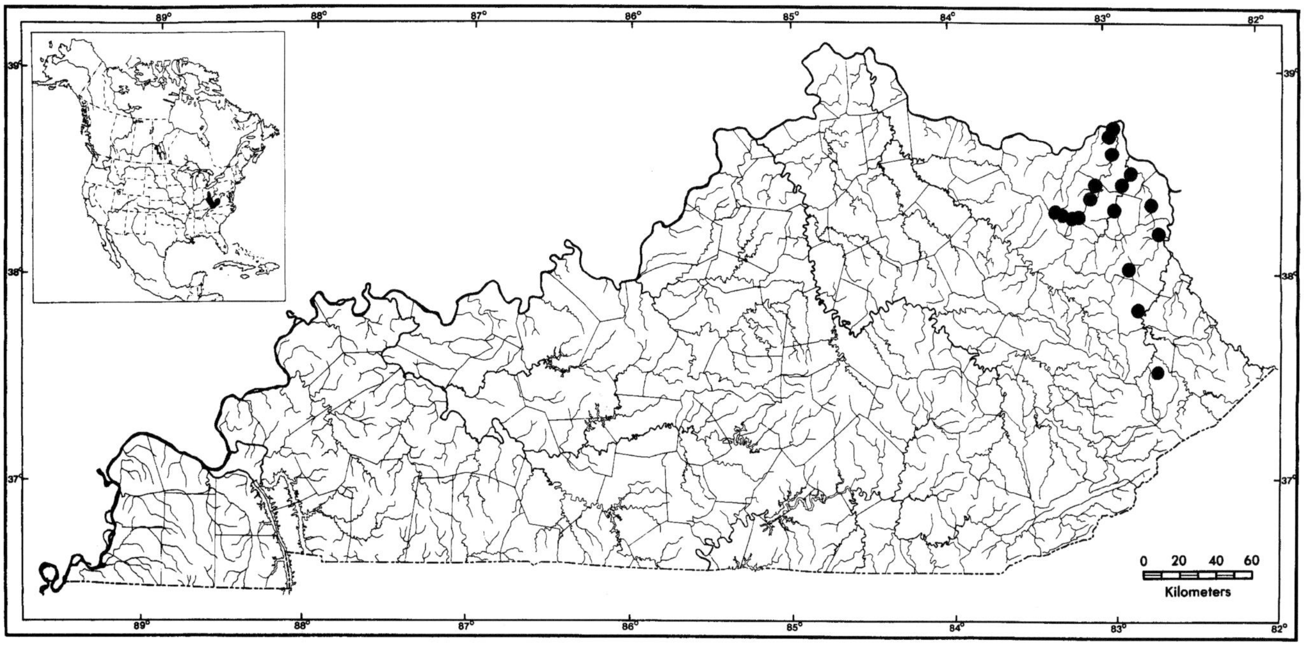

Fig. 82. Distribution of *Cambarus sciotensis* in Kentucky.

Mesial process bulbous at base, narrows to tip. Annulus ventralis of female subrhomboidal, with distinctly narrow transverse ridge; deep trough through cephalic half; deep C–shaped fossa and S–shaped sinus.

MAXIMUM SIZE – The maximum known size for the species is 105.0 mm total length.

MOST LIKE – This species is most like *Cambarus bartonii cavatus.* They have similar coloration, and where their ranges overlap they may be collected together. Character differences between the two species include the shape of the rostrum and the chelae morphology. In *C. sciotensis* the rostral margins are noticeably thickened and nearly straight and the overall appearance of the rostrum is subquadrate, while in *C. bartonii cavatus* the rostral margins are somewhat thickened, and they are rounded anteriorly and the rostrum is much less subquadrate in shape. In *C. bartonii cavatus* the rostrum is usually greatly excavated and spoon shaped, while in *C. sciotensis* the rostrum is usually much more flattened. *Cambarus sciotensis* also has strong longitudinal ridges on the fingers of the chelae, while in *C. bartonii cavatus* the ridges are weak.

DISTRIBUTION AND HABITAT – (Fig. 82) According to Jezerinac et al. (1995), this species is found in central Ohio in the Scioto and Little Scioto rivers and in direct tributaries of the Ohio river in south-central Ohio. In West Virginia and Virginia it is restricted to the Big Sandy, Guyandotte, and New rivers. In Kentucky, *Cambarus sciotensis* is restricted to the Big and Little Sandy River drainages in the far eastern part of the state.

Cambarus sciotensis usually occurs under large flat rocks in medium to large creeks and rivers with cobble and boulder substrates. Roell and Orth (1993) found this species in both pools and riffles in the New River.

LIFE HISTORY – Form I males have been collected from Kentucky in May, October, and November. The lack of form I males from other months is most likely due to a lack of field efforts. No ovigerous females have been collected from the state. In West Virginia Jezerinac et al. (1995) found form I males during March, April, June, July, August, and September, and females with young in late August and early September. They found a positive relationship between size of the female and the total number of eggs she carried. The number of eggs ranged from about 90 to about 200 per female; the smallest female with eggs had a total carapace length of 33 mm. Egg diameters ranged from 2.6 to 2.7 mm. Jezerinac et al. (1995) indicated that most males molted to form I in late June with copulation taking place during June and early July. The females were ovigerous during July and August, and the eggs hatched in late August to early September. Both Jezerinac et al. (1995) and Roell and Orth (1993) have suggested a four-year life span for *C. sciotensis*.

CRAYFISH ASSOCIATES – We have collected *Cambarus sciotensis* with *C. bartonii cavatus*, *C. robustus*, *Orconectes cristavarius*, and *O. sanbornii.*

Cambarus sphenoides Hobbs
Triangleclaw Crayfish

Fig. 83

LIFE COLORS – (Fig. 83) Carapace uniform green to light brown in color, often with slightly darker mottling. Chelae with similar color and pattern as carapace, but usually lighter in color; abdomen of smaller individuals may have pair of distinct dorsal stripes from end of carapace to base of telson and uropods; legs lighter straw or light green in color. Venter cream to whitish.

CHARACTERS – (Fig. 84) Rostrum flattened, only slightly excavated; margins thickened with slight thinning into acumen; margins nearly straight and converging apically; acumen short and triangular with small tubercle at tip. In dorsal view, carapace bullet shaped; in lateral view posterior dorsal portion flat, anterior portion slightly vaulted and rounded. Areola moderately wide, usually 4.6 to 6.6 times longer than wide at its narrowest point; room for four to five punctations, sparsely punctate. Postorbital ridge flattened, ending with rounded tubercle. Hepatic area covered with large round tubercles, branchiostegals covered with small tubercles; cervical spines absent, small tubercle may be present; branchiostegal spines well developed with pointed tubercle; suborbital angle well developed. Antennal scale greatest width distal to midpoint. Chelae triangular in outline, very heavy in appearance. Dorsal surfaces of fingers of chelae with well-developed longitudinal ridges. Mesial margin of palm of chela with two complete rows of tubercles, seven to eight large, serrate tubercles in mesialmost row; second row with five to six smaller tubercles; surface of palm usually covered with few to many very small tubercles. Two to four subpalmar tubercles present. Epistomal zygoma strongly arched.

Central projection of gonopod of form I male corneous, recurved more than 115° to 130° to shaft, with subapical notch. Mesial process bulbous basally, tapering apically to nipplelike structure. Annulus ventralis of female subrhomboidal, with deep C–shaped fossa and S-shaped sinus.

MAXIMUM SIZE – The maximum known size for the species is 89.7 mm total length.

MOST LIKE – With triangular-shaped chelae and two complete rows of seven or fewer tubercles along the mesial margin of the palm, *Cambarus sphenoides* most resembles *C. graysoni* and *C. striatus*. *Cambarus graysoni* differs from *C. sphenoides* in having a weakly arched epistomal zygoma. *Cambarus striatus* differs from *C. sphenoides* in having an obliterated suborbital angle and a narrower areola. In *C. sphenoides* the suborbital angle is well developed and the areola has room for more than two punctations.

DISTRIBUTION AND HABITAT – (Fig. 85) This species is nearly endemic to the Cumberland Plateau, and is found in the Tennessee, Cumberland, and Kentucky River drainages in Kentucky and eastern Tennessee (Bouchard 1978; Hobbs 1989). In Kentucky, *Cambarus sphenoides* has a similar distribution as *Cambarus distans*. It is sporadic in the upper Cumberland River drainage and also gets into the small headwater creeks of

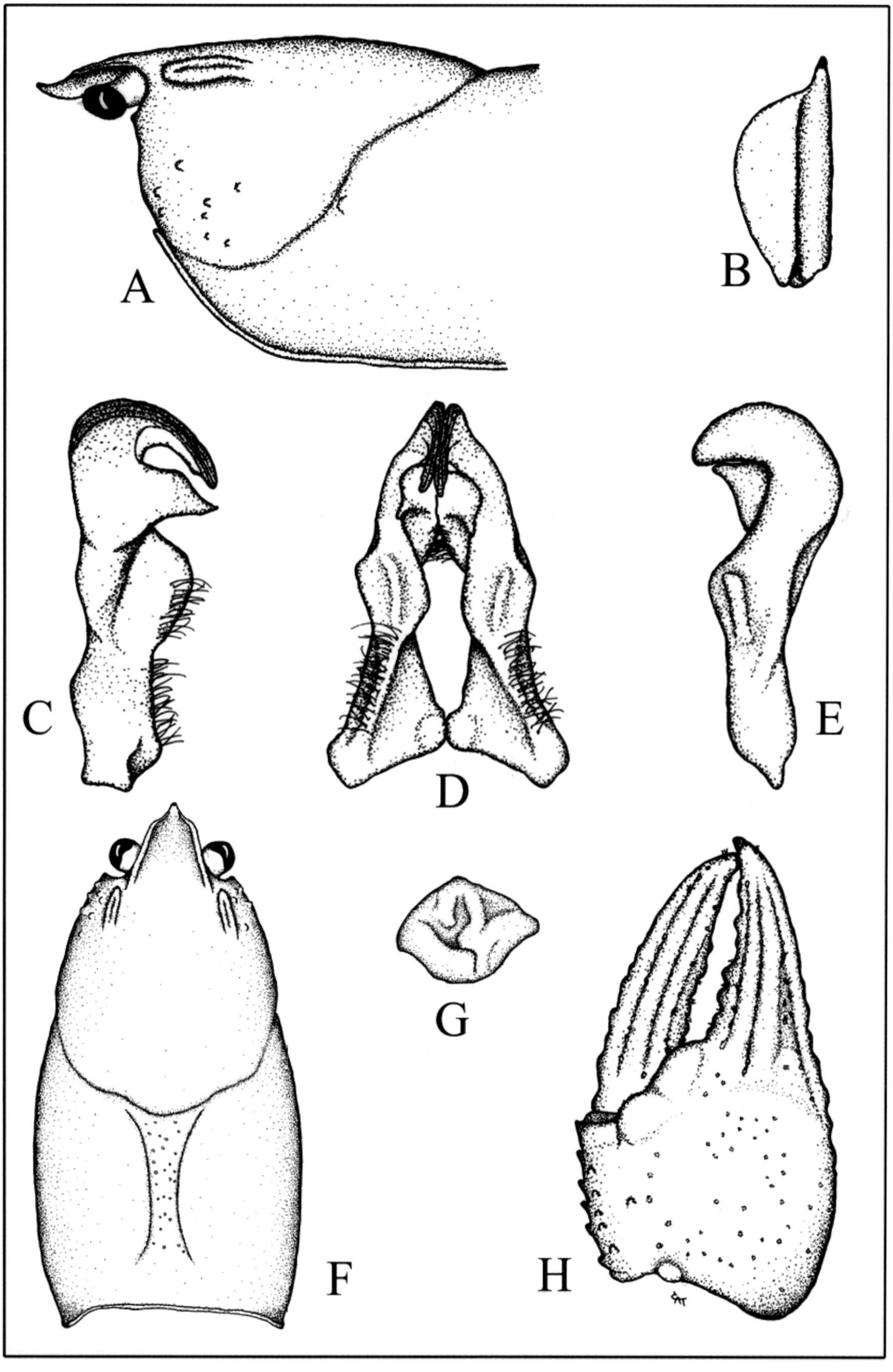

Fig. 84. *Cambarus sphenoides*: A, lateral view of carapace; B, dorsal view of right antennal scale; C, mesial view of form I gonopod; D, ventral view of form I gonopods; E, lateral view of form II gonopod; F, dorsal view of carapace; G, ventral view of annulus ventralis; H, dorsal view of right chela.

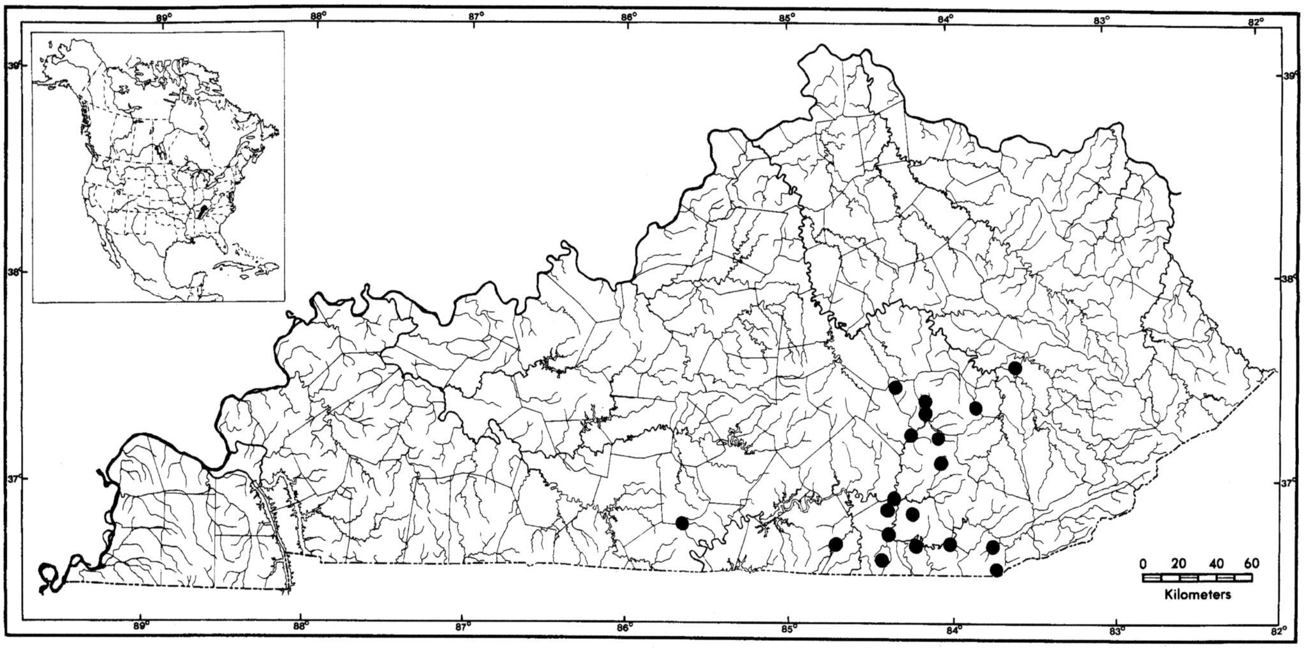

Fig. 85. Distribution of *Cambarus sphenoides* in Kentucky.

the Kentucky River drainage. It is not as common as *C. distans*; however, the two species are often collected together.

Cambarus sphenoides is a secondary burrower, and because of this, if sampling during dryer months it is very easy to overlook. We have collected it most commonly while sampling smaller creeks with a seine, where it occurs under large, flat boulders or occasionally among woody debris. Along these creeks, marginal burrows are usually common, but are very difficult to excavate because of the rocky substrate through which the burrows traverse.

LIFE HISTORY – Form I males were collected in Kentucky in March, April, and May. The absence of form I males from other months is most likely due to the lack of field efforts in other months in the range of *C. sphenoides*. Ovigerous females were not collected by us, not present in museum holdings, nor reported in the literature from other states. Nothing else is known about the life history of this species.

CRAYFISH ASSOCIATES – *Cambarus sphenoides* has been collected with *C. cumberlandensis*, *C. distans*, and *Orconectes cristavarius*.

Cambarus striatus Hay
Ambiguous Crayfish

Fig. 86

LIFE COLORS – (Fig. 86) Because so few live individuals were collected by us, the following color notes were taken from Bouchard (1978). Specimens range from uniform green or green–brown to brown, often with mottling. Mandibular muscle scars on carapace with lighter vermiculated pattern. Abdomen in younger specimens with two distinct dark brown stripes that often begin on carapace and go length of abdomen to base of telson. In older specimens stripes may not be evident. Legs, including chelae, somewhat lighter in color than carapace. Distal ends of chelae red, orange, yellow, or cream.

CHARACTERS – (Fig. 87) Rostrum moderately excavated; margins moderately thickened to acumen, almost straight and slightly converging, rounded anteriorly to acumen; acumen very short, with small apical tubercle. In dorsal view, carapace bullet shape; in lateral view anterior portion moderately vaulted and rounded. Areola very narrow, sometimes obliterated; length up to 9.1 times the width (Bouchard 1978), with room for one or occasionally two punctations. Postorbital ridge flattened without spines or tubercles at end. Hepatic area and branchiostegals covered with many small tubercles; cervical spines absent, sometimes with small tubercle; branchiostegal spines well developed; suborbital angle obsolete, broadly rounded. Antennal scale greatest width distal to midpoint. Chelae triangular in outline. Dorsal surfaces of fingers of chelae with well-developed longitudinal ridges. Mesial margin of palm of chela with two complete rows of strong tubercles; both rows with six to eight tubercles. Two subpalmar tubercles present. Epistomal zygoma strongly arched.

Central projection of gonopod of form I male corneous, recurved 95° to 155° to shaft; most populations lack subapical notch, but distinct notch present in some populations (Bouchard 1978). Mesial process bulbous at base, tapering to one or more subacute to acute tips. Annulus ventralis of female asymmetrical, subquadrangular in outline, with median trough through cephalic half, and sinus running through elevated caudal half.

MAXIMUM SIZE – The maximum known size for the species is 107.8 mm total length.

MOST LIKE – *Cambarus striatus* is most like *C. graysoni*. The coloration and general appearance of the two species are very similar. The main differences we have found in the two species is in the suborbital angle and the degree of curvature of the epistomal zygoma. In *C. striatus* the suborbital angle is obsolete and the area is broadly rounded, while in *C. graysoni* there is a distinct and well-developed suborbital angle. Secondly, *C. striatus* has a strongly arched epistomal zygoma, while in *C. graysoni* the epistomal zygoma is moderately arched to almost straight in some specimens (Bouchard 1978). *Cambarus striatus* also resembles *C. sphenoides* in chelae shape and color; *C. sphenoides* differs from *C. striatus* by possessing an acute suborbital angle and a wider areola.

DISTRIBUTION AND HABITAT – (Fig. 88) This species is widespread throughout the southeastern United States from Kentucky south to Mississippi and east to northwestern Georgia (Bouchard 1978; Hobbs 1989). In Kentucky the species has a disjunct and sporadic distribution. It is known from the lower Green River drainage, and from three isolated localities in the middle Kentucky River drainage. The reasons for its disjunct distribution in the state is unknown.

This species has been variably described as a creek dweller, a primary burrower, and a secondary burrower. In more northern populations it apparently is more commonly found in creeks and rivers and may be a secondary burrower, while the more southern populations are primary burrowers (Hobbs 1981). In Kentucky we have

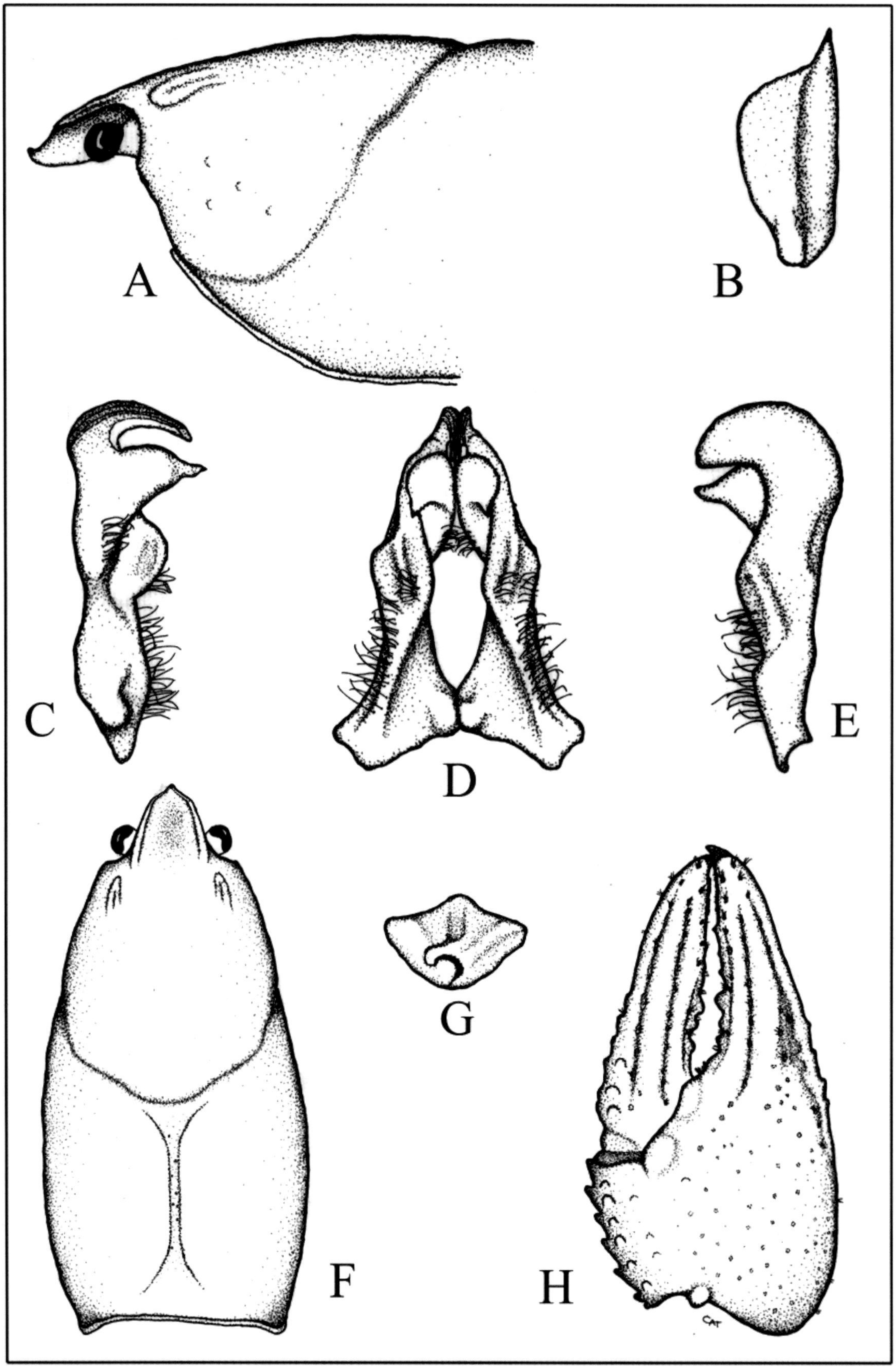

Fig. 87. *Cambarus striatus*: A, lateral view of carapace; B, dorsal view of right antennal scale; C, mesial view of form I gonopod; D, ventral view of form I gonopods; E, lateral view of form II gonopod; F, dorsal view of carapace; G, ventral view of annulus ventralis; H, dorsal view of right chela.

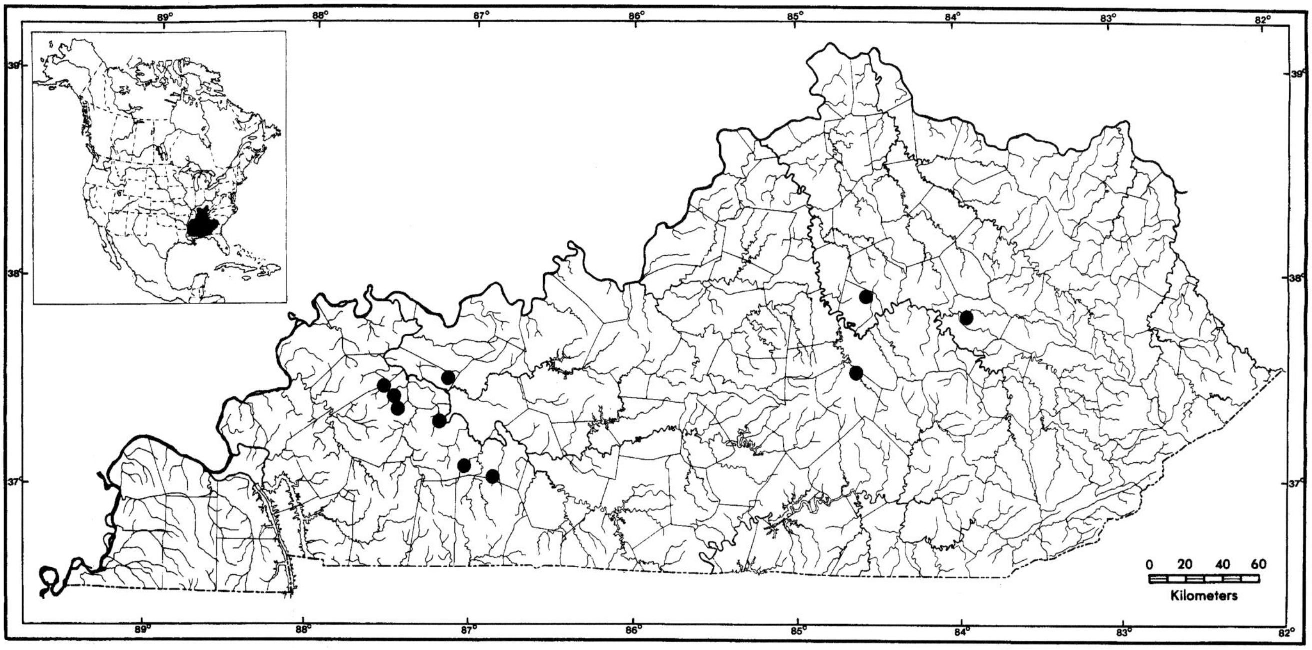

Fig. 88. Distribution of *Cambarus striatus* in Kentucky.

found it only in creeks and rivers of all sizes under large rocks or in submerged debris such as roots and downed trees. In these areas it may be a secondary burrower; however, we have not collected it from burrows.

LIFE HISTORY – Almost nothing is known about the life history of this species. Hobbs (1981) reported form I males from all months except June and August. Bouchard (1978) collected a single female with young on October 9, 1977. He did not report the number of young collected with this female. We have not collected form I males or ovigerous females in Kentucky.

CRAYFISH ASSOCIATES – *Cambarus striatus* has been collected with the following species: *C. tenebrosus, Fallicambarus fodiens, Orconectes immunis, O. juvenilis, O. rafinesquei,* and *O. ronaldi.*

Cambarus tenebrosus Hay
Cavespring Crayfish

Fig. 89A: normal color form.

Fig. 89B: spotted color form.

LIFE COLORS – (Figs. 89A, B) Most populations of this species range from uniform golden yellow to brown to green. Some populations with distinct mottled pattern. In these specimens base color is as described above, but mottling much darker and ranges from rust red to almost black. Mottling most prominent on carapace but may also be on chelae and abdomen. Abdomen may be almost uniform in color, but in mottled individuals dorsum of abdomen may have a pair of broken narrow bands. Legs, other than chelipeds, usually much lighter in color ranging from light brown or green to cream color. Occasionally cave specimens may be somewhat lighter in color but never colorless.

CHARACTERS – (Fig. 90) Rostrum moderately excavated; margins thickened and straight, converging to acumen; acumen short, triangular, sides not thickened, ending in small tubercle. Eyes small. Carapace in dorsal view very wide, almost rectangular in outline. Areola moderately wide, usually 6.5 to 12.0 times longer than wide at its narrowest point, with room for two to four punctations. Postorbital ridge flattened with apical tubercle or small spine. Hepatic area with small tubercles on ventral half; cervical spines poorly to well developed, always present; branchiostegal spines well developed; suborbital angle acute, tipped with sharp tubercle or spine. Antennal scale widest distal to midpoint. Chelae elongated and smooth, rectangular in outline. Dorsal surfaces of fingers of chelae with poorly developed longitudinal ridges. Palm of chelae long. Mesial margin of palm of chela with two complete rows of tubercles; mesialmost row with 7 to 10 low, rounded tubercles; second row with 6 to 8 smaller tubercles. One to three subpalmar tubercles present. Epistomal zygoma moderately arched.

Central projection of gonopod of form I male corneous, recurved slightly more than 90° to shaft, with well-developed subapical notch. Mesial process bulbous at base, tapering to tip. Annulus ventralis of female subrhomboidal with medial longitudinal trough through cephalic half, deep C-shaped fossa and S-shaped sinus.

MAXIMUM SIZE – The maximum known size for the species is 119.05 mm total length.

MOST LIKE – *Cambarus tenebrosus* in general appearance and color most closely resembles *C. bartonii cavatus*. The two are often collected together in the central part of the state. *Cambarus bartonii cavatus* can be distinguished from *C. tenebrosus* by possessing a single complete row of less than eight tubercles along the mesial margin of the palm of the chela, while *C. tenebrosus* has two complete rows of tubercles along the mesial margin of the palm of the chelae with the mesialmost row having eight or more tubercles. In Kentucky *C. rusticiformis* is the closest relative to *C. tenebrosus*, and they may be collected together. The two can easily be separated. First, the margins of the rostrum of *Cambarus rusticiformis* are concave and terminate in spines, while in *C. tenebrosus* the rostral margins are nearly straight and strongly converge into the acumen. Second, *C. rusticiformis* has a single complete mesial row of tubercles on the palm of the chelae, while *C. tenebrosus* has two complete rows.

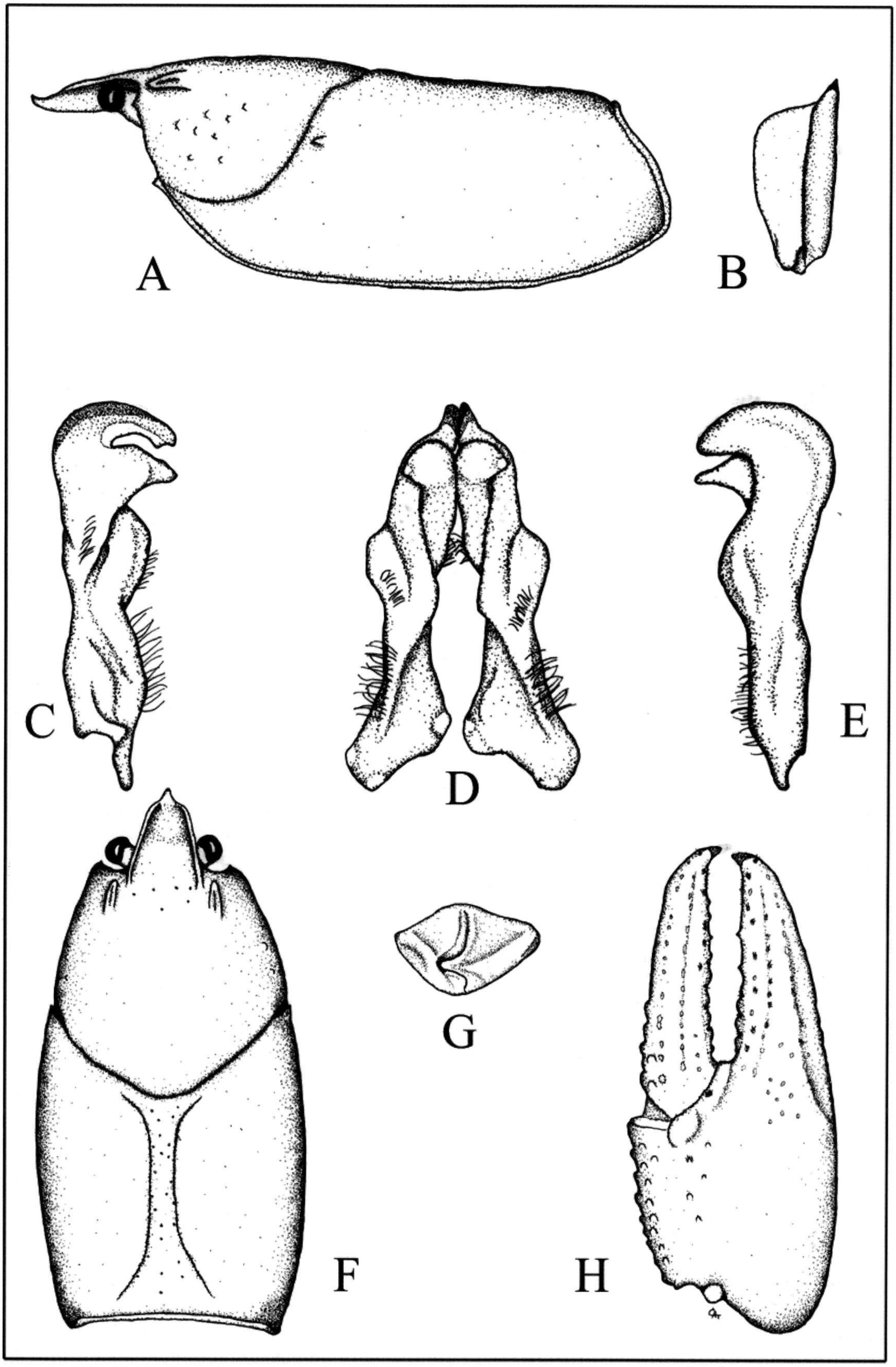

Fig. 90. *Cambarus tenebrosus*: A, lateral view of carapace; B, dorsal view of right antennal scale; C, mesial view of form I gonopod; D, ventral view of form I gonopods; E, lateral view of form II gonopod; F, dorsal view of carapace; G, ventral view of annulus ventralis; H, dorsal view of right chela.

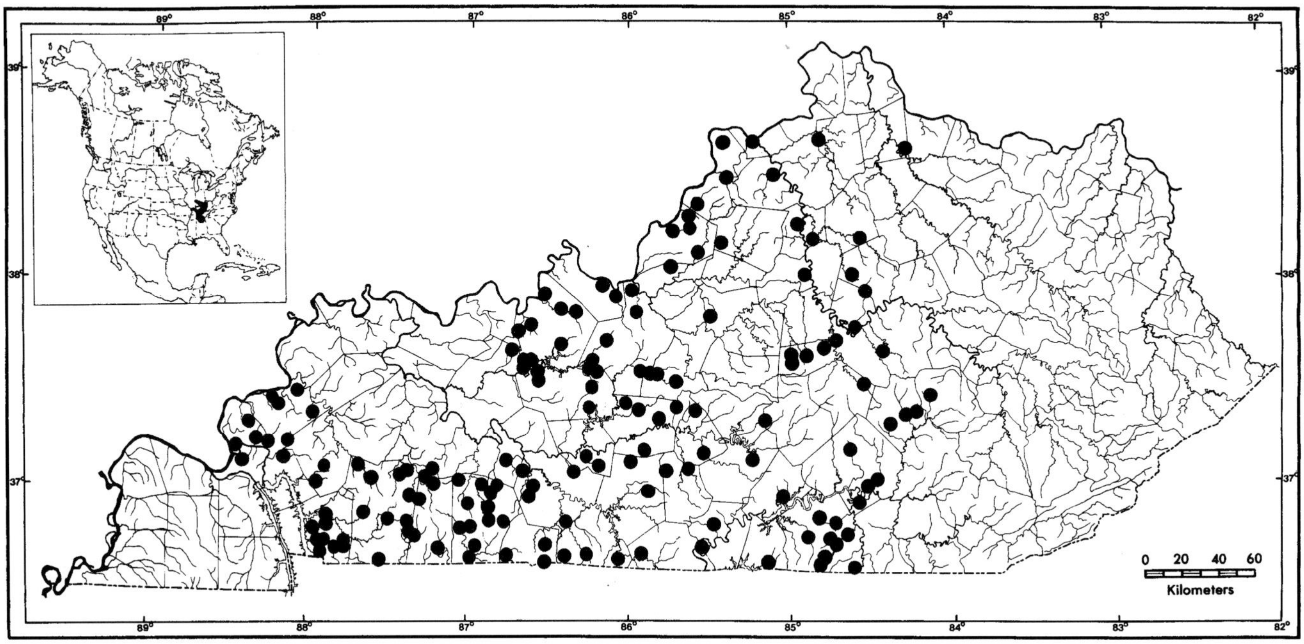

Fig. 91. Distribution of *Cambarus tenebrosus* in Kentucky.

DISTRIBUTION AND HABITAT – (Fig. 91) *Cambarus tenebrosus* is distributed east of the Mississippi River from the Tennessee River drainage in Alabama north to the Vermilion River drainage, a Wabash River tributary, in east–central Illinois (Taylor 1997). In Kentucky it is one of the most widespread species in the state. It is widely distributed from the Licking River drainage west to the lower Cumberland River drainage. It is not found in the lower parts of the Green and Tennessee River drainages, or in the Mississippi Embayment.

Cambarus tenebrosus is commonly associated with karst topography and is often collected from subterranean creeks in the twilight and dark zones of caves. It is also associated with springs, especially springs that originate in caves. Lastly, it is also commonly found in small to large creeks and rivers that are spring fed or partially spring fed. Prins (1968) thought that temperature played an important role in limiting the numbers of *C. tenebrosus* in the lower sections of Doe Run, Meade County, Kentucky. In epigean creeks and springs *C. tenebrosus* is most commonly found under large, flat boulders; however, in caves it is often seen in open water. In Doe Run, Prins (1968) reported that *C. tenebrosus* was also associated with dense moss mats and algal beds under which it lived.

LIFE HISTORY – In our collecting we have not found any ovigerous females. Form I males have been collected in Kentucky in January, March, April, May, October, and November. Prins (1968) studied the ecology of *Cambarus tenebrosus* in Doe Run. He found that mating occurs during the winter and spring. Females oviposit during June and July, and the young are hatched in "secretive places in deeper water" or in burrows in July and August. Seven ovigerous females were observed with eggs ranging from 186 to 266 in number. Egg size averaged approximately 3.0 mm in diameter. Ovigerous females secreted themselves away under dense algal mats or in marginal burrows in the bank. He found that the young after hatching stayed in the burrows until October or November. He estimated that sexual maturity took 20 to 22 months to reach for both sexes, and usually occurred when an individual attained a total length of about 40 mm. He indicated that females probably reproduced twice, and individuals of this species probably lived for 36 to 38 months. Stomach content analysis indicated that *C. tenebrosus* fed primarily on vascular plants, filamentous algae, and detritus.

CRAYFISH ASSOCIATES – *Cambarus tenebrosus* has been collected with the following species: *Barbicambarus cornutus*, *C. cumberlandensis*, *C.*

5/20/05 mouth of Grahamton Cave, Camp Carlson, Meade Co. Ky.

diogenes, *C.* sp. A, *C. graysoni*, *C. ortmanni*, *C. rusticiformis*, *Orconectes barrenensis*, *O. compressus*, *O. margorectus*, *O. immunis*, *O. placidus*, *O. putnami*, *O. rafinesquei*, *O. ronaldi*, *O. rusticus*, and *O. tricuspis*.

TAXONOMIC NOTE – In addition to *Cambarus rusticiformis* and *C. tenebrosus*, Hobbs (1989) also recognized two other species, *C. ornatus* Rhoades and *C. laevis* Faxon, belonging to the subgenus *Erebicambarus* as occurring in Kentucky. Taylor (1997) analyzed morphological variation in *C. tenebrosus*, *C. laevis*, *C. ornatus* and an additional species, *C. cahni* Rhoades, known from a single cave in northwestern Alabama. His findings indicated that the latter three nominal species were junior synonyms of *C. tenebrosus*. We have followed the findings of Taylor (1997) and report only *C. tenebrosus* for the state.

Cambarus thomai Jezerinac
Little Brown Mudbug

Fig. 92

LIFE COLORS – (Fig. 92) Individuals uniform dark green to brown. Tips of fingers of chelae and rostral margins orange to red, rostral margins occasionally more yellow to cream. Ventral surface of crayfish uniform cream to straw color.

CHARACTERS – (Fig. 93) Rostrum deeply excavated; margins thickened, more or less straight, converging apically, no marginal spines. Acumen very short, triangular, not clearly defined, small tubercle at tip. In dorsal view carapace slender and bullet shaped; in lateral view dorsally rounded. Areola obliterated. Postorbital ridge flattened with no terminal tubercles or spines. Hepatic area with few scattered round tubercles, branchiostegal with small tubercles on sides; cervical spines absent, but with small tubercle in area; branchiostegal spines absent; suborbital angle acute, with no tubercle or spine at tip. Antennal scale widest at midpoint. Chelae heavy, elongated and triangular in appearance. Dorsal surfaces of dactyl of chela with strong longitudinal ridges; ridges on propodus not as prominent. Mesial margin of palm of chela with six to eight tear-shaped, serrate tubercles; numerous tubercles cover entire dorsal surface of palm. Zero or one subpalmar tubercle. Epistomal zygoma strongly arched.

Central projection of gonopod of form I male corneous and short, recurved about 90° to shaft, without subapical notch. Mesial process bulbous at base narrows apically to point. Annulus ventralis of female subcircular with very deep central fossa and short S–shaped sinus.

MAXIMUM SIZE – The maximum known size for the species is 106.0 mm total length.

MOST LIKE – By possessing an obliterated areola and a broad concavity at the base of the opposable margins of the dactyl, *C. thomai* most closely resembles *C.* sp. A. The two have quite different color patterns. Whereas *C. thomai* tends to be more uniform in color, *C.* sp. A is very colorful with bright reds, oranges, and yellows on the chelae and on all of the abdominal tergites. The rostrum of *C. thomai* tends to be straight and directed slightly downward, while the rostrum of *C.* sp. A is broadly rounded from the eye forward to the tip and strongly directed downward at the tip. *Cambarus thomai* also resembles *C. diogenes* and *C. ludovicianus*; however, they are rarely collected together. *Cambarus thomai* differs from these two species in that the length of the dactyl of the chelae in *C. thomai* is less than 1.8 times the length of the palm while in the other two species the length of the dactyl of the chelae is more than 1.9 times the length of the palm (Jezerinac 1993). In addition, *C. diogenes* and *C. ludovicianus* have a distinct triangular space on the palm of the chelae where there are no tubercles; *C. thomai* lacks this space and tubercles completely cover the palm of the chelae.

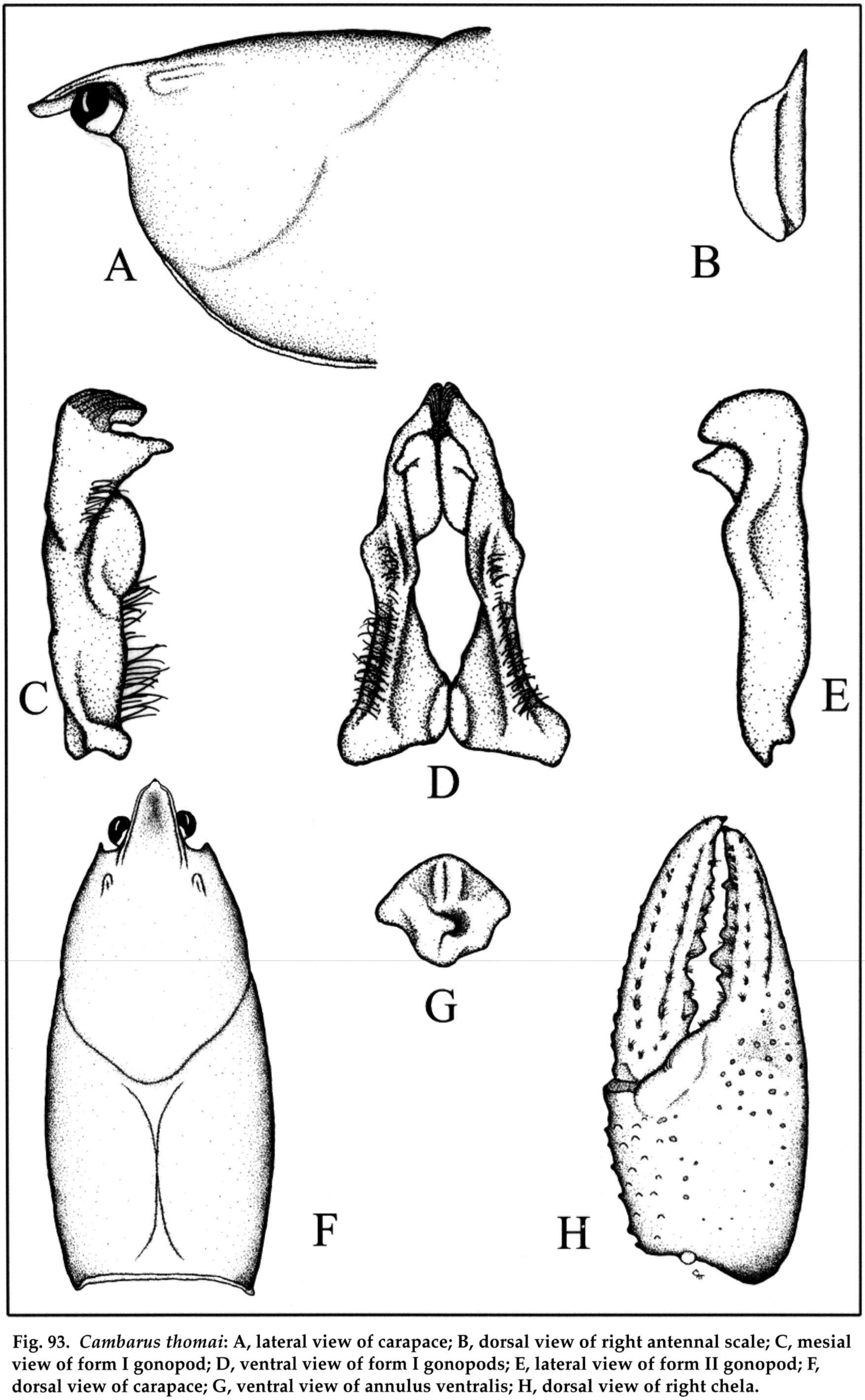

Fig. 93. *Cambarus thomai*: A, lateral view of carapace; B, dorsal view of right antennal scale; C, mesial view of form I gonopod; D, ventral view of form I gonopods; E, lateral view of form II gonopod; F, dorsal view of carapace; G, ventral view of annulus ventralis; H, dorsal view of right chela.

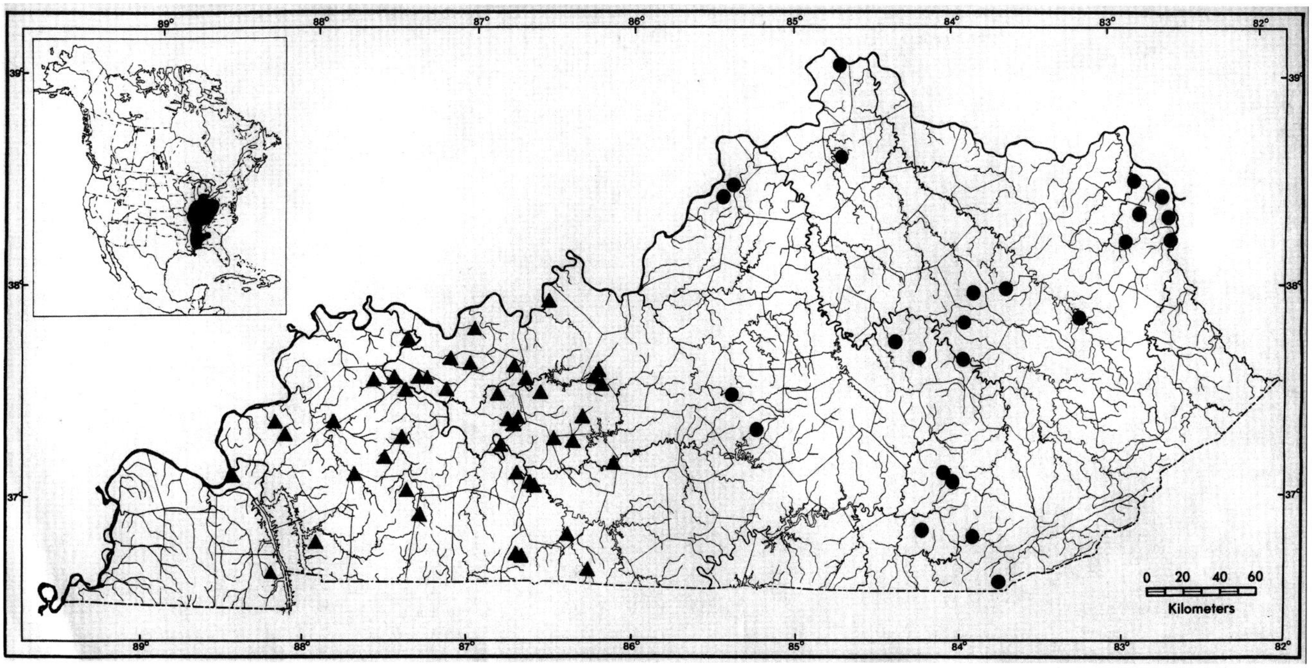

Fig. 94. Distribution of *Cambarus thomai* and *C.* sp. A in Kentucky: solid circles = *C. thomai*; solid triangles = *C.* sp. A. Total North American range is that of *C. thomai* and *C.* sp. A.

DISTRIBUTION AND HABITAT – (Fig. 94) The range of *Cambarus thomai* includes most of Ohio, southwestern Pennsylvania, western West Virginia, and eastern Kentucky (Jezerinac 1993; Jezerinac et al. 1995). In Kentucky *C. thomai* occurs sporadically from the Big and Little Sandy River drainages in the far eastern part of the state west to the uppermost Salt and Green River drainages and south to the upper Cumberland River drainage.

Jezerinac et al. (1995) reported that *Cambarus thomai* was a primary and secondary burrower that is common in wet roadside ditches. We have found this species in similar habitat but also in burrows along the margins of creeks. The single ovigerous female we collected was found in a dense brush pile within a first order creek. Jezerinac et al. (1995) reported similar findings of ovigerous females in creeks that were lined with burrows. Jezerinac et al. (1995) describe in detail the burrow construction of this species.

LIFE HISTORY – Form I males have been collected in Kentucky during January and August. Jezerinac (1993) and Jezerinac et al. (1995) reported form I males from March through October. Jezerinac et al. (1995) reported the smallest form I male that they collected had a total carapace length of 26.6 mm, which suggested to them that breeding males must be at least two years old. Jezerinac (1993) reported ovigerous females in March, April, May, and June and females with attached young in May and June. Jezerinac et al. (1995) reported collecting ovigerous females in March and April. They also reported that a female had 430 eggs attached to the abdomen while another had 197 eggs, and that the smallest known ovigerous female had a total length of 62 mm. We collected an ovigerous female with a total length of 86.4 mm on March 24, 1997, which had 184 eggs attached. The eggs had a mean diameter of 2.0 mm.

CRAYFISH ASSOCIATES – *Cambarus thomai* has been collected near burrows of *C. deweesae* and *C. ortmanni*.

Cambarus veteranus Faxon
Big Sandy Crayfish

Fig. 95

LIFE COLORS – (Fig. 95) Carapace, chelae, and abdomen uniform dark green to brown. Rostral margins and large tubercles on chelae bright orange to deep red. Legs and venter somewhat lighter, cream to whitish in color.

CHARACTERS – (Fig. 96) Rostrum moderately excavated, narrow; margins moderately thickened, greatly converging and slightly rounded at base of acumen. Acumen triangular with corneous tubercle at tip. In dorsal view, carapace very wide at branchiostegals, narrows somewhat anteriorly to head; in lateral view dorsoventrally flattened. Areola very wide, usually 3.7 to 5.6 times longer than wide at its narrowest point, room for 8 to 10 punctations. Postorbital ridge flattened, with a tubercle and well-developed corneous spine at tip. Hepatic area with moderately dense covering of large round tubercles; branchiostegals laterally covered with small round tubercles; cervical spines well developed; branchiostegal spines well developed; suborbital angle absent. Antennal scale greatest width distal to midpoint. Chelae triangular in outline, with long fingers. Dorsal surfaces of fingers of chelae without strong longitudinal ridges. Mesial margin of palm of chela with two rows of moderately large, rounded tubercles; mesialmost row with 9 to 10 tubercles; second row with 8 to 10 smaller tubercles. No subpalmar tubercles. Epistomal zygoma strongly arched.

Central projection of gonopod of form I male corneous, recurved 90° to 105° to shaft, with strong subapical notch. Mesial process bulbous at base and apically ends in one or two points. Annulus ventralis of female subrhomboidal, asymmetrical to symmetrical in outline with median trough through cephalic half, and a deep C-shaped fossa and S-shaped sinus.

MAXIMUM SIZE – The maximum known size for the species is 96.4 mm total length.

MOST LIKE – *C. veteranus* is most likely to be confused with *C. robustus* since both are large crayfish with elongated chelae and deep dorsoventral depressions at the base of the propodus. However, it can be separated from *C. robustus* by possessing a narrow triangular rostrum and by lacking a suborbital angle.

DISTRIBUTION AND HABITAT – (Fig. 97) *Cambarus veteranus* has a relatively limited distribution. It is known from the upper Guyandotte and Bluestone rivers within the New River drainage in West Virginia and the Big Sandy River drainage in Virginia and Kentucky (Hobbs 1989; Jezerinac et al. 1995). In Kentucky it is known only from the upper portions of the Big Sandy River drainage in Floyd and Pike counties. A single collection from the Red River in Estill Co. (Kentucky River drainage) from August 1967 is in the EKU Crayfish Collection. Subsequent collections in the area have been fruitless; therefore, we consider it to be a dubius record.

Cambarus veteranus is most commonly found under very large flat boulders in both riffles and pools of medium-sized creeks and rivers.

LIFE HISTORY – We have collected form I males in July in Kentucky and Jezerinac et al. (1995) reported them in West Virginia during April, July, and August. No ovigerous females have ever been collected, nor has copulation been observed (Jezerinac et al. 1995). Nothing else is known about its life history.

Jezerinac et al. (1995) suggested that the limited distribution of the species may be due in part to coal mining in the areas from which it is known. They also suggested that it may not be able to compete well with *Cambarus sciotensis*, which is more common and widespread in the same drainages from which *C. veteranus* is known.

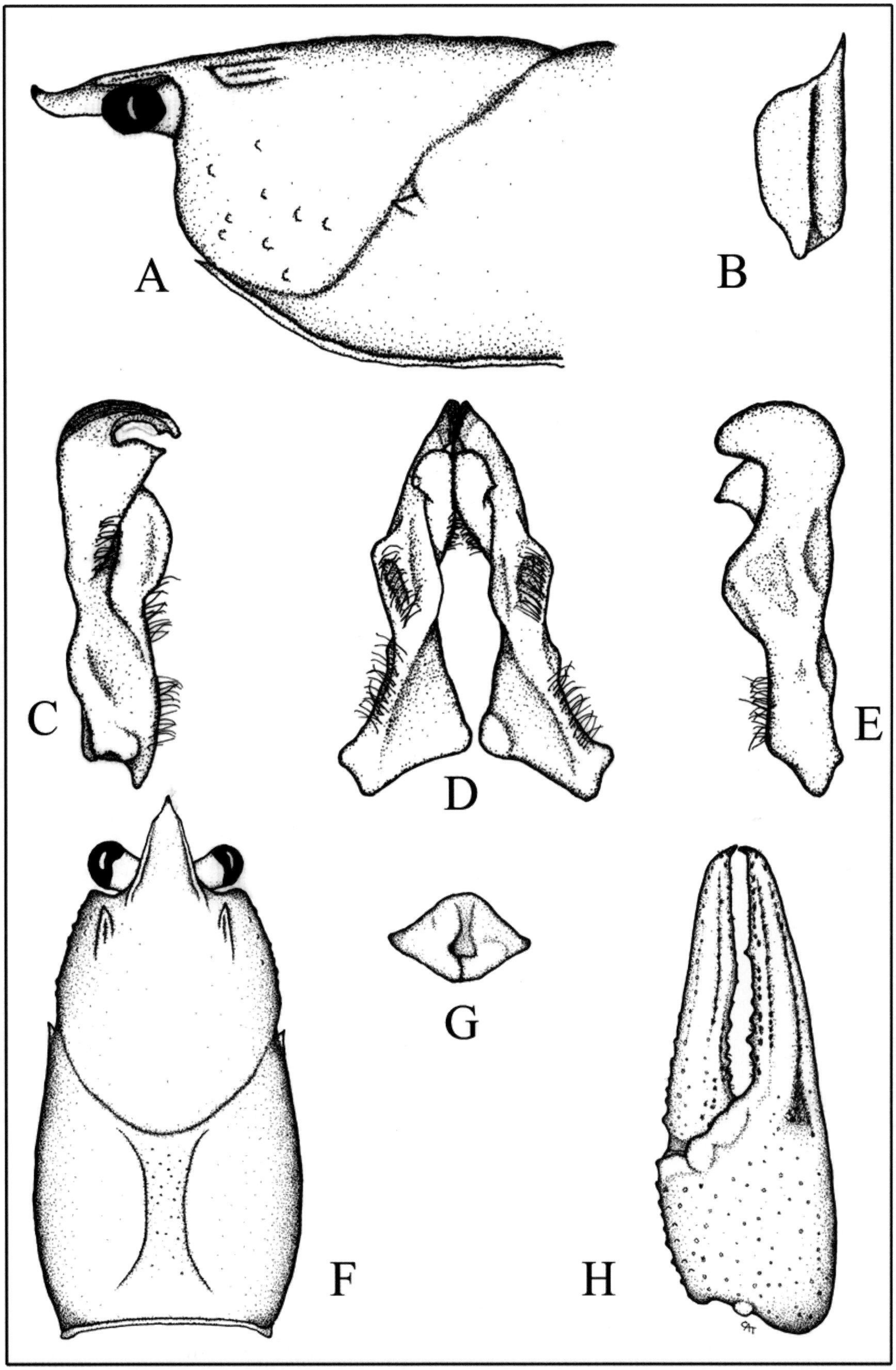

Fig. 96. *Cambarus veteranus*: A, lateral view of carapace; B, dorsal view of right antennal scale; C, mesial view of form I gonopod; D, ventral view of form I gonopods; E, lateral view of form II gonopod; F, dorsal view of carapace; G, ventral view of annulus ventralis; H, dorsal view of right chela.

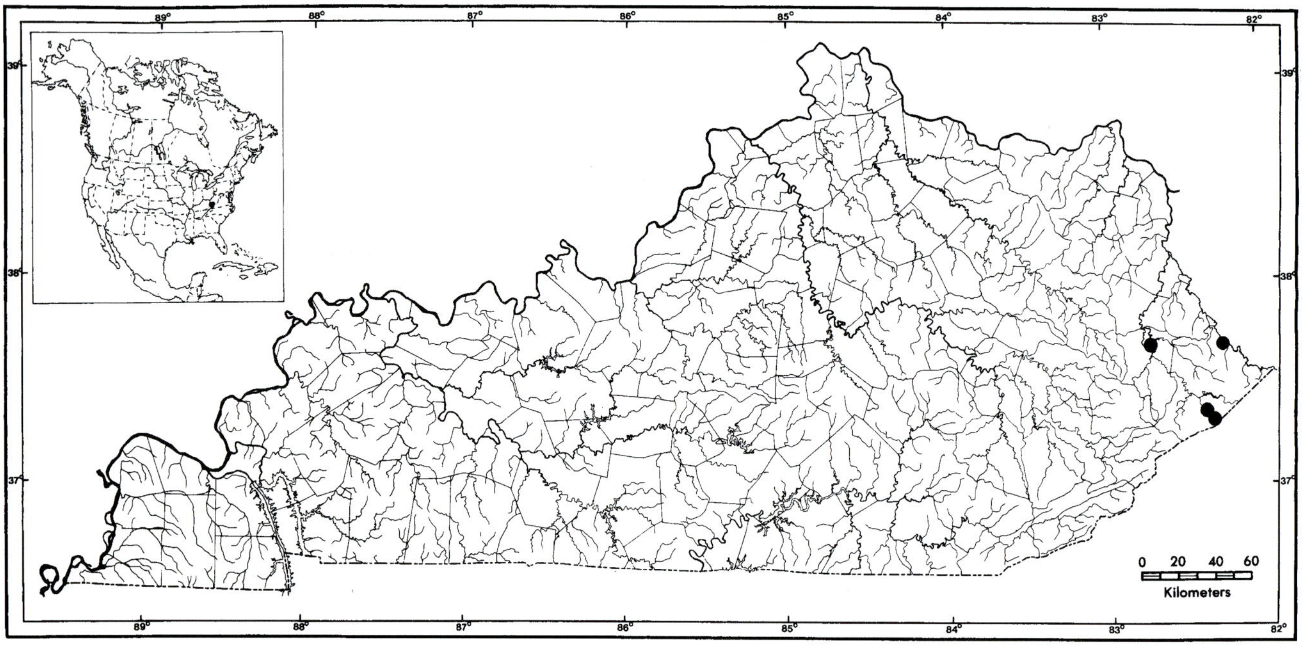

Fig. 97. Distribution of *Cambarus veteranus* in Kentucky.

CRAYFISH ASSOCIATES – *Cambarus veteranus* has been collected with the following species: *C. bartonii cavatus*, *C. sciotensis*, *C. robustus*, and *Orconectes cristavarius*.

CONSERVATION STATUS – AFS Threatened, KSNPC Special Concern.

Cambarus sp. A
Paintedhand Mudbug

Fig. 98

LIFE COLORS – (Fig. 98) Overall base color olive green to tan brown. Carapace, chelae, and abdomen with same base color. Conspicuous darker, round vermiculations on carapace at mandibular muscles. Rostral margins, large tubercles on chelae, posterior margins of all abdominal segments, telson, and uropods all trimmed in bright red to burgundy. Legs cream color, with joints highlighted in red. Venter cream to whitish.

CHARACTERS – (Fig. 99) Rostrum deeply excavated, more so distally; broad at base, gradually narrowing distally; margins slightly concave and thickened up to acumen; acumen short and triangular, small tubercle at tip. Carapace in dorsal view bullet shaped, entire carapace laterally compressed, gradually narrows anteriorly; in lateral view both sections of carapace broadly rounded. Areola obliterated. Postorbital ridge very low, no tubercle or spine at tip. Hepatic area sparsely covered with small tubercles, branchiostegals without tubercles, but surface rugose; cervical spines and branchiostegal spines absent; suborbital angle acute, but no tubercle or spine on end. Antennal scale greatest width distal to midpoint. Chelae subtriangular in outline; dactyl with broad notch at base, when fingers closed with large round opening at base. Dorsal surfaces of fingers of chelae with weakly developed longitudinal ridges; sometimes well devel-

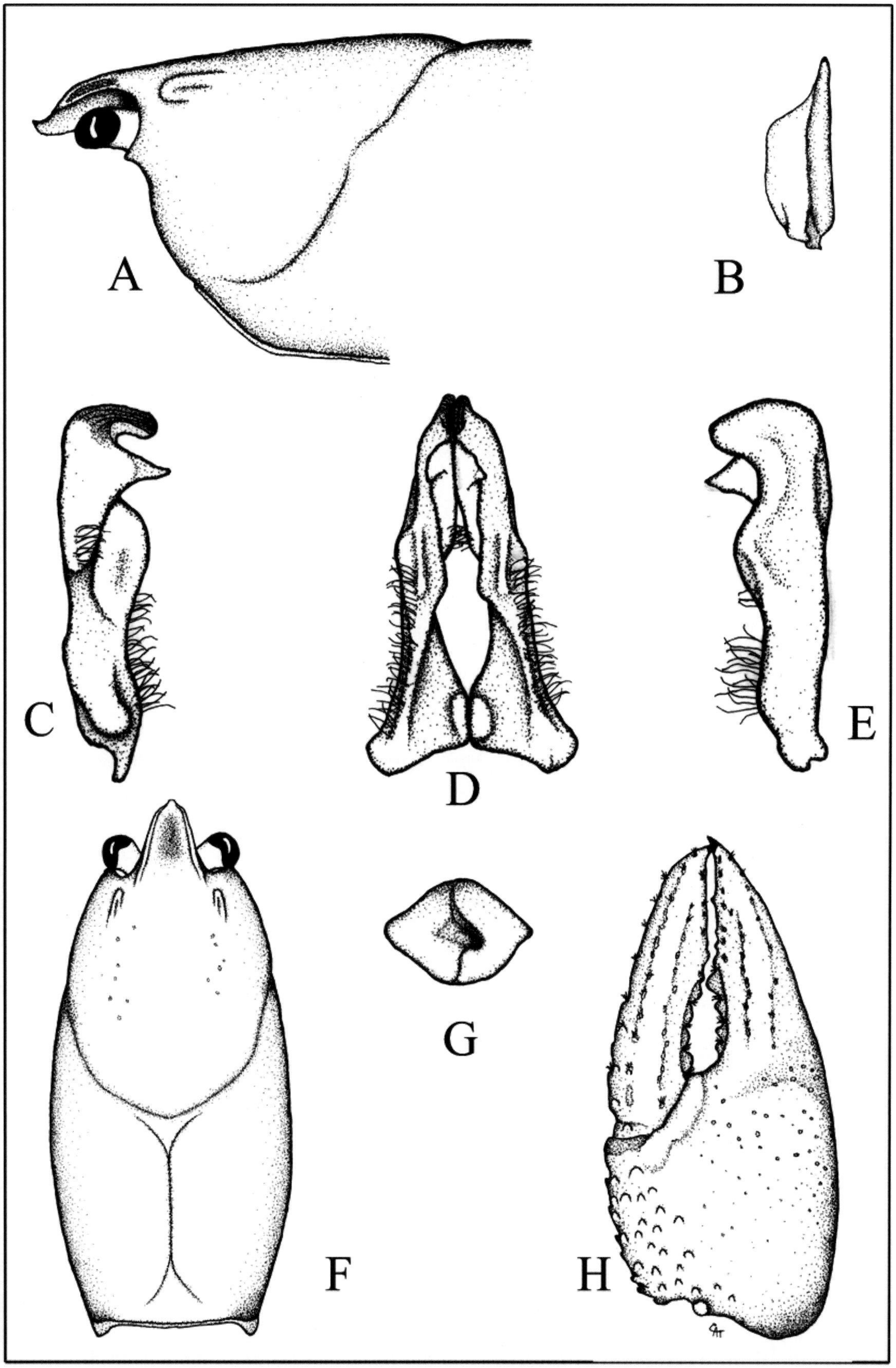

Fig. 99. *Cambarus* sp. A: A, lateral view of carapace; B, dorsal view of right antennal scale; C, mesial view of form I gonopod; D, ventral view of form I gonopods; E, lateral view of form II gonopod; F, dorsal view of carapace; G, ventral view of annulus ventralis; H, dorsal view of right chela.

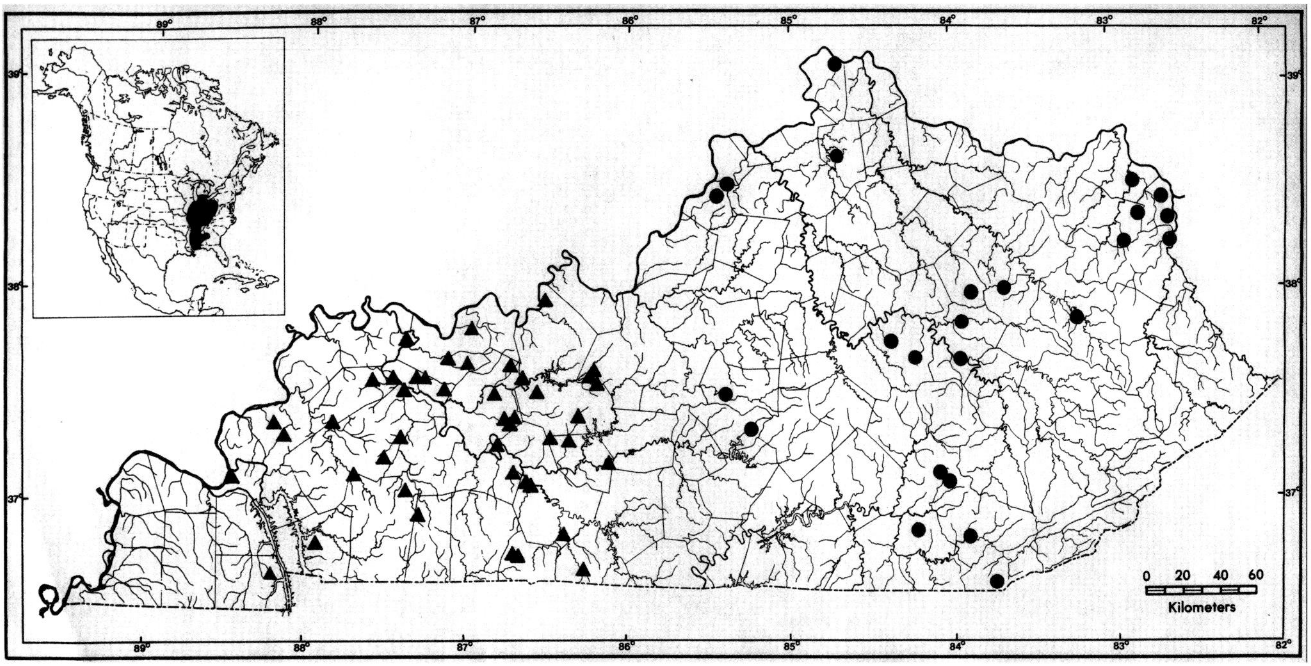

Fig. 94. Distribution of *Cambarus thomai* and *C.* sp. A in Kentucky: solid circles = *C. thomai*; solid triangles = *C.* sp. A. Total North American range is that of *C. thomai and C.* sp. A.

oped in form I males. Palm of chela covered with many tubercles, not in distinct rows, except for mesialmost row; mesialmost row with six to eight tubercles, basal tubercles large and serrate, distally are smaller and flattened; zero to two subpalmar tubercles. Epistomal zygoma strongly arched.

Central projection of gonopod of form I male corneous, short, and usually recurved approximately 90° to shaft, with no distinct subapical notch. Mesial process fingerlike, gradually tapers to end. Annulus ventralis of female round to subrhomboidal with deep central fossa, relatively long straight sinus.

MAXIMUM SIZE – The maximum known size for the species is 110.2 mm total length.

MOST LIKE – *Cambarus* sp. A most resembles *C. thomai, C. diogenes,* and *C. ludovicianus*. In life, C. sp. A differs from *C. thomai* in coloration. *Cambarus thomai* is more or less uniform in color from dark green to brown, while C. sp. A is very colorful. The most morphological character to distinguish the two species is the shape of the rostrum in a lateral view. In C. sp. A the rostrum is distinctly rounded from the eye to its end, and directed ventrally; in C. *thomai* the rostrum is more or less straight its entire length and just slightly directed ventrally. *Cambarus* sp. A differs from C. *diogenes* and C. *ludovicianus* in that the dactyl length is less than 1.8 times the length of the palm, while in the other two species the length of the dactyl is greater than 1.9 times the length of the palm. It also differs from the latter two species in that they usually have a distinct gap in the tubercles covering the palm of the chela.

DISTRIBUTION AND HABITAT – (Fig. 94) *Cambarus* sp. A ranges from southern Michigan, western Ohio, Indiana, most of Illinois, west to eastern Missouri, and south to the Gulf of Mexico in Mississippi, Alabama, and far western Florida (Jezerinac 1993). It is widely distributed in the western part of Kentucky from Louisville to the Mississippi River embayment.

This species is almost always found in burrows and most often lives as a secondary burrower; however, it may also be found in fields without adjacent surface waters. In western Kentucky we have found it most frequently in burrows along the margins of creeks. In the spring they are somewhat easy to collect since they like to be in lateral channels or chambers close to the surface. Large adults are also frequently encountered in spring months among woody debris or under large cobble in creeks.

LIFE HISTORY – Little is known about the life history of this species. We collected form I males in February, March, April, and May in Kentucky and ovigerous females in April and May. A single ovigerous female with a total length of 78.3 mm was collected on May 3, 2001. She carried 72 eggs with a mean diameter of 2.7 mm.

CRAYFISH ASSOCIATES – *Cambarus* sp. A has been collected with the following species: *Cambarus diogenes, C. graysoni, C. striatus, C. tenebrosus, Fallicambarus fodiens, Orconectes immunis, O. putnami, O. rafinesquei, O. ronaldi*, and *O. rusticus*.

TAXONOMIC NOTES – This is an undescribed species, currently being described by Roger Thoma. Jezerinac (1993) referred to this species as Species A; however, he did not provide any diagnosis distinguishing it from *Cambarus thomai*. Until the formal description of this species, these two taxa are currently recognized as *C. thomai*. As such we have plotted the ranges of both on a single distribution map.

Fallicambarus fodiens **Photo by M. Redmer**

Fallicambarus Hobbs

The genus *Fallicambarus* contains 16 species and occurs primarily along the Atlantic and Gulf Coastal plains of the southern and eastern United States. The single species that occurs in Kentucky, *F. fodiens*, occurs as far north as southern Ontario, Canada. All members of the genus are primary burrowers and resemble burrowing members of the genus *Cambarus* in gonopod and overall carapace shape. The most notable difference between members of the two genera is the deep notch at the base of the opposable margin of the dactyl found in members of *Fallicambarus*.

Fallicambarus fodiens (Cottle)
Digger Crayfish

Fig. 100

LIFE COLORS – (Fig. 100) Carapace, chelae, and abdomen uniform light to dark brown, sometimes almost grey in color; legs and venter of body may be significantly lighter, from straw to cream to almost white. Abdomen sometimes with darker brown longitudinal stripes that go full length of abdomen; stripes sometimes extend onto carapace.

CHARACTERS – (Fig. 101) Rostrum moderately excavated, bent ventrally; margins slightly thickened, straight, but slightly converging near tip, no rostral spines. Acumen short, margins thinner than rostral margins. Carapace in dorsal view bullet shaped, rounded and narrowing anteriorly; in lateral view rounded dorsally, especially anteriorly. Areola obliterated. Hepatic area with few tubercles; branchiostegals covered laterally with pinpoint small round tubercles; cervical spines absent; branchiostegal spines represented by very small tubercles; suborbital angle obsolete. Antennal scale widest distal to midpoint; scale length short, shorter than or just reaching tip of acumen; no spines on basal sclerites of antennae. Chelae almost oval or egg-shaped in outline; dorsoventrally flattened especially at base of fingers; in lateral view chela curved and sickle shape; often with dense patch of long setae at base of propodus; with very distinct basal notch on dactyl, gap present at area of notch when fingers closed. Dorsal surfaces of fingers of chelae without well-developed longitudinal ridges. Mesial margin of palm of chela with two distinct rows of tubercles; mesialmost row with six to eight tear-shaped and serrate tubercles; second row with four to six tubercles; three to seven large tear-shaped subpalmar tubercles. Epistomal zygoma strongly arched.

Central projection of gonopod of form I male corneous, very short and thick, recurved approximately 90° to shaft, with subapical notch. Mesial process hard and tubular at base, apically soft, twisted, and spatulalike distally. Annulus ventralis of female subcircular in outline, slightly asymmetrical with very deep medial longitudinal trough through cephalic half, deep C–shaped fossa and S-shaped sinus.

MAXIMUM SIZE – The maximum known size for the species is 76.4 mm total length.

MOST LIKE – *Fallicambarus fodiens* most closely resembles *Cambarus diogenes* and its relatives *C. ludovicianus*, *C.* sp. A, and *C. thomai*, especially when the specimens are preserved and have lost their color. All of these species have an obliterated areola and a notch at the base of the dactyl of the chelae; however, in *F. fodiens* this notch is deeper and lacks tubercles along its distal margin. *Fallicambarus fodiens* can also be separated from the previously named species in that it lacks a suborbital angle. This character works well even in small individuals.

DISTRIBUTION AND HABITAT – (Fig. 102) *Fallicambarus fodiens* has a very large but disjunct range (Hobbs and Robison 1989; Hobbs 1989; Jezerinac et al. 1995). The species is found on the Atlantic Coastal Plain from Maryland south to eastern Georgia and in the Mississippi River Embayment and Gulf drainages from Texas east to western Florida and Alabama, and north along the Mississippi drainage to Illinois, Indiana, and Ohio. It also extends farther north into the Great Lakes drainage, Michigan, and southern Ontario. Jezerinac and Stocker (1987) and Jezerinac et al. (1995) reported a third apparently disjunct population along the Ohio River in west-central West Virginia. In Kentucky, *F. fodiens* is known from the northern half of the state from the Salt River drainage west to the Mississippi Embayment. It seems to be the most common between the Tennessee and Mississippi rivers.

This species is a primary burrower making burrows in ephemeral wetlands, wooded flood-

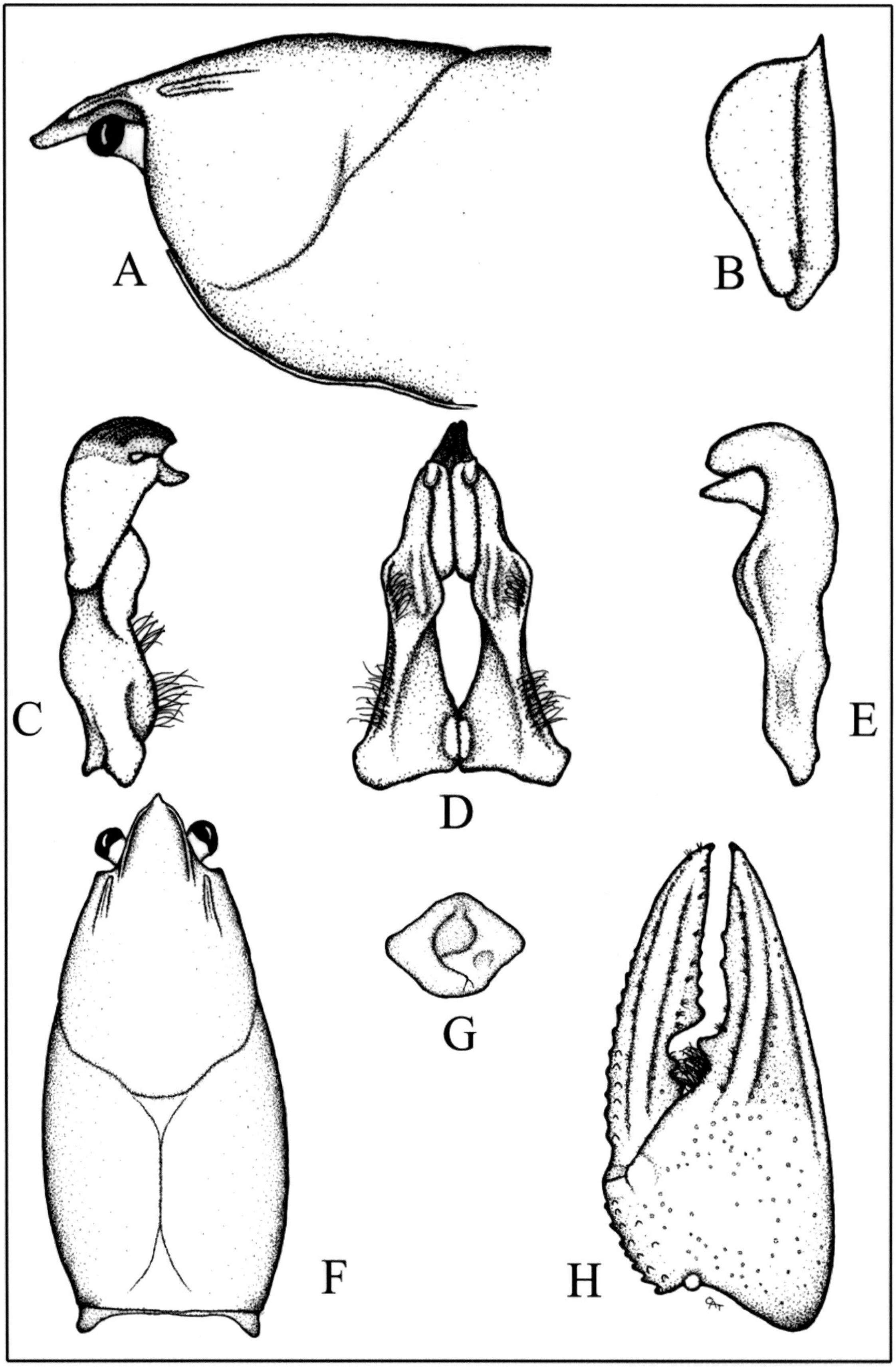

Fig. 101. *Fallicambarus fodiens*: A, lateral view of carapace; B, dorsal view of right antennal scale; C, mesial view of form I gonopod; D, ventral view of form I gonopods; E, lateral view of form II gonopod; F, dorsal view of carapace; G, ventral view of annulus ventralis; H, dorsal view of right chela.

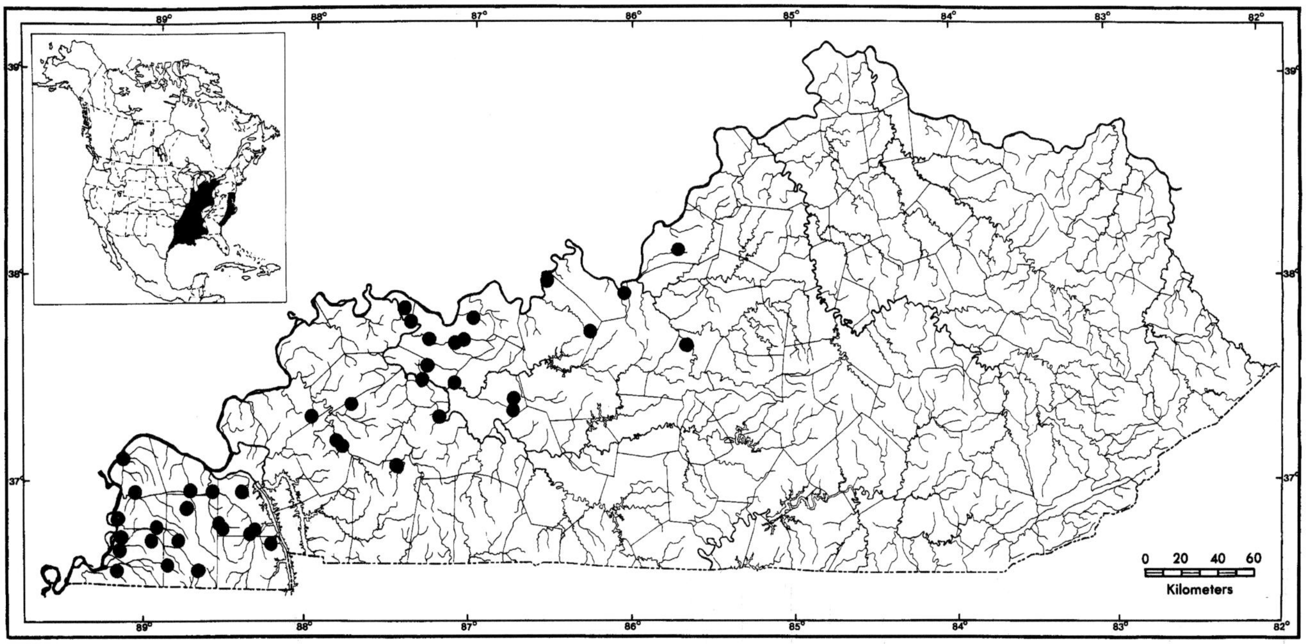

Fig. 102. Distribution of *Fallicambarus fodiens* in Kentucky.

plains, and low-lying fields. Page (1985), Jezerinac et al. (1995), and Pflieger (1996) all reported it from similar habitats. Because this species is a primary burrower, it is often difficult to collect and hence very easy to overlook.

LIFE HISTORY – Form I males have been collected in Kentucky in March, April, May, and October. A single ovigerous female was collected in the state on March 9, 2004. Most of what is known about the life history of the species comes from populations in other states. Creaser (1931) reported copulation in the fall and early spring. Jezerinac et al. (1995) reported form I males during the months of March and April in West Virginia, and Hobbs and Robison (1989) reported them from February through May and August through November in Arkansas. Bovbjerg (1952), Page (1985), Pearse (1910), Cummins (1921), and Creaser 1931) have found ovigerous females from February through April. We collected a single female with a total length of 93.3 mm carrying 239 young. Mean total length of young was 10.8 mm. Hay (1896), Williamson (1907), Crocker and Barr (1968), and Jezerinac et al. (1995) have reported females with young from the end of February through May. Brown (1955) reported collecting a female with attached young from a burrow in southern Illinois in October. Pflieger (1996), based on collecting very small young in February in Missouri, also surmised that some young may hatch in the fall or early winter at the latitude of southeastern Missouri. Page (1985) reported three females with eggs having 48, 68, and 162 eggs, respectively. He also reported collecting seven females in February and March with young. Mean number of young for those females was 132.1. Those data from other more northern latitudes suggest that egg laying occurs from late winter to spring and that young hatch between early spring and early fall. Until more data are collected from Kentucky populations, it is premature to speculate when peak reproductive activities occur. A size frequency distribution from a population in southern Illinois suggested that this population reaches sexual maturity and dies in its second year (Page 1985). Page (1985) also found in this population that all form I males had greater than 40-mm carapace length and males smaller than this were all form II. He also found all females greater than 30-mm carapace length were in their second year and carried eggs. The sex ratio of the population was nearly 1:1.

CRAYFISH ASSOCIATES – We have collected *Fallicambarus fodiens* with the following species: *Cambarus diogenes, C.* sp. A, *Orconectes immunis, O. palmeri palmeri, Procambarus acutus, P. clarkii,* and *P. viaeviridis.*

Orconectes rusticus **Photo by C. Lukhaup**

Orconectes Cope

The genus *Orconectes* contains approximately 88 species and subspecies and is one of the most diverse crayfish genera in the world. Members of the genus occur natively in the central and eastern United States and Canada. Approximately 70% of its members occur only in the Central Highlands of the United States. With a few exceptions, *Orconectes* species inhabit creeks and rivers of all sizes and very rarely burrow. Those that do burrow are tertiary burrowers and do so only during drought conditions. Three members of the genus that inhabit Kentucky occur in subterranean creeks. Males possess two terminal elements on the gonopod and have hooks on the ischia of the third, and occasionally third and fourth, pair of pereiopods. Kentucky is home to 23 species of *Orconectes*, 7 of which are not found outside of the state.

Orconectes australis packardi Rhoades
Appalachian Cave Crayfish

Fig. 103

LIFE COLORS – (Fig. 103) Albinistic; dorsal, ventral, and lateral surfaces of chelae, carapace, and abdomen white to cream to light yellow in color. Larger individuals may have several shades of cream and light yellow, with brighter yellow highlighting various grooves and joints of the individual.

CHARACTERS – (Fig. 104) Eyes greatly reduced and lacking pigment. Rostrum moderately excavated, distally margins straight and slightly converging, proximally slightly convex; margins terminating in spines at base of long acumen. Acumen longer than width of rostrum at marginal spines. Carapace dorsoventrally flattened. At least one cervical spine or tubercle present on each side; if multiple cervical spines or tubercles present, spine or tubercle size gradually decreases dorsally. All spines on carapace sharply pointed. Areola wide, 4.2 to 5.9 times longer than wide at its narrowest point, with room for three to five punctations. Antennal scale widest at midpoint. Chelae long and thin; mesial margin of palm with 10 to 14 tubercles; dorsal surfaces of palms with several irregular rows of tubercles. Fingers of chelae smooth or with weak longitudinal ridges. Opposable margin of propodus with enlarged fourth and ninth tubercles, opposable margin of dactyl usually with enlarged fifth tubercle.

Gonopod of form I male terminating in two short elements: a straight, pointed and almost triangular corneous central projection and a slightly longer and thinner pointed, noncorneous mesial process. Mesial process arched caudodistally. Small angular shoulder present on caudal surface of gonopod at base of central projection. Hobbs and Barr (1972) reported the presence of very short and rudimentary caudal and cephalic processes between central projection and mesial process on several individuals from Pulaski County. Annulus ventralis of female subrhomboidal, lacking fossa; possessing centrally located and smoothly rounded longitudinal ridge and short, curving sinus running through elevated knob at center of caudal edge.

MAXIMUM SIZE – The maximum known size for the species is 68.0 mm total length.

MOST LIKE – Three species of Kentucky crayfishes are albinistic and inhabit caves: *Orconectes australis packardi*, *O. inermis inermis*, and *O. pellucidus*. Males of *O. australis packardi* differ from males of *O. inermis inermis* and *O. pellucidus* in possessing a mesial process that is of equal length or only slightly longer than the central projection and a small angular shoulder at the base of the central projection.

DISTRIBUTION AND HABITAT – (Fig. 105) *Orconectes australis packardi* appears to be endemic to the state of Kentucky (Hobbs 1989). Within the state it occurs in subterranean passages of the middle Cumberland River drainage in McCreary, Pulaski, Rockcastle, and Wayne counties. It has been recorded from 16 different caves in that region. Hobbs and Barr (1972) reported that in the southeastern portion of the species' range in Kentucky and in northern Tennessee, *O. a. packardi* may interbreed with the nominate subspecies, *O. a. australis*. However, no subsequent authors have reported specimens possessing "pure" *O. a. packardi* characters from the state of Tennessee.

Orconectes australis packardi occurs exclusively in subterranean creeks. Within those creeks the species is encountered along the edges of pools and in shallow lotic reaches. *Orconectes a. parkardi* has also on occasion been observed walking along cave floors out of water. Unlike most other crayfishes, *O. a. parkardi* is not frequently encountered under loose rocks or debris; it is instead usually found wondering along creek beds or clinging to cave walls in pooled areas. This is most likely due to the lack of large predators in subterranean creeks from which *O. a. packardi* is known.

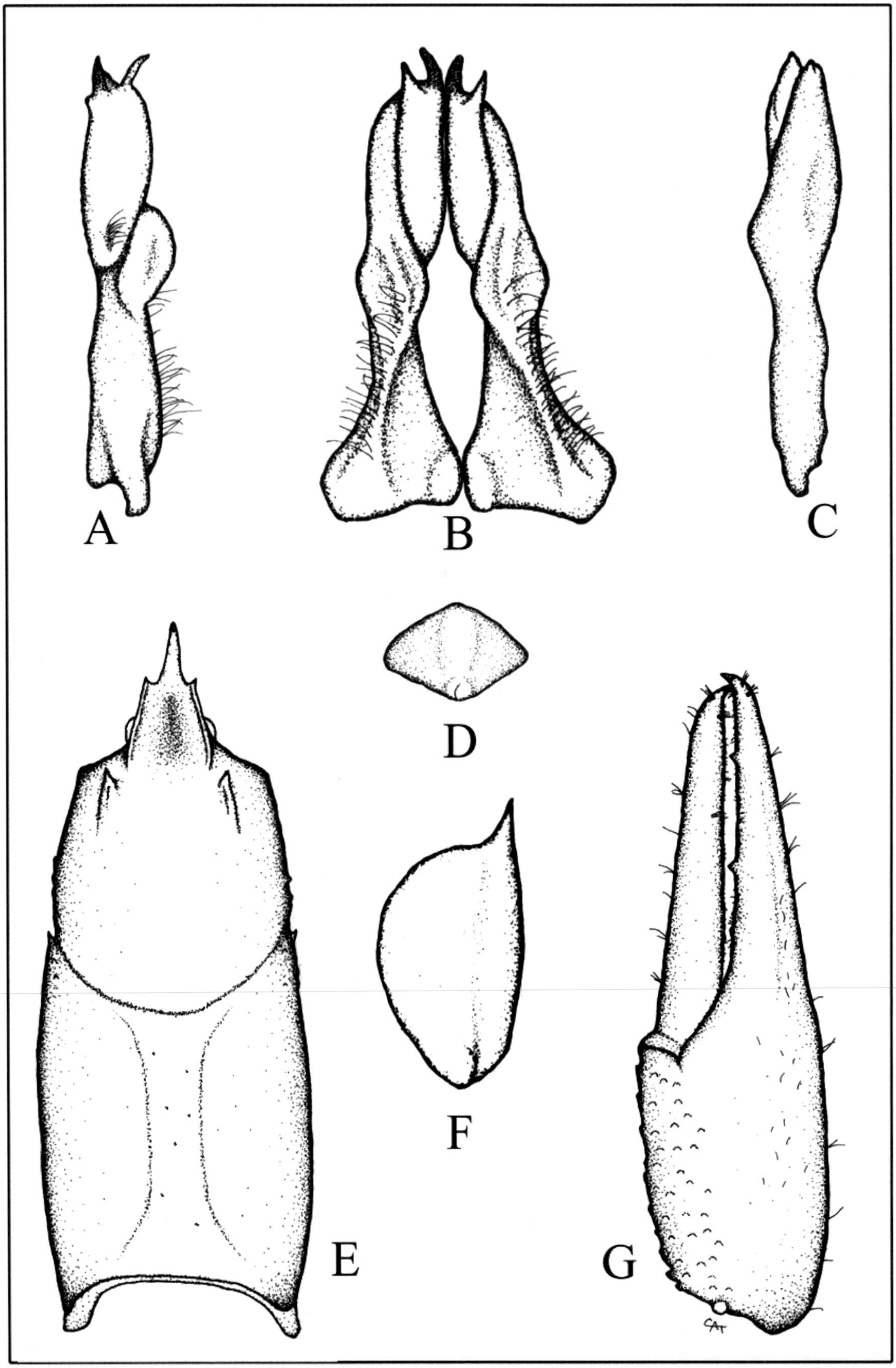

Fig. 104. *Orconectes australis packardi*: A, mesial view of form I gonopod; B, ventral view of form I gonopods; C, lateral view of form II gonopod; D, ventral view of annulus ventralis; E, dorsal view of carapace; F, dorsal view of right antennal scale; G, dorsal view of right chela.

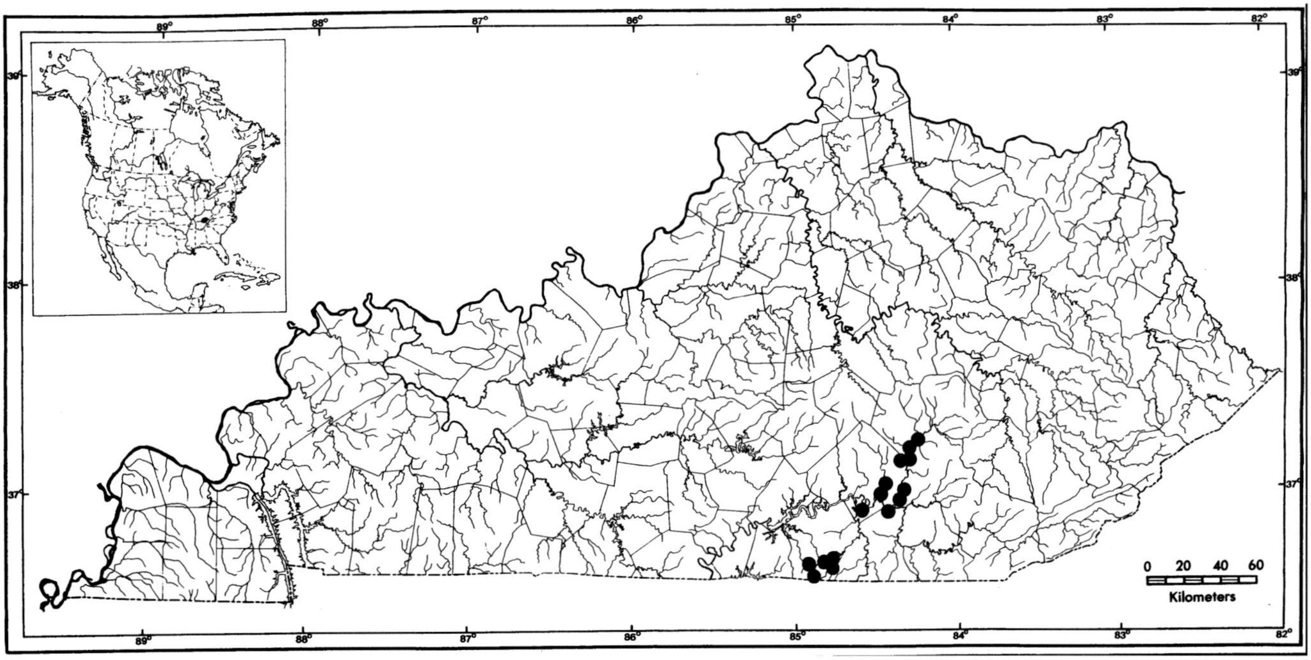

Fig. 105. Distribution of *Orconectes australis packardi* in Kentucky.

LIFE HISTORY – Very little is known about the life history of *Orconectes australis packardi*. Form I males have been collected in January, February, March, June, August, September, October, and November. No ovigerous females have been collected; however, a female carrying young was collected in January of 1967 (Hobbs and Barr 1972). Cooper (1975) conducted an extensive study of the biology of the closely related *O. australis australis* in a northern Alabama cave. In that study, Cooper found that individuals of *O. a. australis* were capable of moving up to 290 m per day and that growth in the species was very slow, averaging 0.2 mm per month. Using growth rates and individual size measurements, Cooper (1975) estimated longevity in *O. a. australis* to range from 40 to 50 years. He found a near 1:1 sex ratio and form I males in roughly equal numbers in all months of the year. He did not collect any ovigerous females and found only two females carrying young, one in January and one in October.

CRAYFISH ASSOCIATES – No other crayfish species have been collected from habitats containing *Orconectes australis packardi*.

CONSERVATION STATUS – AFS Threatened; KSNPC Threatened.

Orconectes barrenensis Rhoades
Barren River Crayfish

Fig. 106A: narrow gaping form.

Fig. 106B: wide gaping form.

LIFE COLORS – (Figs. 106A, B) Base color of dorsal and lateral surfaces of chelae, carapace, and abdomen light brown. Dorsal and lateral surfaces of carapace and abdomen densely mottled with light to dark brown patches of varying size and shape. Lateral surfaces of abdomen occasionally lighter in color than other surfaces of carapace. Carapace with wide U-shaped dark saddle centered at dorsal surface of caudal margin, extending forward on lateral surfaces to approximately midpoint. Dorsal and lateral surfaces of carapace also with two large dark patches, one on either side of areola, just posterior to cervical groove. Fingers of chelae with orange or red tips. Venter cream to whitish.

CHARACTERS – (Fig. 107) Rostrum narrow and moderately excavated; distally margins straight and converging, proximally margins convex; margins terminating in spines at base of acumen. Acumen equal to or slightly longer than width of rostrum at marginal spines. Weak median carina occasionally present. Carapace slightly dorsoventrally flattened. Areola wide, 4.3 to 7.0 times longer than wide at its narrowest point, with room for three to five punctations. Cervical spines absent. Antennal scale widest distal to midpoint. Incisor region of mandible dentate, occasional specimen with straight edge. Chelae large, with short fingers; dorsal surfaces of palms usually covered with numerous shallow punctations. Dorsal surfaces of fingers of chelae with weakly developed longitudinal ridges. Wide gap between bases of opposable margins of fingers of chela, gap usually narrower in Nolin River drainage specimens or in females or form II males. Mesial margin of palm of chela with two rows of weak tubercles, seven or eight tubercles in mesialmost row.

Gonopod of form I male terminating in two long, thin elements: a thin, corneous central projection slightly arched caudodistally; and a shorter, straight, noncorneous mesial process. Tip of central projection usually overhanging mesial process. Caudal surface of gonopod with or without smoothly rounded shoulder at base of central projection. Annulus ventralis of female subrhomboidal; with median trough running through cephalic half and two caudally directed protuberances overhanging central fossa; fossa very wide laterally and with sinus running from center to caudal margin.

MAXIMUM SIZE – The maximum known size for the species is 62.0 mm total length.

MOST LIKE – Numerous species of crayfishes in Kentucky possess form I gonopods with two long, thin terminal elements. Of those species, *Orconectes barrenensis* is most similar to *O. placidus* and *O. durelli* in that all three have a mandible with a dentate incisor region, lack a strong angular shoulder, and have a gonopod with an arched central projection less than 50% of the total length of gonopod. *Orconectes barrenensis* differs from both *O. placidus* and *O. durelli* by lacking cervical spines. *Orconectes barrenensis* also differs from *O. placidus* by possessing chelae with shorter fingers, the dactyl of *O. barrenensis* being less than two times the length of the mesial margin of the palm.

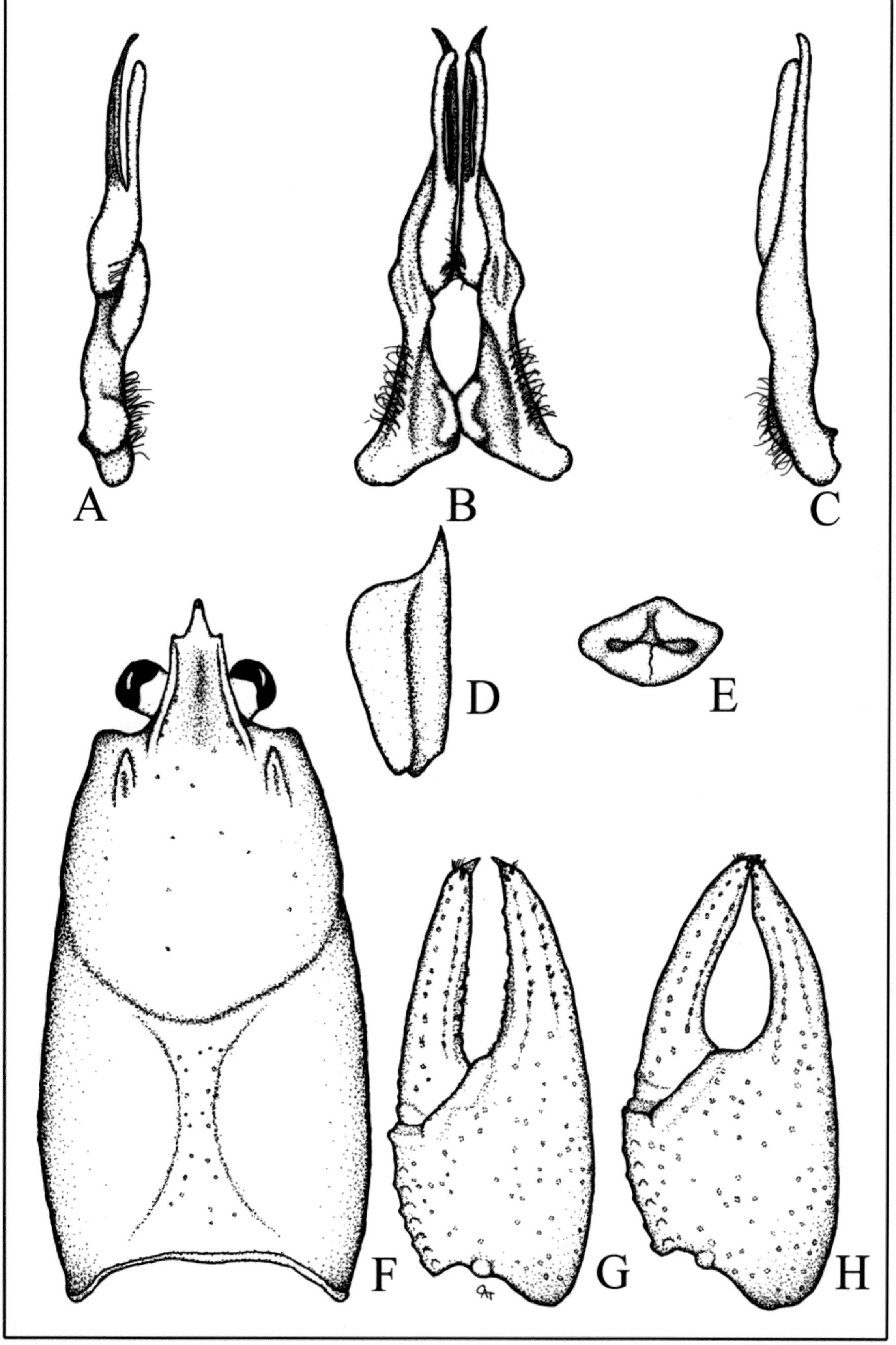

Fig. 107. *Orconectes barrenensis*: A, mesial view of form I gonopod; B, ventral view of form I gonopods; C, lateral view of form II gonopod; D, dorsal view of right antennal scale; E, ventral view of annulus ventralis; F, dorsal view of carapace; G, dorsal view of right chela showing narrow gaping form; H, dorsal view of right chela showing wide gaping form.

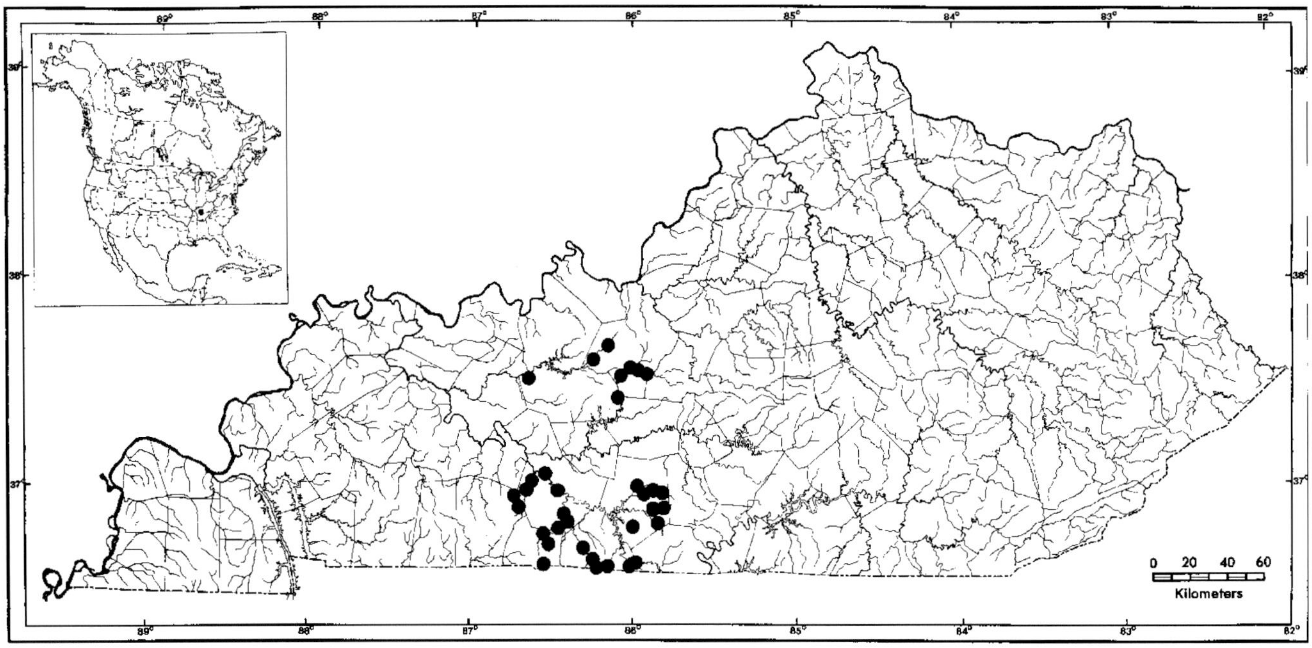

Fig. 108. Distribution of *Orconectes barrenensis* in Kentucky.

DISTRIBUTION AND HABITAT – (Fig. 108) *Orconectes barrenensis* is endemic to the Green River drainage of north-central Tennessee and central Kentucky (Hobbs 1989). Within Kentucky, the species has a disjunct distribution and occurs in three different Green River drainages. It is widespread and common in the Barren River drainage and Nolin River drainage upstream of Nolin Lake. It also occurs sporadically in the upper Rough River drainage in Grayson and Hardin counties (Lawson 2003).

Orconectes barrenensis is most common in clear small to large-sized creeks ranging in width from 3 to 15 m with bottom substrates of gravel and cobble. The species is also known to occur in areas of the Barren River, where widths can exceed 20 m. Within these creeks and rivers, *O. barrenensis* is usually encountered under gravel and small cobble in riffle areas.

LIFE HISTORY – Form I males have been collected from February to July and October to November. A lack of field efforts may explain the absence of form I males in other months of the year. No ovigerous females or females with young have been collected. Nothing else is known about the life history of this species.

CRAYFISH ASSOCIATES – The following species have been collected from habitats containing *Orconectes barrenensis*: *Barbicambarus cornutus*, *Cambarus graysoni*, *C. rusticiformis*, *C. tenebrosus*, *O. compressus*, *O. durelli*, *O. putnami*, *O. rafinesquei*, and *O. rusticus*.

Orconectes bisectus Rhoades
Crittenden Crayfish

Fig. 109

LIFE COLORS – (Fig. 109) Base color of dorsal and lateral surfaces of chelae, carapace, and abdomen light to dark brown. Dorsal surfaces of carapace anterior to cervical groove, caudal margin of carapace, and dorsal surface of first abdominal segment with darker brown patches. Fingers of chelae with orange tips and subdistal black bands. Large knobs at base of dactyl orange. Venter cream to whitish.

CHARACTERS – (Fig. 110) Rostrum moderately excavated; distally margins straight, proximally margins slightly convex; margins terminating in small spines or tubercles at base of long acumen. Acumen longer than width of rostrum at marginal spines. Carapace dorsoventrally flattened, more so in larger individuals. Areola of moderate width, 6.4 to 11.0 times longer than wide at its narrowest point, with room for three to five punctations. Cervical spines present. Antennal scale widest at midpoint. Chelae large with moderately long fingers, dorsal surfaces of palms usually covered with shallow punctations and short setae. Dorsal surfaces of fingers of chelae with well-developed longitudinal ridges and one row of deep punctations that coalesce; ridges more strongly developed on larger individuals. Mesial margin of palm of chela and dactyl with strong tubercles; palm with one row and several additional scattered tubercles lateral to mesial row; 8 to 10 tubercles in mesial row. Males with longer chelae than females of the same carapace length.

Gonopod of form I male terminating in two short terminal elements: a corneous straight central projection with a truncated tip and a shorter, noncorneous mesial process. Mesial processes diverge laterally from one another when viewed ventrally. Annulus ventralis of female subcircular; large, centrally located fossa; wide median trough running from the cephalic to the caudal margin; sinus running from fossa to caudal margin.

MAXIMUM SIZE – The maximum known size for the species is 75.7 mm total length.

MOST LIKE – *Orconectes bisectus* is most similar to other members of the genus that possess short, straight terminal elements on the form I gonopod. Those species are *O. jeffersoni, O. margorectus, O. rafinesquei, O. sanbornii*, and *O. tricuspis*. The combination of a central projection with a truncated tip and mesial prossesses that diverge from one another laterally make the form I male gonopod of *Orconectes bisectus* unique in comparison to the other five species. *O. bisectus* also differs from *O. margorectus* by possessing a mandible with a dentate incisor region and from *O. jeffersoni* by possessing terminal elements of the gonopod that are less than 25% of the total length of the gonopod.

DISTRIBUTION AND HABITAT – (Fig. 111) *Orconectes bisectus* is endemic to the state of Kentucky. It has an extremely limited distribution in the lower Ohio River drainage in that it occurs only in Ohio River tributaries in the eastern portion of Crittenden County.

Orconectes bisectus is most common in small to medium-sized creeks ranging in width from 2 to 10 m with bottom substrates of bedrock, cobble, and some mud. Individuals are usually encountered under cobble in areas with flow or among woody debris along creek edges.

LIFE HISTORY – Very little is known about the life history of *Orconectes bisectus*. Form I males have been collected in January, March, April, and May. Ovigerous females have been collected in March and May. The absence of form I males and ovigerous females in other months of the year is more likely due to a lack of field efforts rather than being a reflection of actual reproductive

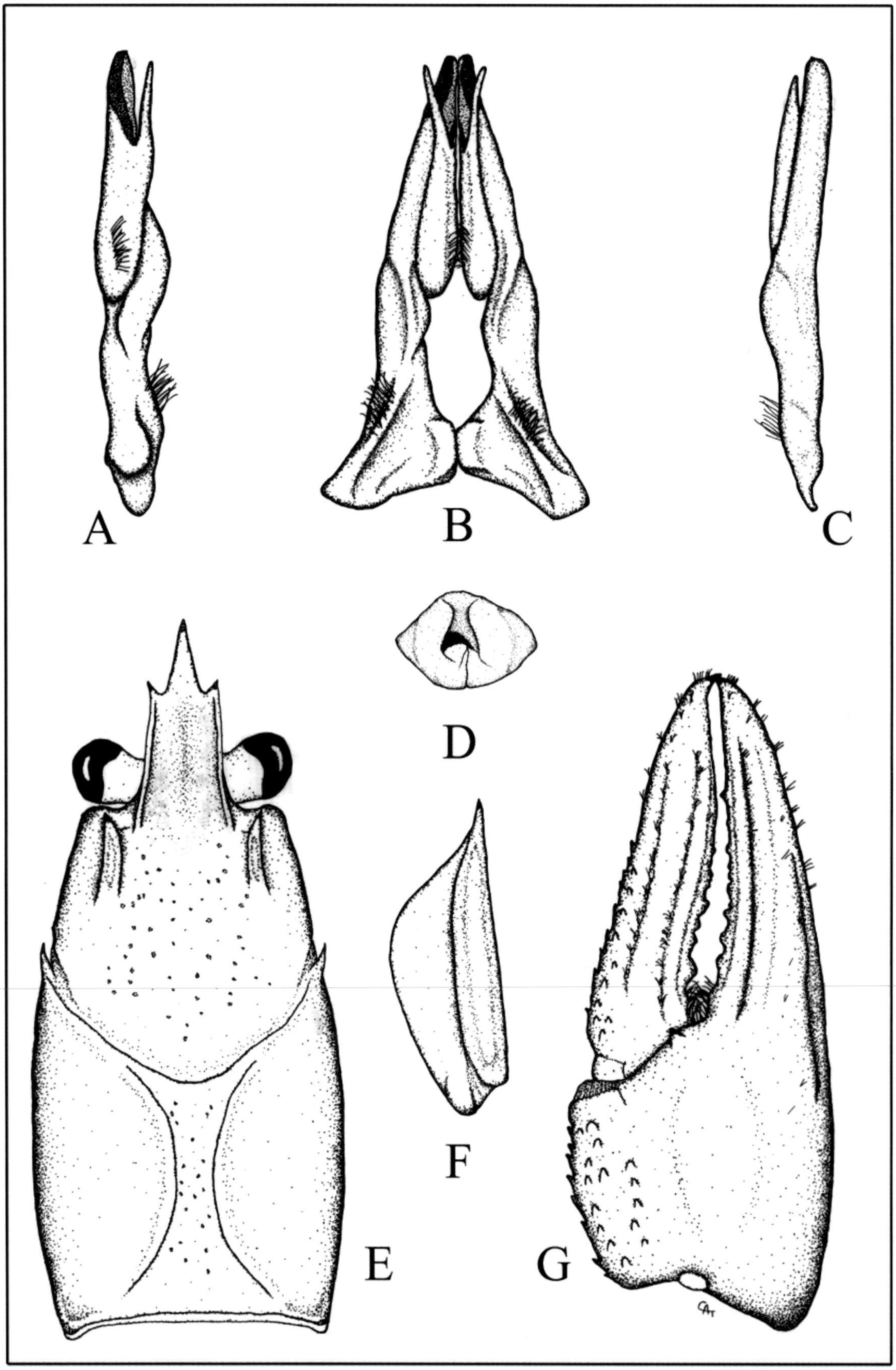

Fig. 110. *Orconectes bisectus*: A, mesial view of form I gonopod; B, ventral view of form I gonopods; C, lateral view of form II gonopod; D, ventral view of annulus ventralis; E, dorsal view of carapace; F, dorsal view of right antennal scale; G, dorsal view of right chela.

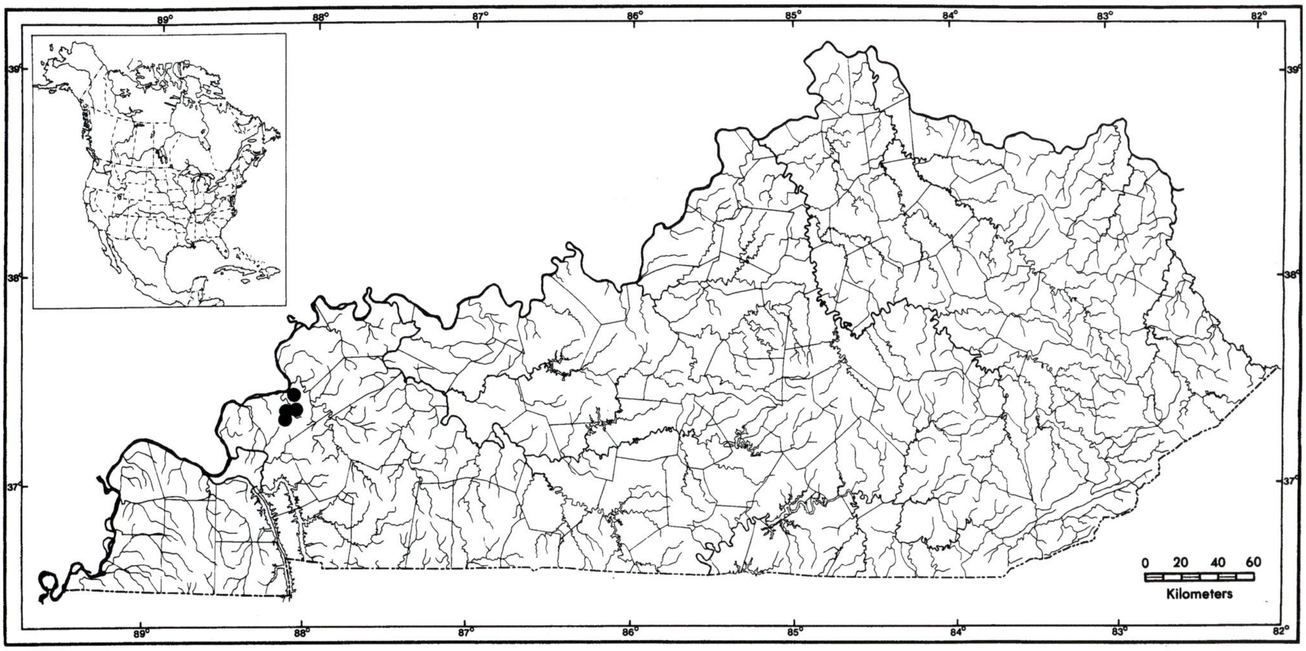

Fig. 111. Distribution of *Orconectes bisectus* in Kentucky.

cycles. Without additional collections of this species from summer and fall months, it is premature to speculate on the timing of copulation and egg-laying activities. Two ovigerous females collected in March measured 57.7 mm and 60.1 mm in total length and carried 134 and 146 eggs, respectively. Average egg diameter for both females was 2.0 mm.

CRAYFISH ASSOCIATES – The following species have been collected from habitats containing *Orconectes bisectus*: *Cambarus diogenes*, *C.* sp. A, and *O. immunis*.

CONSERVATION STATUS – AFS Endangered; KSNPC Threatened.

Orconectes burri Taylor and Sabaj
Blood River Crayfish

Fig. 112

LIFE COLORS – (Fig. 112) Dorsal and lateral surfaces of chelae, carapace, and abdomen with light brown to tan base color and densely mottled with light to dark brown patches of varying size. Dorsal surface of carapace with wide dark brown patch just anterior to cervical groove and wide U-shaped dark saddle centered at caudal margin and extending anteriorly to midlength of lateral surfaces of carapace. Fingers of chelae with orange tips and wide subdistal black bands. Caudal edge of chela with centrally located large dark blue to black knob. Venter cream to whitish.

CHARACTERS – (Fig. 113) Rostrum moderately excavated; distally margins straight and slightly converging, proximally margins slightly convex; margins terminating in spines or tubercles at base of short acumen. Acumen length slightly less than width of rostrum at marginal spines. Weak median carina present. Carapace not dorsoven-

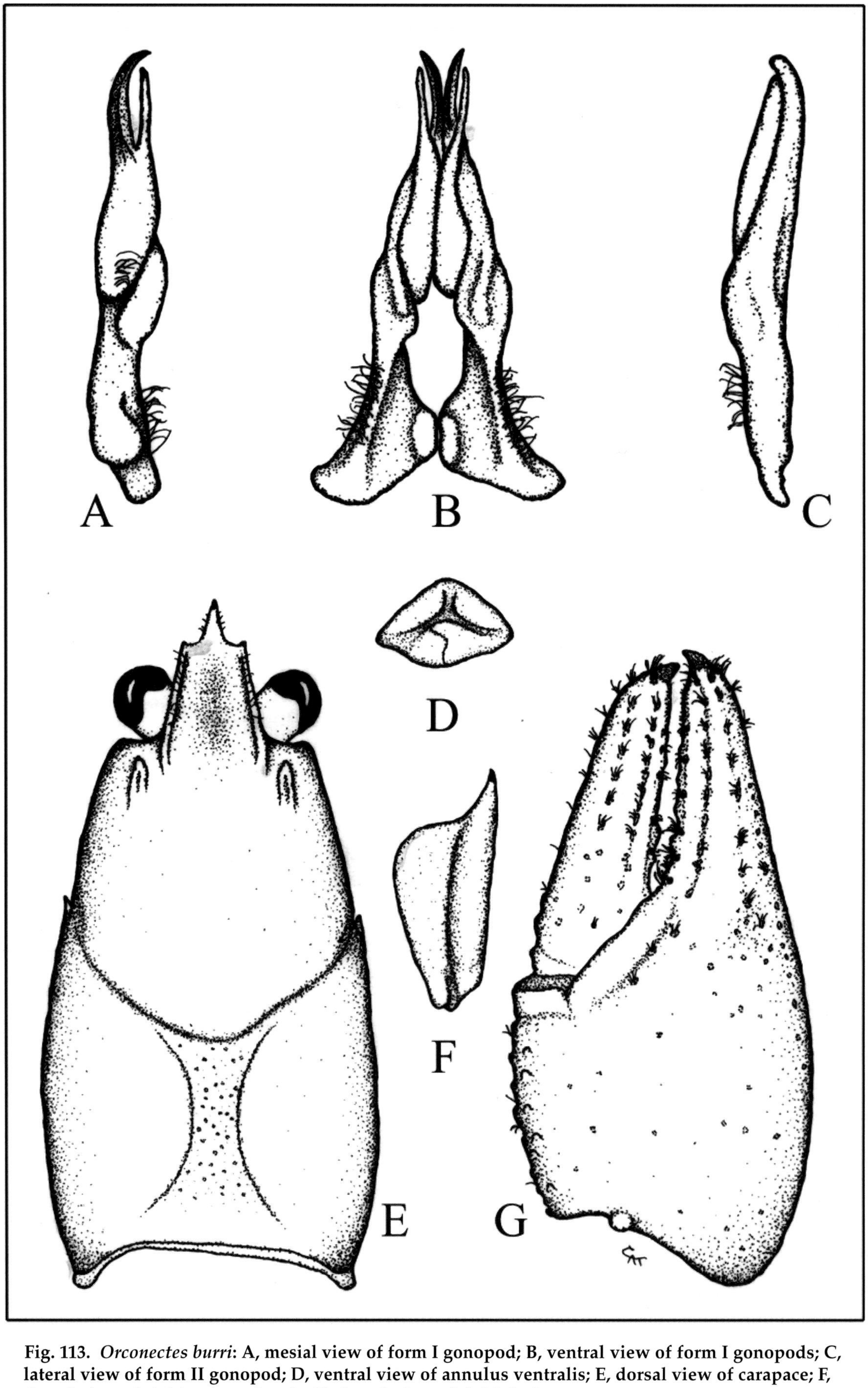

Fig. 113. *Orconectes burri*: A, mesial view of form I gonopod; B, ventral view of form I gonopods; C, lateral view of form II gonopod; D, ventral view of annulus ventralis; E, dorsal view of carapace; F, dorsal view of right antennal scale; G, dorsal view of right chela.

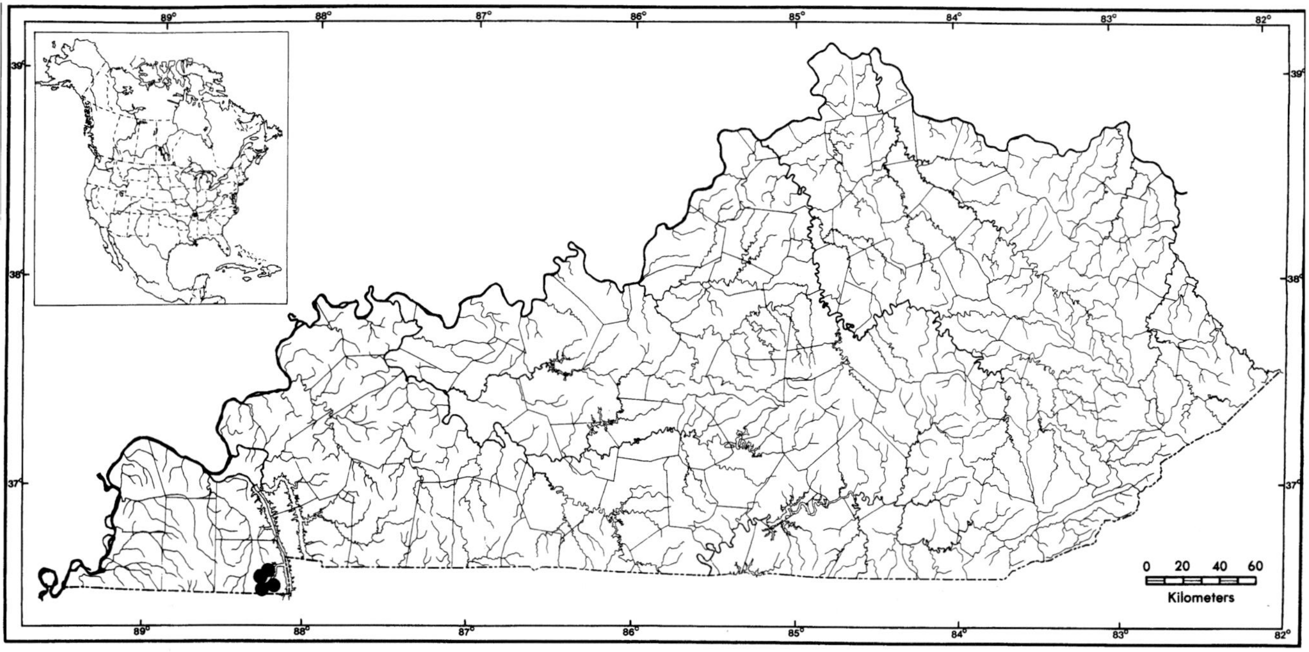

Fig. 114. Distribution of *Orconectes burri* in Kentucky.

trally flattened, subcircular in cross section. Areola wide, 4.4 to 7.5 times longer than wide at its narrowest point, with room for five to eight punctations. Cervical spines present. Antennal scale widest distal to midpoint. Chelae stout with short, thick fingers; dorsal surfaces of palms usually covered with numerous shallow punctations. Dorsal surfaces of fingers of chelae with well-developed longitudinal ridges. Mesial margin of palm of chela with two rows of tubercles and small tufts of setae, eight or nine tubercles in mesialmost row.

Gonopod of form I male terminating in two short elements: a thin, caudodistally arched, corneous central projection and a shorter, straight, noncorneous mesial process. Tip of central projection bent at 90° angle to main shaft and overhanging mesial process. Annulus ventralis of female subrhomboidal; median trough running between two strongly elevated protuberances at center of cephalic margin; centrally located fossa; sinus running from left (or occasionally right) corner of fossa to caudal margin.

MAXIMUM SIZE – The maximum known size for the species is 64.5 mm total length.

MOST LIKE – The form I gonopod of *Orconectes burri* is superficially similar to other members of the genus such as *O. bisectus, O. jeffersoni, O. margorectus, O. rafinesquei, O. sanbornii*, and *O. tricuspis* in that it has short, more or less straight terminal elements. *Orconectes burri* differs from those species by possessing a central projection with a tip bent at a 90° angle to the main shaft of the gonopod and which overhangs the mesial process. The strongly mottled color pattern of *O. burri* is also atypical for most Kentucky species of *Orconectes*.

DISTRIBUTION AND HABITAT – (Fig. 114) *Orconectes burri* is endemic to the Blood River drainage, a Tennessee River tributary in western Kentucky and northwest Tennessee (Taylor and Sabaj 1998). The Blood River is a relatively small tributary of the Tennessee River, with its headwaters originating in Henry County, Tennessee. Within Kentucky the Blood River drainage and the range of *O. burri* are contained entirely within the boundaries of Calloway County.

Orconectes burri is most common in small to medium-sized creeks ranging in width from 3 to 10 m with bottom substrates of gravel and sand. Within these creeks, the species is usually encountered in woody debris piles or woody vegetation root masses along creek banks.

LIFE HISTORY – Very little is known about the life history of *Orconectes burri*. The species was only recently discovered and described (Taylor and Sabaj 1998) and collections of the species are sparse. Form I males have been collected in

March, April, May, and October but ovigerous females have not been collected. In their description of the species, Taylor and Sabaj (1998) noted that most collections of *O. burri* contained two distinct year classes and suggested that the species has a two-year life cycle.

CRAYFISH ASSOCIATES – *Cambarus diogenes* and *Procambarus acutus* are the only species known to have been collected from habitats containing *Orconectes burri*.

CONSERVATION STATUS – KSNPC Threatened.

Orconectes compressus (Faxon)
Slender Crayfish

Fig. 115

LIFE COLORS – (Fig. 115) Base color of dorsal and lateral surfaces of chelae, carapace, and abdomen light tan or light rust; abdomen usually lighter in color than carapace. Dorsal and lateral surfaces with or without numerous randomly scattered small dark brown spots or patches. Dorsal surface of carapace with wide dark brown patch just posterior to cervical groove and wide U–shaped dark saddle centered at caudal margin and extending anteriorly to midlength of lateral surfaces of carapace. Fingers of chelae with orange or light tan tips. Venter cream to whitish.

CHARACTERS – (Fig. 116) Small adult size, rarely exceeding 44 mm in total length. Rostrum narrow and moderately excavated; margins straight, converging anteriorly, terminating in spines or tubercles at base of acumen. Acumen equal to or slightly longer than width of rostrum at marginal spines. Strong median carina present. Carapace strongly compressed laterally. Areola wide, 2.9 to 5.0 times longer than wide at its narrowest point, with room for 7 to 10 punctations. Cervical spines absent. Antennal scale widest at midpoint. Chelae stout and smooth, with short fingers and long palm; chelae covered with shallow punctations. Dorsal surfaces of fingers of chelae lacking longitudinal ridges. Gap between opposable margins of fingers of chelae at their bases, gap more pronounced in males. Mesial margin of palm of chela with single row of six to eight smoothly rounded tubercles.

Gonopod of form I male terminating in two thin, caudodistally recurved elements: a sharply pointed, corneous central projection and a noncorneous mesial process of approximately equal length. Distal end of mesial process dorsoventrally flattened and ladlelike; bent at 90° angle to main shaft. Annulus ventralis of female subrhomboidal; large, deep, sinuate fossa; sinus running from caudal corner of fossa to caudal margin of annulus. Lateral margins of annulus ventralis elevated.

MAXIMUM SIZE – The maximum known size for the species is 51.0 mm total length.

MOST LIKE – The laterally compressed carapace of *Orconectes compressus* is unique among Kentucky crayfishes and that character alone should distinguish this species. With two strongly recurved elements, the form I gonopod of *O. compressus* is similar to those of *O. immunis* and *O. palmeri palmeri*. *Orconectes immunis* differs from *O. compressus* in having strongly developed longitudinal ridges on the fingers of its chelae and a notch at the base of the opposable margin of the dactyl. *Orconectes compressus* differs from *O. palmeri palmeri* in possessing an open areola.

DISTRIBUTION AND HABITAT – (Fig. 117) *Orconectes compressus* occurs in the middle and lower Tennessee River drainage of northern Alabama, extreme northeastern Mississippi, and western Tennessee; the middle and lower Cumberland River drainage in Tennessee and

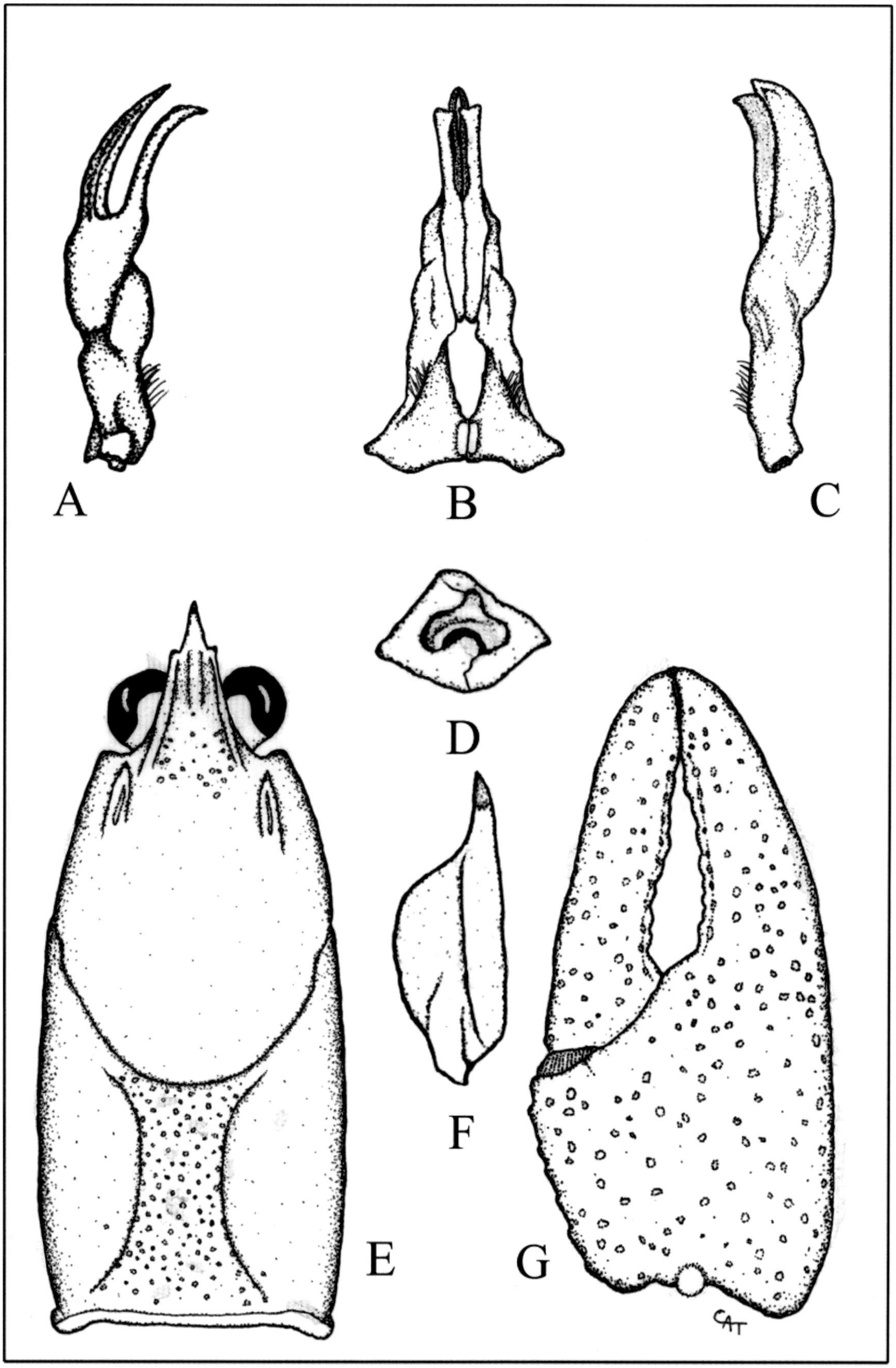

Fig. 116. *Orconectes compressus*: A, mesial view of form I gonopod; B, ventral view of form I gonopods; C, lateral view of form II gonopod; D, ventral view of annulus ventralis; E, dorsal view of carapace; F, dorsal view of right antennal scale; G, dorsal view of right chela.

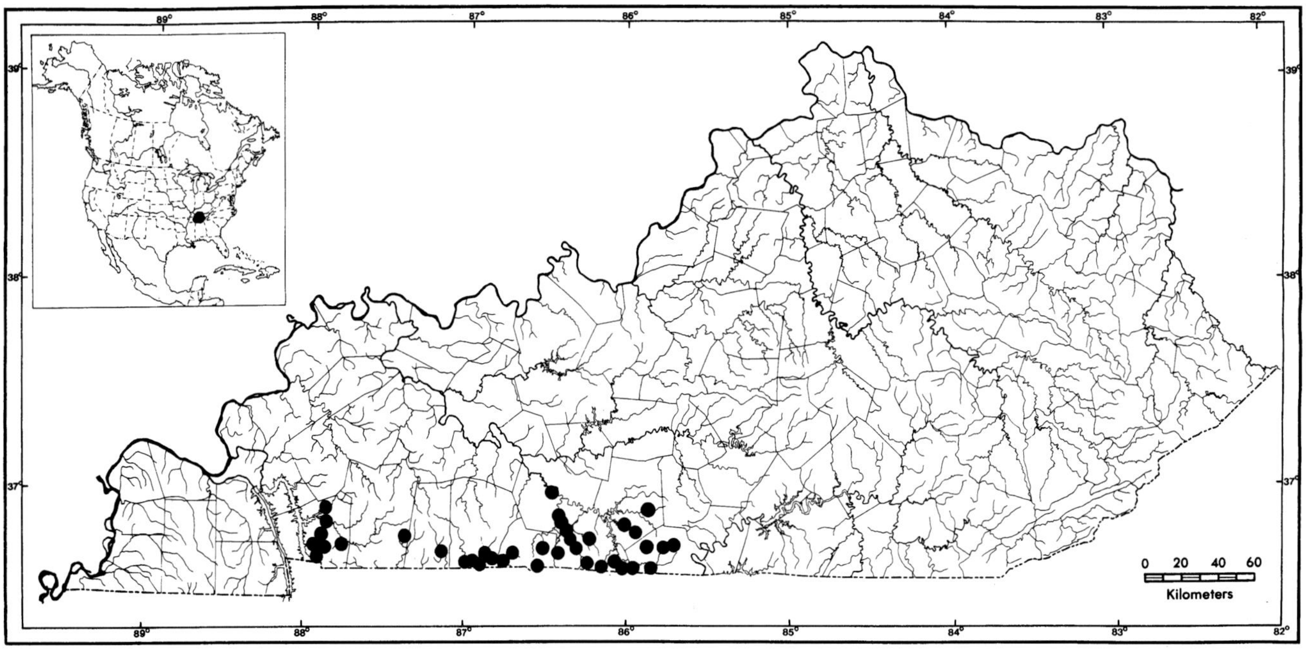

Fig. 117. Distribution of *Orconectes compressus* in Kentucky.

Kentucky; and the Barren River drainage of Tennessee and Kentucky (Hobbs 1989). Within Kentucky the species occurs commonly in west- and south-flowing Cumberland River tributaries from Trigg to Simpson counties and in the Barren River drainage upstream of the Gasper River.

Orconectes compressus is most common in clear, small to medium-sized creeks ranging in width from 2 to 10 m with abundant gravel and cobble substrates. Within these creeks, the species is usually encountered in shallow riffles with a smaller-sized gravel substrate. The species can also be found in deeper runs and occasionally in pools. Because of its small adult body size and the fact that other sympatric species of crayfishes are usually too large to find refuge under the gravel pieces, *O. compressus* is frequently the only species found in the shallow riffles of creeks and can occur in very high densities there.

LIFE HISTORY – Numerous collections of *Orconectes compressus* have been made in Kentucky and Tennessee from all months of the year. Within those collections, form I males occur in all months but January, June, July, and August. Ovigerous females have been collected only in March and April. These data suggest that mating activities occur in either fall or early spring and that egg laying occurs only in the spring. Three ovigerous females collected in March measured 43.1 mm, 36.7 mm, and 29.3 mm total length and carried 57, 45, and 38 eggs, respectively, thus supporting the observation that fecundity in crayfishes is positively correlated to female body size. Egg size was slightly smaller in *O. compressus* than in most other *Orconectes*, averaging 1.8 mm in diameter. As reported by Bouchard (1972), the species will dig short tunnels in gravel substrates, even under nondrought conditions.

CRAYFISH ASSOCIATES – The following species have been collected from habitats containing *Orconectes compressus*: *Barbicambarus cornutus, Cambarus graysoni, C. tenebrosus, O. barrenensis, O. durelli, O. placidus, O. putnami,* and *O. tricuspis.*

Orconectes cristavarius Taylor
Spiny Stream Crayfish

Fig. 118

LIFE COLORS – (Fig. 118) Base color of dorsal and lateral surfaces of chelae, carapace, and abdomen variable, tan to brown to olive green. Dorsum of first abdominal segment occasionally with dark brown transverse bar and carapace with dark brown U-shaped saddle centered at caudal margin and extending forward to midlength of lateral surfaces of carapace. Posterior edges of abdominal segments, dorsal knob at base of dactyl, and lateral edges of chelae red or orange. Dorsum of chelae with small dark flecks. Fingers of chelae with orange tips followed proximately by thin black bands. Venter cream to whitish.

CHARACTERS – (Fig. 119) Rostrum moderately excavated; distally margins straight and slightly converging, proximally margins slightly convex; margins terminating in spines at base of acumen. Acumen equal to or slightly longer than width of rostrum at marginal spines. Median carina present or absent. Carapace slightly compressed dorsoventrally. Areola of moderate width and variable, 4.2 to 11.3 times longer than wide at its narrowest point, with room for two to six punctations. Cervical spines present. Antennal scale widest at midpoint. Chelae large and with two distinct forms: longer finger, shorter palm form and shorter finger, longer palm form (see below). Mesial margin of palm of both forms with two rows of tubercles, 7 to 11 tubercles in mesialmost row. Dorsal and lateral surfaces of chelae with small tufts of setae; dorsal surfaces of fingers with well-defined longitudinal ridges. Fifth or sixth tubercle of opposable margin of dactyl usually enlarged.

Gonopod of form I male terminating in two long, thin elements: a sharply pointed corneous central projection that is caudodistally arched at distal end and a shorter, straight noncorneous mesial process. Caudal surface of gonopod with strong angular shoulder at base of central projection. Annulus ventralis of female subcircular; with median trough running through cephalic half, two caudally directed protuberances overhanging centrally located fossa; sinus running from left corner of fossa to caudal edge.

As indicated above, variation in chela shape and median carina exists in *O. cristavarius* and this variation shows a geographic pattern. The length of the palm region of the chela increases as one moves in northeasterly direction; specimens from the upper Cumberland River drainage have shorter palm regions, specimens from the Kentucky and Licking River drainages have intermediate palm lengths, while specimens from the Big Sandy River drainage have longer palm regions. Specimens from the upper Cumberland River drainage usually have strong carinas, specimens from the Licking and Big Sandy drainages usually do not possess carinas, while geographic intermediate Kentucky River drainage populations are highly variable and contain individuals with and without carinas.

MAXIMUM SIZE – The maximum known size for the species is 84.0 mm total length.

MOST LIKE – Numerous species of crayfishes in Kentucky possess form I gonopods with two long, thin terminal elements. Of those species, *Orconectes cristavarius* is most similar to *O. juvenilis* in that both possess a strong angular shoulder, cervical spines, and gonopod with an arched central projection less than 50% of the total length of gonopod. *Orconectes cristavarius* differs from *O. juvenilis* in possessing a mandible with a dentate incisor region.

DISTRIBUTION AND HABITAT – (Fig. 120) *Orconectes cristavarius* is a relatively wide-ranging species, occurring in upland creeks of the Ohio River basin from the upper Cumberland River drainage in southeastern Kentucky, eastward to the Elk River drainage in West Virginia, upper New River drainage in western North Carolina,

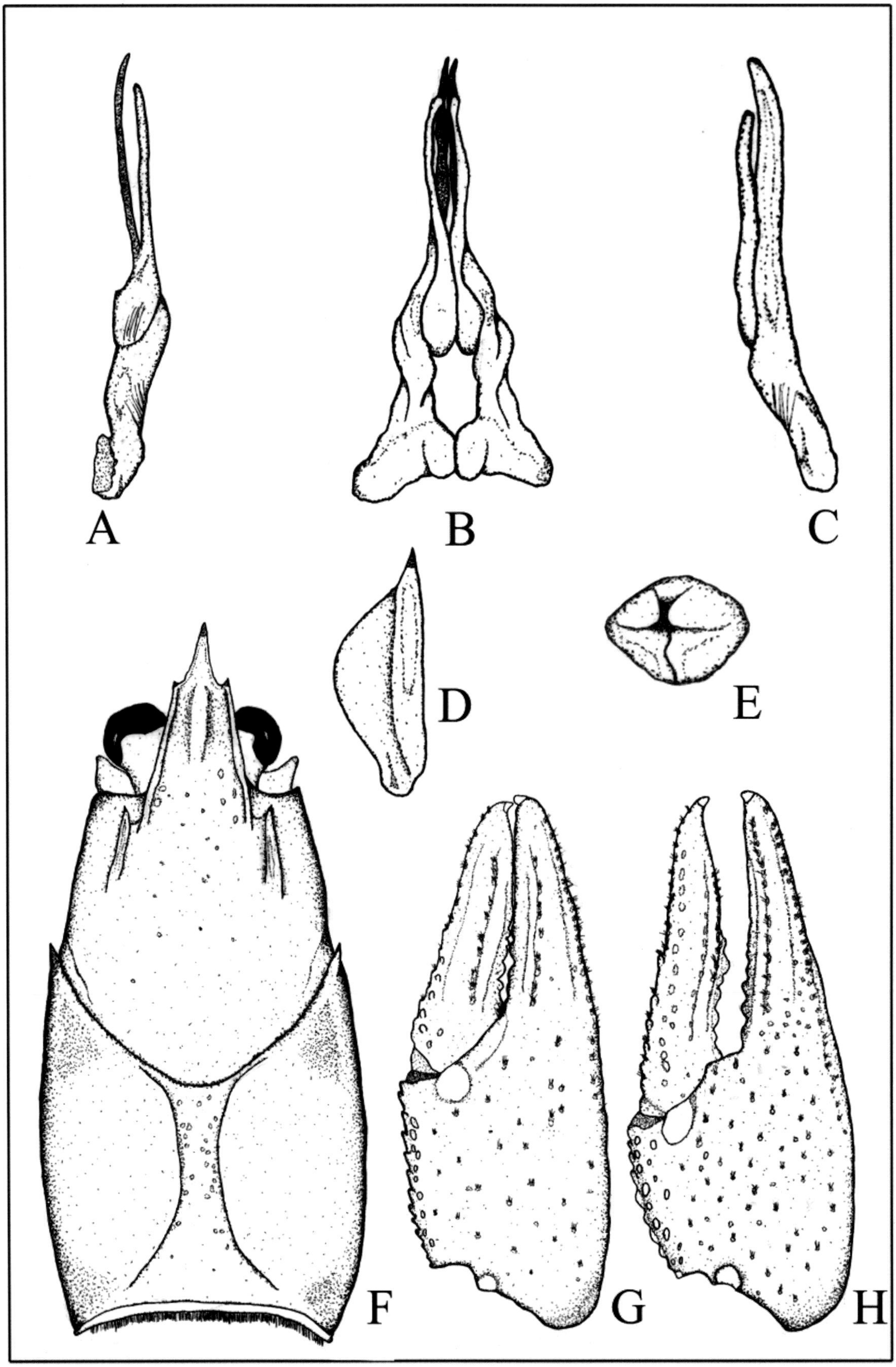

Fig. 119. *Orconectes cristavarius*: A, mesial view of form I gonopod; B, ventral view of form I gonopods; C, lateral view of form II gonopod; D, dorsal view of right antennal scale; E, ventral view of annulus ventralis; F, dorsal view of carapace; G, dorsal view of northeasterly form right chela; H, dorsal view of southwesterly form right chela.

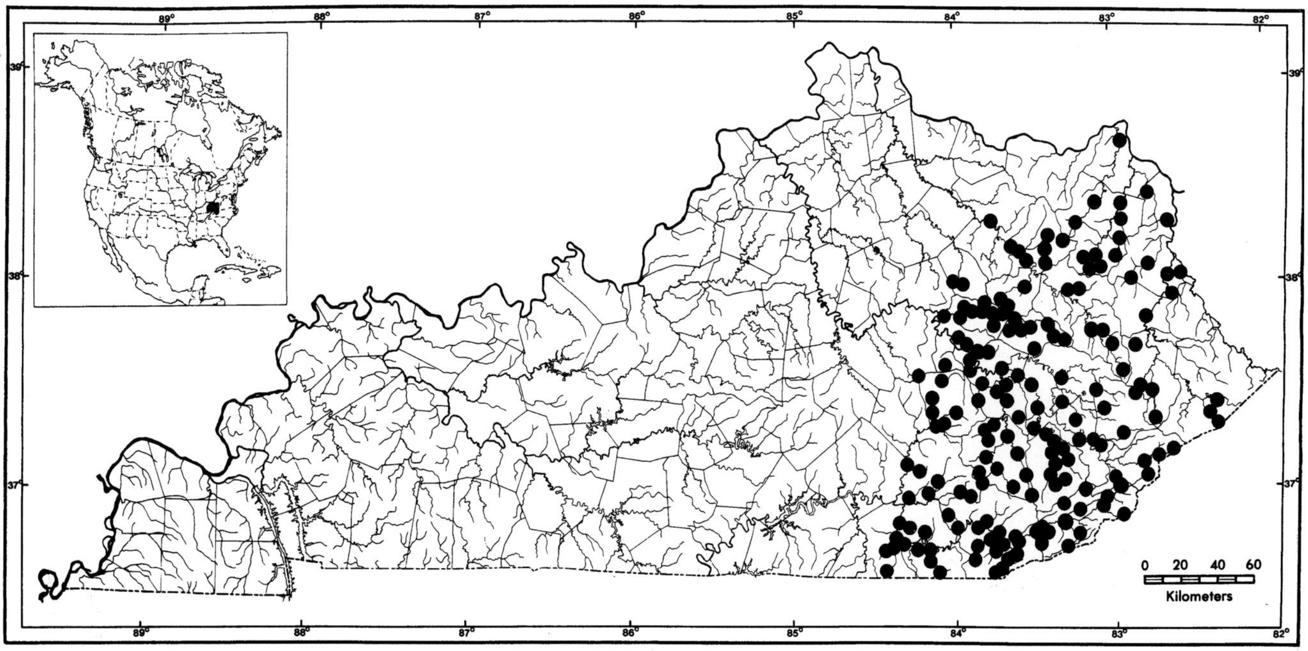

Fig. 120. Distribution of *Orconectes cristavarius* in Kentucky.

and upper New and Tennessee River drainages in Virginia (Taylor 2000). Within Kentucky, the species is widespread and common in the upper Cumberland, Kentucky, and Licking River drainages and in the Big Sandy drainage. The species also occurs in the Little Sandy River and Tygarts Creek drainages of eastern Kentucky.

Orconectes cristavarius occurs commonly in almost every size creek or river except for large rivers. Substrates at sites from which *O. cristavarius* was collected were predominately gravel, cobble, and boulders. Jezerinac et al. (1995) reported the species (as *O. spinosus*) from creeks with bedrock substrates. The species is usually encountered under cobble or large pieces of gravel in shallow riffles or runs but can also be found in woody debris or in undercut creek banks.

LIFE HISTORY – Several hundred collections of *Orconectes cristavarius* have been made in Kentucky and neighboring states from all months of the year. Within those collections, form I males occur in all months. Taylor (2000) reported that form I males were common in the months of March, July, August, September, and October and uncommon in January, May, June, and December. Ovigerous females have been collected only in March and April. The absence of ovigerous females in hundreds of collections from other months suggests that egg laying occurs only in the spring. The presence of form I males year-round suggests that copulation and gamete transfer may occur at any time. Jezerinac et al. (1995) reported observing copulation on numerous occasions from mid-July to early October. Taylor (2000) reported 43 eggs from a 37-mm total length female and 74 eggs from a 42.6-mm female. Egg size averaged 2.0 mm in diameter.

CRAYFISH ASSOCIATES – The following species have been collected from habitats containing *Orconectes cristavarius*: *Cambarus bartonii cavatus, C. buntingi, C. cumberlandensis, C. robustus, C. sciotensis, C. sphenoides, C. veteranus, O. juvenilis,* and *O. sanbornii.*

TAXONOMIC NOTES – Until its description by Taylor (2000), several names had been applied to populations of *Orconectes cristavarius* in eastern Kentucky and neighboring West Virginia. Rhoades (1944) referred to Kentucky populations of the species as *O. juvenilis,* while Jezerinac et al. (1995) assigned West Virginia populations to *O. spinosus*.

Orconectes durelli Bouchard and Bouchard
Saddle Crayfish

Fig. 121

LIFE COLORS – (Fig. 121) Base color of dorsal and lateral surfaces of chelae, carapace, and abdomen olive green to light brown and densely mottled with light to dark brown patches of varying size and shape. Carapace with a wide U-shaped dark saddle centered at the dorsal surface of caudal margin, extending forward on lateral surfaces of carapace to cervical groove. Light cream horizontal stripe directly below dark extensions of saddle on lateral surfaces. Fingers of chelae with orange or red tips, proximal borders with wide black bands. Diffuse light cream bands occasionally found proximal to black bands. Venter cream to whitish. In southern portions of the range (i.e., Duck, Tennessee River drainages in Tennessee), specimens are not mottled; each abdominal segment with thin red posterior margin; saddle more prominent.

CHARACTERS – (Fig. 122) Rostrum moderately excavated; distally margins straight and slightly converging, proximally margins straight or slightly convex; margins terminating in spines at base of long acumen. Acumen slightly longer than width of rostrum at marginal spines. Weak median carina occasionally present. Carapace slightly flattened dorsoventrally. Areola width variable, 4.4 to 13.0 times longer than wide at its narrowest point, with room for two to six punctations. Cervical spines present. Antennal scale widest distal to midpoint. Chelae stout with short fingers, dorsal surfaces of palms usually covered with numerous shallow punctations. Dorsal surfaces of fingers of chelae with weakly developed longitudinal ridges. Mesial margin of palm of chela with two rows of weak tubercles, 8 to 11 tubercles in mesialmost row.

Gonopod of form I male terminating in two long, thin elements: a thin, corneous central projection that is slightly arched caudodistally and a shorter, straight, noncorneous mesial process. Tip of central projection bent at a 45° angle to shaft and overhanging mesial process. Annulus ventralis of female subrhomboidal; with median trough running through cephalic half, two caudally directed protuberances overhanging centrally located fossa; sinus running from center of fossa to caudal margin.

MAXIMUM SIZE – The maximum known size for the species is 72.6 mm total length.

MOST LIKE – Numerous species of crayfishes in Kentucky possess form I gonopods with two long, thin terminal elements. Of those species *Orconectes durelli* is most similar to *O. placidus* in that both have a mandible with a dentate incisor region, lack a strong angular shoulder, and have a gonopod with an arched central projection less than 50% of the total length of gonopod. *Orconectes durelli* differs from *O. placidus* in possessing stocky chelae with shorter fingers; the dactyl is less than two times the length of the mesial margin of the palm. The form I gonopod of *O. durelli* is also similar to that of *O. barrenensis* and the two species do occur together at sites in the Barren River drainage. The two can easily be separated using cervical spines; *O. durelli* has them while *O. barrenensis* does not.

DISTRIBUTION AND HABITAT – (Fig. 123) *Orconectes durelli* occurs in the middle and lower Tennessee, middle and lower Cumberland, and upper Barren River drainages of central Tennessee and southern Kentucky (Bouchard and Bouchard 1995). Within Kentucky the species has a disjunct distribution and occurs in three different drainages. It is known to occur in tributaries of the Wolf River in Clinton County, tributaries of the upper Barren River drainage in Allen and Monroe counties, and in Wildcat Creek, a tributary of the lower Tennessee River in Calloway County.

Orconectes durelli is most common in clear, small to large-sized creeks ranging in width from 3 to 15 m with bottom substrates of gravel and cobble. Within these creeks, the species is usually

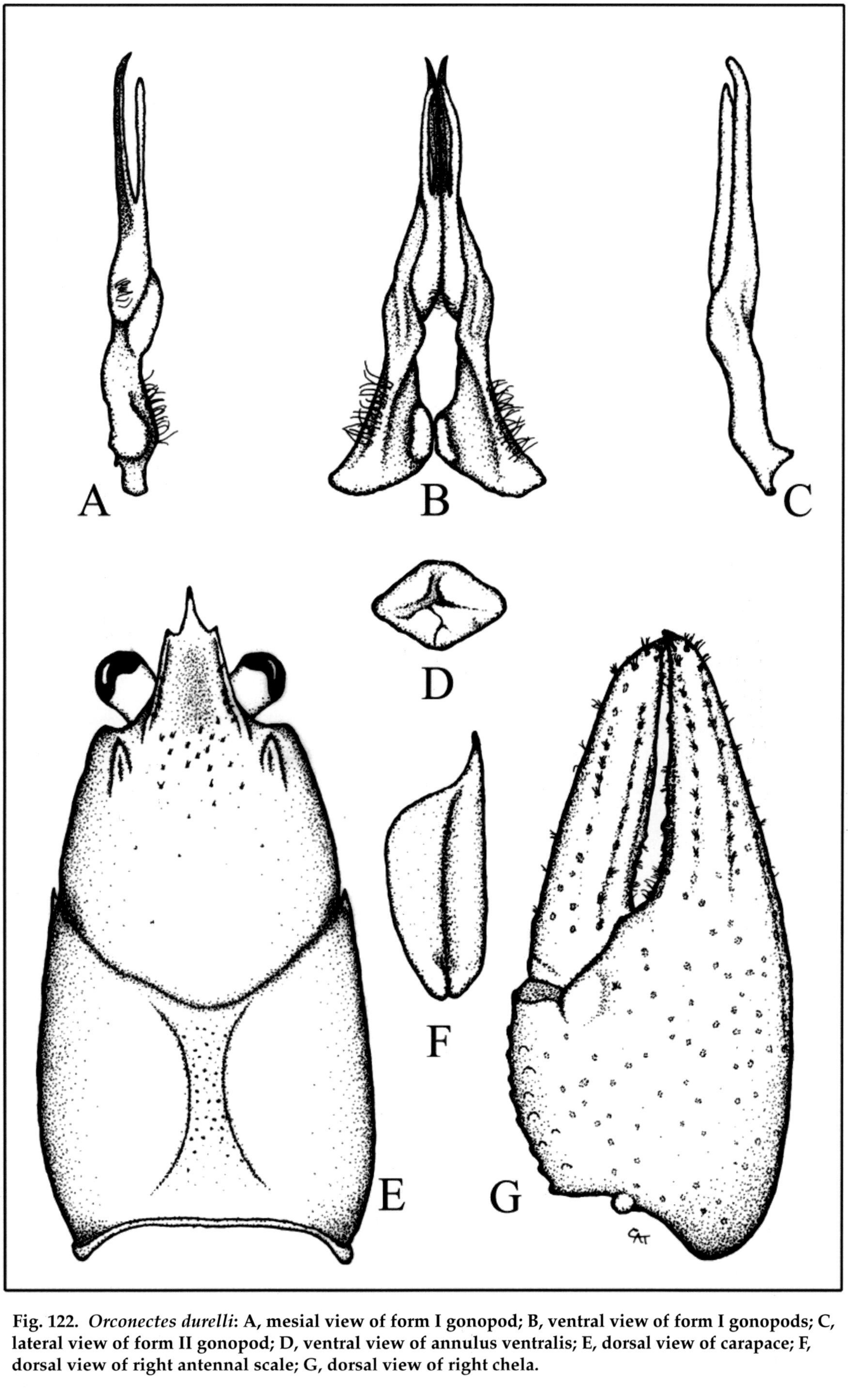

Fig. 122. *Orconectes durelli*: A, mesial view of form I gonopod; B, ventral view of form I gonopods; C, lateral view of form II gonopod; D, ventral view of annulus ventralis; E, dorsal view of carapace; F, dorsal view of right antennal scale; G, dorsal view of right chela.

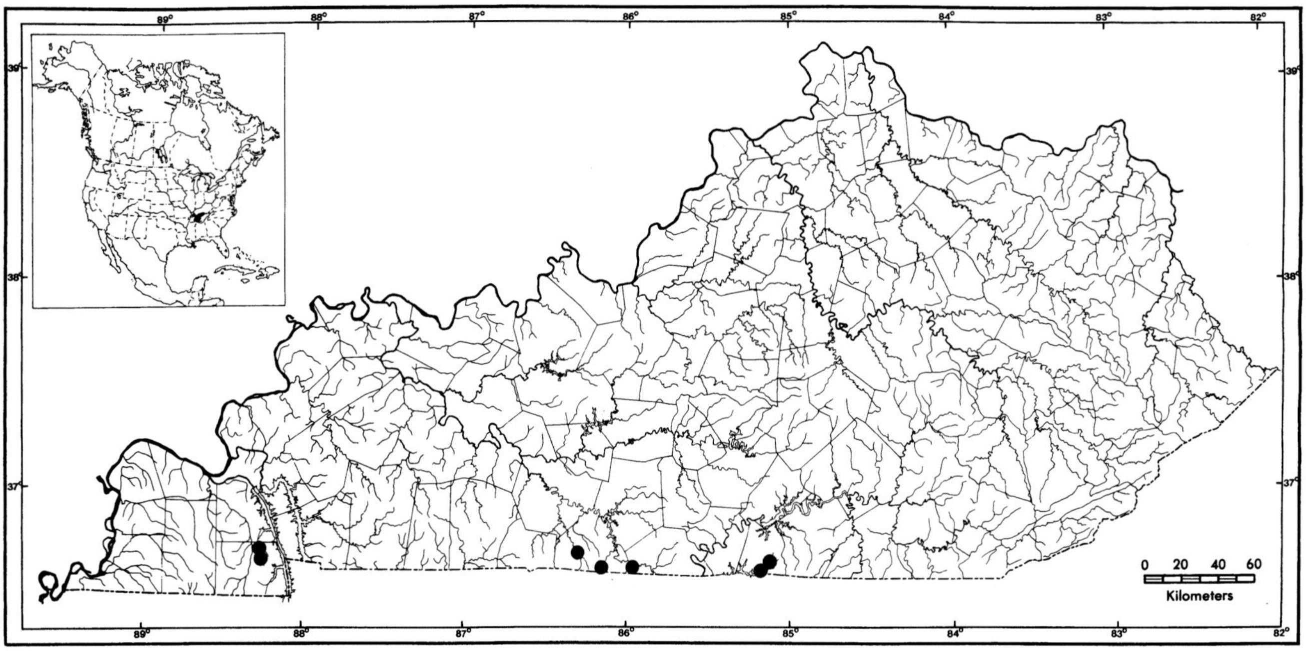

Fig. 123. Distribution of *Orconectes durelli* in Kentucky.

encountered under gravel and small cobble in riffle areas. The species can also be found under woody debris piles or root masses along creek margins or in pools.

LIFE HISTORY – Form I males have been collected in all months but February and December; however, they are much more common in the months of March, April, October, and November. Ovigerous females have been collected only in March and April and females carrying young have been collected in April. The lack of ovigerous females in summer and fall months suggests that reproductive activities occur in either fall or early spring and that egg laying occurs only in the spring. Four ovigerous females collected in March measured 55.1, 49.5, 53.8, and 41.0 mm in total length and carried 102, 121, 110, and 74 eggs, respectively. Egg size was slightly larger in *Orconectes durelli* than in other *Orconectes,* averaging 2.4 mm in diameter. As reported for *O. compressus, O. durelli* will dig short tunnels or shallow burrows under loose gravel during nondrought conditions. Nothing else is known about the life history of this species.

CRAYFISH ASSOCIATES – The following species have been collected from habitats containing *Orconectes durelli*: *Cambarus tenebrosus, O. barrenensis, O. compressus, O. placidus, O. putnami,* and *Procambarus acutus.*

Orconectes immunis (Hagen)
Calico Crayfish

Fig. 124

LIFE COLORS – (Fig. 124) Dorsal and lateral surfaces of chelae, carapace, and abdomen with dark gray or olive green base color. Lighter gray or tan stripe of varying widths running down midline of areola and abdomen. Lateral surfaces of abdomen lighter gray or tan. Dorsal surfaces of chelae occasionally mottled with dark spots of varying size. Fingers of chelae with orange or red tips. Venter cream to whitish.

CHARACTERS – (Fig. 125) Rostrum moderately excavated; margins slightly convex and converging anteriorly; margins terminating in weak spines or tubercles at base of acumen or terminating abruptly without spines or tubercles. Acumen length equal to or shorter than width of rostrum at marginal spines. Carapace slightly compressed laterally. Areola narrow, 10.6 to 25.0 times longer than wide at its narrowest point, with room for one to three punctations. Cervical spines present. Antennal scale widest point proximal to midpoint. Chelae large and with long, thin fingers; dorsal surfaces of fingers of chelae with longitudinal ridges. Shallow notch immediately posterior to enlarged tubercle at base of opposable margin of dactyl; second and fourth or fifth tubercle on opposable margin of propodus enlarged. Tufts of long setae usually present at base of propodus. Mesial margin of palm of chela and dactyl with pointed, anteriorly directed tubercles. Mesial margin of palm with two rows for tubercles; mesialmost row with six to eight tubercles.

Gonopod of form I male terminating in two thin, caudodistally recurved elements: a sharply pointed, corneous central projection recurved at approximately 90° to shaft and a noncorneous mesial process of about equal length. Distal end of mesial process dorsoventrally flattened and ladlelike; recurved at 90° to shaft. Annulus ventralis of female ovoid; with narrow angular trough through cephalic margin and a large, deep, sinuate fossa; fossa directed to either right or left side; sinus running from lateral margin of fossa to caudal margin of annulus.

MAXIMUM SIZE – The maximum known size for the species is 102.9 mm total length.

MOST LIKE – With two long, strongly recurved elements, the form I gonopod of *Orconectes immunis* is similar to two other Kentucky species, *O. compressus* and *O. palmeri palmeri*. *Orconectes immunis* differs from *O. compressus* in having strongly developed longitudinal ridges on the fingers of its chelae and a notch at the base of the opposable margin of the dactyl. *Orconectes immunis* is also not as strongly compressed laterally as *O. compressus*. *Orconectes immunis* differs from *O. palmeri palmeri* in possessing an open areola. With a shallow notch found near the base of the dactyl the chelae of *O. immunis* may superficially resemble *Fallicambarus fodiens, Cambarus diogenes, C. ludovicianus, C. thomai,* and *C.* sp. A. *Orconectes immunis* differs from all of those species in possessing an open areola and thin terminal elements on the gonopod.

DISTRIBUTION AND HABITAT – (Fig. 126) *Orconectes immunis* is a wide-ranging species and occurs from southern Quebec and New England westward across the upper Midwest to Wyoming and eastern Colorado and south to extreme northwestern Tennessee (Hobbs 1989). While Hobbs (1989) also listed the species from Alabama, we have been unable to locate any specimens collected from that state. Within Kentucky, the species occurs commonly across the western half of the state in the lower Ohio, Green, Cumberland, and Tennessee River drainages and in several Mississippi River tributaries. It is most common in the lower Green River drainage.

Orconectes immunis occurs in a wide variety of habitats, from ponds and lakes to flowing creeks

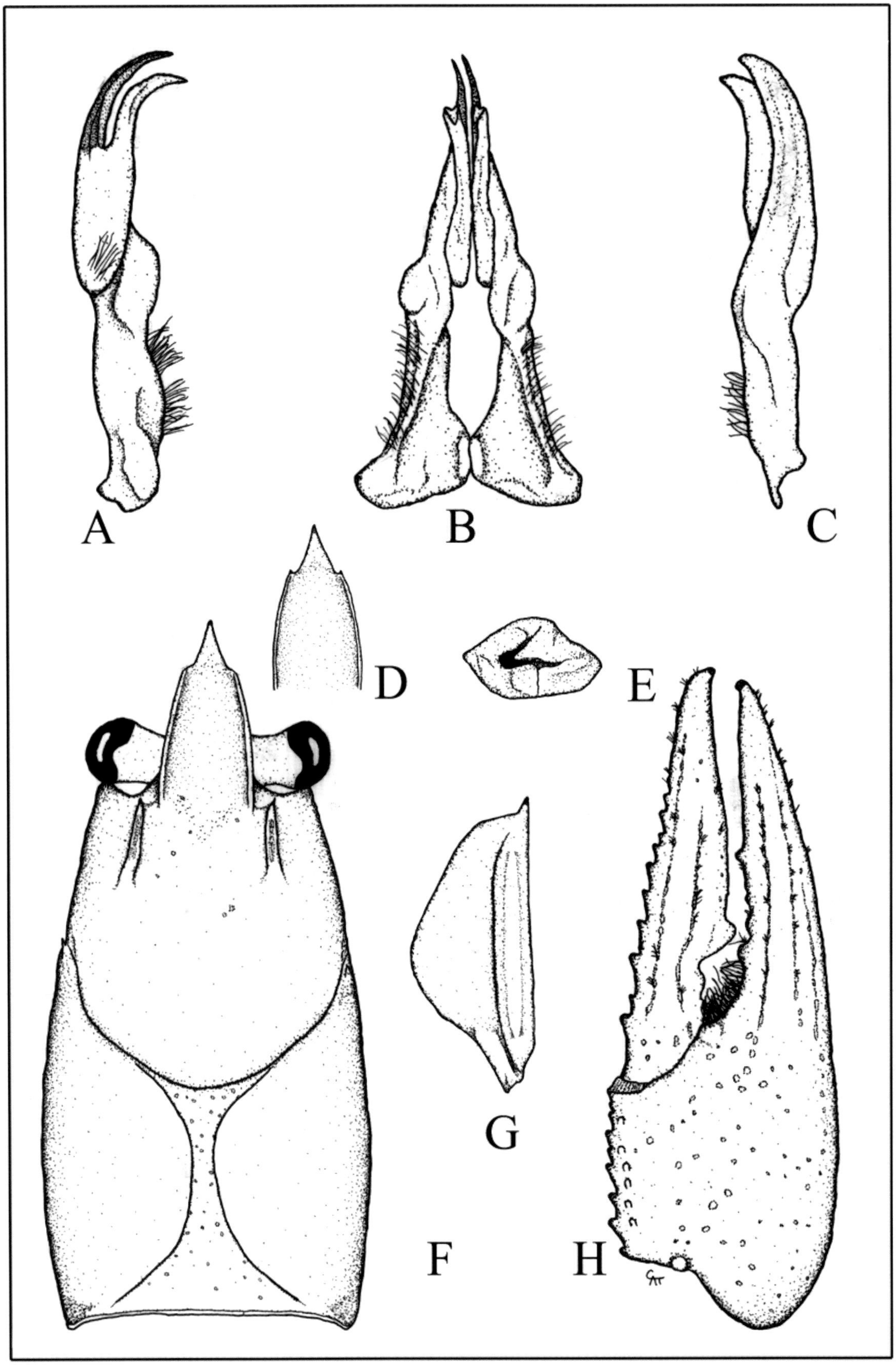

Fig. 125. *Orconectes immunis*: A, mesial view of form I gonopod; B, ventral view of form I gonopods; C, lateral view of form II gonopod; D, ventral view of rostrum with margins terminating in spines; E, ventral view of annulus ventralis; F, dorsal view of carapace; G, dorsal view of right antennal scale; H, dorsal view of right chela.

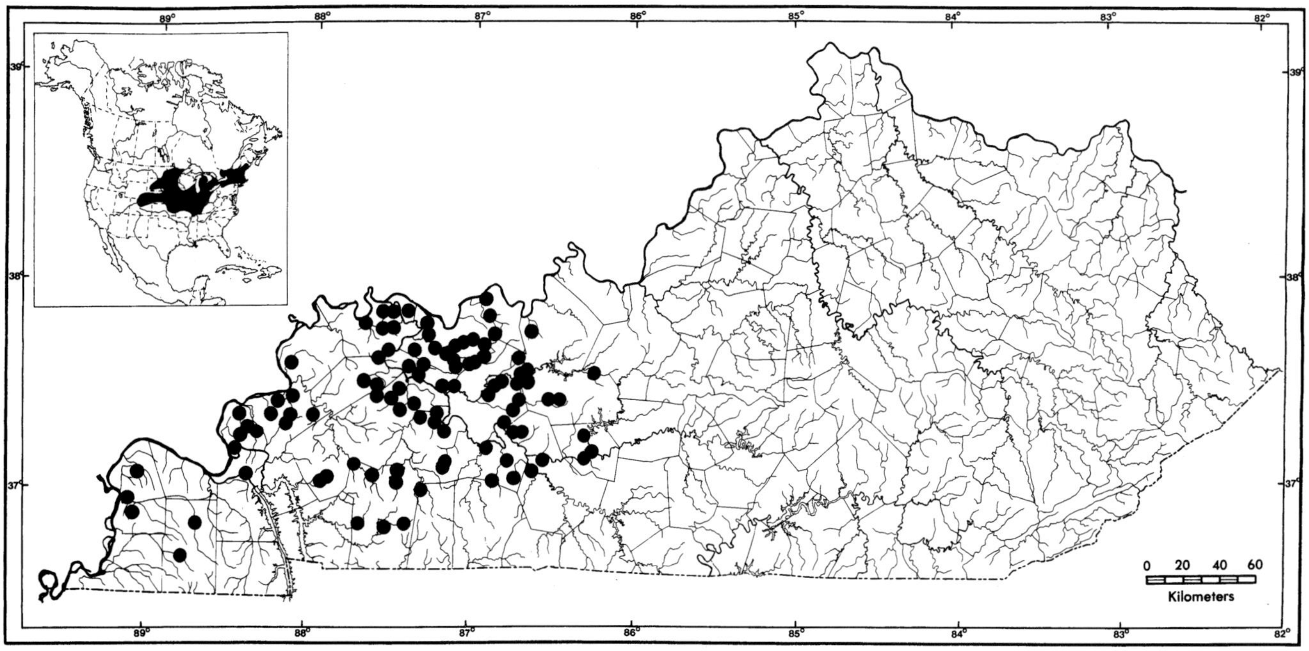

Fig. 126. Distribution of *Orconectes immunis* in Kentucky.

with cobble substrates. However, the species is most commonly found in lower gradient, mud-bottomed creeks and shallow, well-vegetated ditches in the floodplains of larger creeks and rivers. Given its ability to build burrows, *O. immunis* will also inhabit seasonally flooded ponds and lakes. Within these habitats, the species is usually encountered under leaf litter or in woody debris, in growths of aquatic vegetation, or in shallow burrows.

LIFE HISTORY – In Kentucky, form I males have been found from March through October and ovigerous females have been collected only in March and April, suggesting that hatching may occur in late spring. It is unknown when peak mating activity occurs in Kentucky populations. In Missouri, Pflieger (1996) also reported ovigerous females in March and April and counted 12 to 423 eggs on females ranging in size from 45.7 to 68.5 mm total length.

Detailed life history studies have been conducted on *Orconectes immunis* in Iowa (Caldwell and Bovbjerg 1969) and New York (Tack 1941) and a summary of those data is presented below. Mating appears to occur occasionally at all times of the year. However, in New York it occurred most often from mid-June to mid-October. In Iowa eggs were laid in either the fall or spring and eggs generally hatched in May. Adults are nocturnal while juveniles are active throughout the day. Aquatic invertebrates such as amphipods, cladocerans, insects, and vascular plant material make up a large percentage of the species' diet. Algae is also commonly consumed. *Orconectes immunis* will frequently burrow in seasonally wet habitats. Burrows are usually constructed in the dry fall months and are simple horizontal shafts into pond banks or vertical shafts into dry pond or creek bottoms with enlarged chambers at their termini. Burrow depths of up to 35 cm have been reported and their openings may be capped with low chimneys, mounds, or ridges. Burrow openings may be plugged in very dry or cold weather conditions. Longevity in the species is usually two years with high mortality among form I males occurring after the peak mating season in their second year of life.

CRAYFISH ASSOCIATES – The following species have been collected from habitats containing *Orconectes immunis*: *Cambarus diogenes, C. graysoni, C. striatus, C. tenebrosus,* C. sp. A, *Fallicambarus fodiens, O. bisectus, O. kentuckiensis, O. margorectus, O. putnami, O. rafinesquei, O. rusticus, O. tricuspis,* and *Procambarus acutus.*

Orconectes inermis inermis Cope
Ghost Crayfish

Fig. 127

LIFE COLORS – (Fig. 127) Albinistic; dorsal, ventral, and lateral surfaces of chelae, carapace, and abdomen white to cream to light yellow. Larger individuals may have several shades of cream and light yellow, with yellows highlighting various grooves and joints of the individual.

CHARACTERS – (Fig. 128) Eyes greatly reduced and lacking pigment. Rostrum moderately excavated or flat; distally margins straight and slightly converging, proximally margins slightly convex; margins terminating in spines or tubercles at base of acumen. Acumen usually longer than width of rostrum at marginal spines. Carapace dorsoventrally flattened. At least one cervical spine present per side; often several of varying size. Hepatic region highly variable, with 0 to 13 spines of various sizes per side. Most spines on carapace sharply pointed. Areola wide, 4.5 to 6.7 times longer than wide at its narrowest point, with room for three to six punctations. Antennal scale widest at midpoint. Chelae long and thin; mesial margin of palm with 10 to 14 small tubercles, dorsal surfaces of palms with several irregular rows of tubercles. Fingers of chelae smooth or with weak longitudinal ridges. Opposable margin of propodus with enlarged fourth or fifth and eighth or ninth tubercles, opposable margin of dactyl usually with enlarged fifth tubercle.

Gonopod of form I male terminating in two short elements: a slightly curved and truncated corneous central projection and a slightly longer, noncorneous mesial process. Mesial process broad basally, constricting rapidly to form a pointed tip, and arched caudodistally. Annulus ventralis of female oval; lacking fossa; possessing a centrally located and smoothly rounded longitudinal ridge and short sinus running to the caudal edge.

MAXIMUM SIZE – The maximum known size for the species is 79.0 mm total length.

MOST LIKE – Three species of Kentucky crayfishes are albinistic and inhabit caves: *Orconectes australis packardi*, *O. inermis inermis*, and *O. pellucidus*. Males of *O. inermis inermis* differ from males of *O. australis packardi* and *O. pellucidus* in possessing a mesial process that is of equal length or only slightly longer than the central projection and by lacking an angular shoulder at the base of the central projection.

DISTRIBUTION AND HABITAT – (Fig. 129) *Orconectes inermis inermis* occurs in a narrow band of karst extending from Hart and Green counties, Kentucky north to Lawrence County, Indiana (Hobbs and Barr 1972). Hobbs and Barr (1972) reported that at the northern terminus of the species' range in Monroe County, Indiana, *O. inermis inermis* may interbreed with *O. i. testii*. Within Kentucky the species occurs in subterranean passages of the upper Green and Nolin River drainages in Green, Hardin, and Hart counties, and in subterranean passages that drain into Ohio River tributaries in Breckinridge, Hardin, and Meade counties.

Orconectes inermis inermis occurs exclusively in subterranean creeks. Within those creeks the species is most frequently encountered in shallow pools with mud and silt substrates where it is frequently seen out in the open. While the species may occasionally be found in riffle habitats, Hobbs III (1976) suggested that the higher concentrations of food matter such as organic detritus in pooled habitats accounts for the species' preference for such habitats. As mentioned for *O. australis parkardi*, the lack of large predators in subterranean creeks may also account for *O. inermis inermis'* habitat choices.

LIFE HISTORY – Through the work of Jegla (1964, 1969) and Hobbs III (1976, 1978) several aspects of the life history of *Orconectes inermis inermis* have been recorded. All of these data were collected

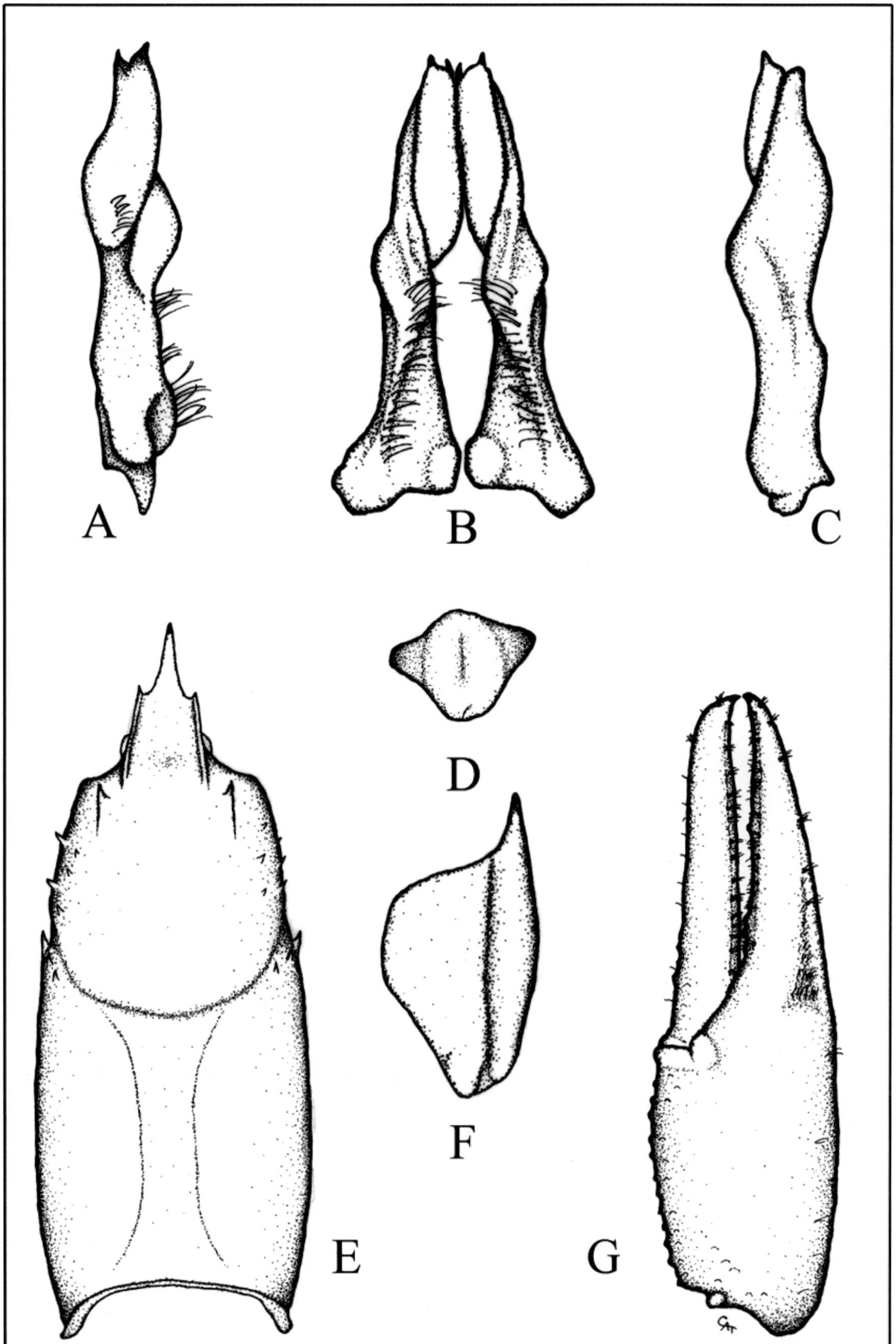

Fig. 128. *Orconectes inermis inermis*: A, mesial view of form I gonopod; B, ventral view of form I gonopods; C, lateral view of form II gonopod; D, ventral view of annulus ventralis; E, dorsal view of carapace; F, dorsal view of right antennal scale; G, dorsal view of right chela.

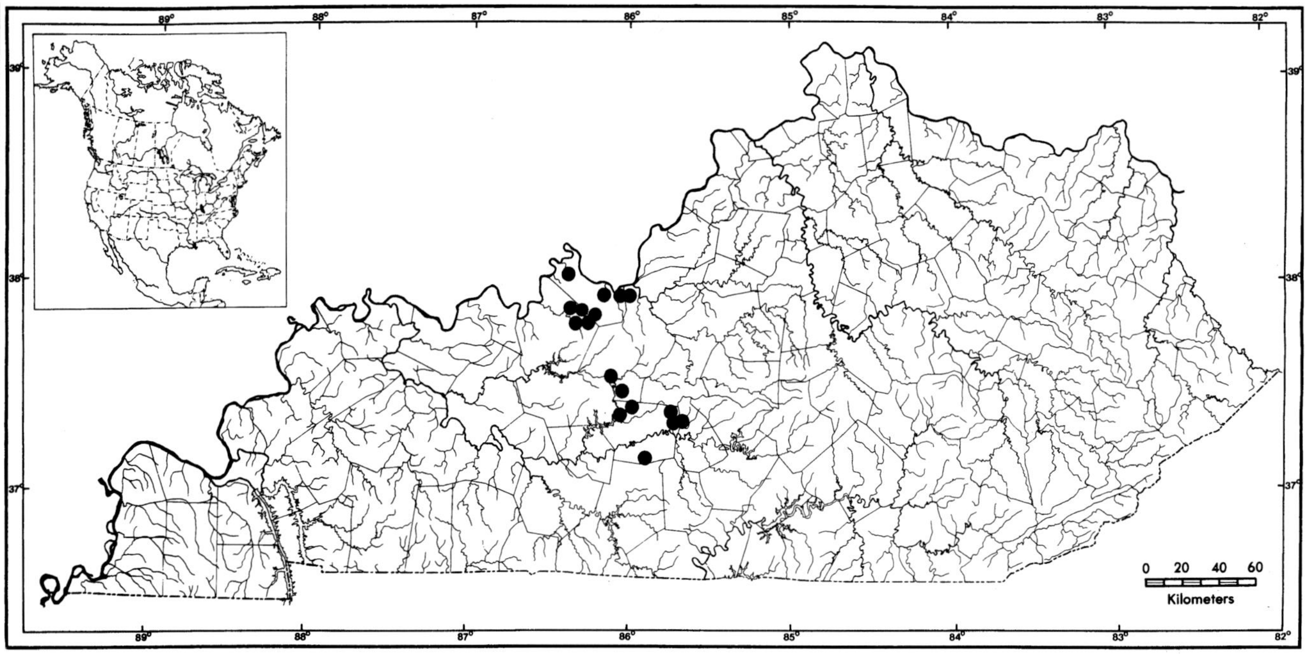

Fig. 129. Distribution of *Orconectes inermis inermis* in Kentucky.

from Indiana populations of the subspecies and are summarized below. Molting in the subspecies was not restricted to definite seasons and form I males were encountered throughout the year. However, in fall and winter months form I males accounted for 77 to 97% of the total male population. Ovigerous females were only encountered during summer months (June and August). Average clutch size was 45 eggs and egg size ranged from 2.0 to 2.5 mm in diameter. Mohr and Poulson (1966) suggested that spring floods trigger ovulation in the species. The timing of copulatory activities has not been recorded. Diet consisted primarily of plant material washed into caves, while other small crustaceans such as isopods and amphipods were also consumed if available. The home-range size appeared to be related to body size, sex, and stage of reproduction: for males, range size averaged approximately 20 linear m while those of females averaged 23 m.

In Kentucky, form I males have been collected in every month but January, March, and April. No ovigerous females have been collected in the state.

CRAYFISH ASSOCIATES – No other crayfish species have been collected from habitats containing *Orconectes inermis inermis*.

CONSERVATION STATUS – KSNPC Special Concern.

5/13/05 – Belt Cave, Hardin Co. Ky.

Orconectes jeffersoni Rhoades
Louisville Crayfish

Fig. 130

LIFE COLORS – (Fig. 130) Base color of dorsal and lateral surfaces of chelae, carapace, and abdomen light brown. Lateral surfaces of the abdomen occasionally lighter brown than base color of carapace. Dorsum of carapace anterior to cervical groove with laterally wide dark brown patch. Carapace with wide U-shaped dark saddle centered at the dorsal surface of caudal margin, extending forward on lateral surfaces of carapace to cervical groove. Darkness of saddle, especially lateral arms, highly variable among individuals. Abdomen with long, dark brown or black wedge extending down center of dorsum. Fingers of chelae with orange tips bordered proximally by subdistal black or dark brown bands. Large knobs at base of dactyl and joints of walking legs orange. Venter cream to whitish.

CHARACTERS – (Fig. 131) Rostrum deeply excavated, distally margins straight and slightly converging, proximally margins slightly convex, margins terminating in small spines or tubercles at base of acumen. Length of acumen equal to width of rostrum at marginal spines. Carapace dorsoventrally flattened, more so in larger individuals. Areola of moderate width, 7.6 to 8.9 times longer than wide at its narrowest point, with room for 2 to 4 punctations. Cervical spines present. Antennal scale widest at midpoint. Incisor region of mandible dentate, occasional specimen with straight edge. Chelae large with moderately long fingers, dorsal surfaces of palms usually covered with shallow punctations. Dorsal surfaces of fingers of chelae with weak longitudinal ridges. Mesial margin of palm of chela and moveable finger with tubercles, palm with two discreet rows of tubercles, several additional scattered tubercles lateral to rows; 7 to 10 tubercles in mesialmost row. Males with slightly longer chelae than females of the same carapace length.

Gonopod of form I male terminating in two straight midlength terminal elements: a corneous central with a pointed and slightly bent tip and a shorter, noncorneous mesial process. Caudal surface of gonopod with or without smoothly rounded shoulder at base of central projection. Annulus ventralis of female subrhomboidal: with centrally located fossa; narrow median trough running from fossa to the cephalic margin, S-shaped sinus running from fossa to caudal margin.

MAXIMUM SIZE – The maximum known size for the species is 75.7 mm total length.

MOST LIKE – *Orconectes jeffersoni* is most similar to other members of the genus that possess short, more or less straight terminal elements on the form I gonopod. Those species are *O. bisectus, O. margorectus, O. rafinesquei, O. sanbornii*, and *O. tricuspis*. *Orconectes jeffersoni* differs from all of those species in possessing longer terminal elements. The central projection of *O. jeffersoni* ranges in length from 29 to 34% (mean = 31%) of the total length of the gonopod, the central projections of *O. bisectus, O. margorectus, O. rafinesquei, O. sanbornii*, and *O. tricuspis* are less than 25% of the total length of the gonopod. *Orconectes jeffersoni* also differs from *O. sanbornii* and *O. tricuspis* by having a mesial process that is not dorsoventrally flattened at the tip.

DISTRIBUTION AND HABITAT – (Fig. 132) *Orconectes jeffersoni* is endemic to the state of Kentucky. It has an extremely limited distribution in the lower Ohio River drainage where it is known only from the Pond, Beargrass, and Harrods Creek drainages in Bullitt, Jefferson, and Oldham counties in north–central Kentucky.

Orconectes jeffersoni occurs in small to medium-sized creeks ranging in width from 2 to 10 m with substrates of bedrock, cobble, and some mud. The species is usually encountered under flat cobble in areas with flow or among woody debris along creek edges. *Orconectes jeffersoni* appears to

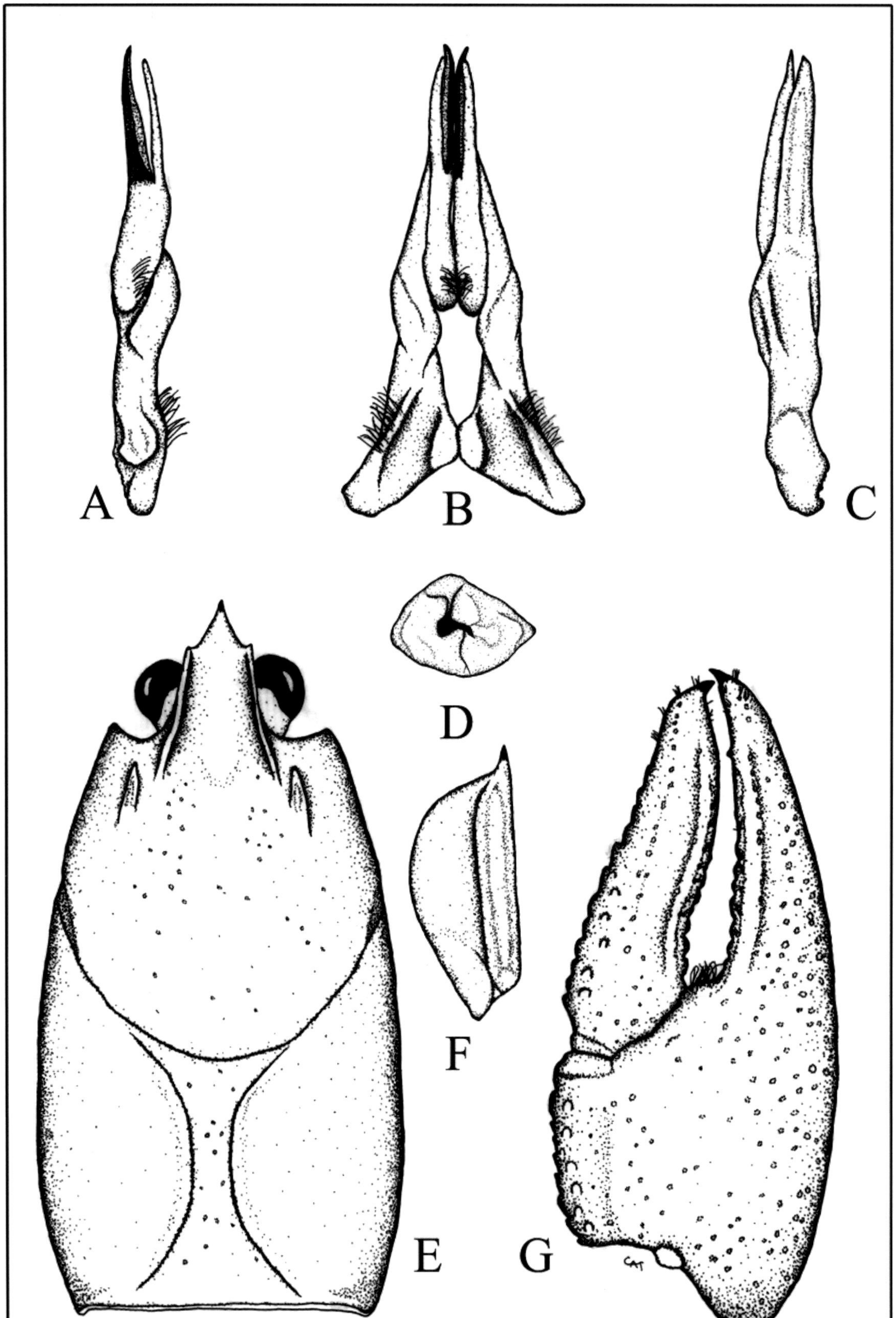

Fig. 131. *Orconectes jeffersoni*: A, mesial view of form I gonopod; B, ventral view of form I gonopods; C, lateral view of form II gonopod; D, ventral view of annulus ventralis; E, dorsal view of carapace; F, dorsal view of right antennal scale; G, dorsal view of right chela.

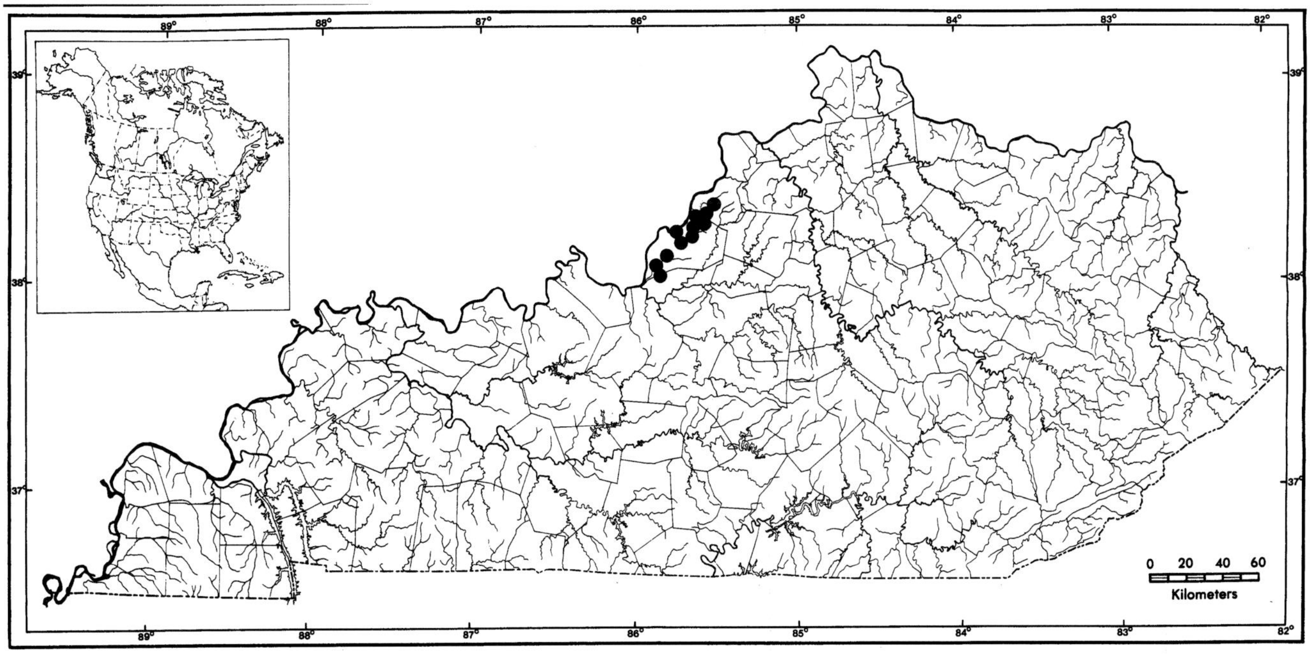

Fig. 132. Distribution of *Orconectes jeffersoni* in Kentucky.

be quite tolerant of habitat alteration. The species can be found commonly in creek reaches flowing through heavily urbanized regions of the Louisville metropolitan area. We found it in creeks flowing through residential neighborhoods, golf courses, and shopping mall parking lots.

LIFE HISTORY – Very little is known about the life history of *Orconectes jeffersoni*. Form I males have been collected in February, April, July, September, October, and November; however, form I males are much more abundant in February, April, October, and November. Ovigerous females have been collected only in April and juveniles less than 30 mm total length have been collected in May. These data indicate that egg hatching occurs in midspring. It is unknown when peak mating occurs.

CRAYFISH ASSOCIATES – Only two species have been collected from habitats containing *Orconectes jeffersoni*, *Cambarus ortmanni* and *C. tenebrosus.*

CONSERVATION STATUS – AFS Endangered; KSNPC Endangered.

5/12/05 at U. of L. Horner Refuge – S. Fk. Harrods
Jefferson Co. Ky.

Orconectes juvenilis (Hagen)
Kentucky River Crayfish

Fig. 133

LIFE COLORS – (Fig. 133) Base color of dorsal and lateral surfaces of chelae and carapace brown to olive green. Abdomen with light brown base color, a long rusty brown or brick red wedge extending down center of dorsum, and orange oval spot on lateral surfaces of each abdominal segment. Fingers of chelae with orange tips bordered proximally by wide black bands. Large knobs at base of dactyl, tubercle lateral to base of palm of chela, and joints of walking legs orange. Venter cream to whitish.

CHARACTERS – (Fig. 134) Rostrum moderately excavated; distally margins straight, proximally margins convex; margins terminating in spines at base of acumen. Acumen equal to or slightly longer than width of rostrum at marginal spines. Median carina absent. Carapace slightly compressed dorsoventrally. Areola of moderate width, 7.0 to 13.0 times longer than wide at its narrowest point, room for two to four punctations. Cervical spines present. Antennal scale widest at midpoint or just distal to midpoint. Incisor region of mandible with straight edge. Chelae large with moderately long fingers, dorsal surfaces of palms usually covered with shallow punctations. Dorsal surfaces of fingers of chelae with weak longitudinal ridges. Mesial margin of palm of chela and dactyl with tubercles, palm with two discreet rows of tubercles, several additional scattered tubercles lateral to rows; seven to nine tubercles in mesialmost row. Slight gap between basal opposable margins of fingers, gap wider in form I males. Males with longer chelae than females of the same carapace length.

Gonopod of form I male terminating in two long, thin elements: a sharply pointed corneous central projection that is slightly arched caudodistally and a shorter, noncorneous mesial process with a cephalically recurved tip. Caudal surface of gonopod with strong angular shoulder at base of central projection. Annulus ventralis of female subrhomdoidal; with median trough running through cephalic half and two caudally directed protuberances overhanging centrally located fossa; fossa laterally wide; sinus extends from center of fossa to caudal edge of annulus.

MAXIMUM SIZE – The maximum known size for the species is 81.5 mm total length.

MOST LIKE – Numerous species of crayfishes in Kentucky possess form I gonopods with two long, thin terminal elements. Of those species *Orconectes juvenilis* is most similar to *O. cristavarius* and *O. rusticus* in that all three possess a strong angular shoulder at the base of the central projection, cervical spines, and a gonopod with an arched central projection less than 50% of the total length of gonopod. *Orconectes juvenilis* differs from *O. cristavarius* in possessing a mandible with a straight-edged incisor region. While *O. rusticus* also has a straight-edged incisor region, *O juvenilis* differs from that species in possessing longer terminal elements. The length of the central projection of *O. juvenilis* is at least 44% of the total length of the gonopod while the length of the central projection of *O. rusticus* is less than 44% of the total length of the gonopod.

DISTRIBUTION AND HABITAT – (Fig. 135). *Orconectes juvenilis* occurs in a relatively narrow stretch of the middle Ohio River drainage and the upper Cumberland River drainage (Taylor 2000). The species is currently known only from Kentucky and southern Indiana. In Kentucky, *O. juvenilis* is widespread and common in the middle and lower Kentucky River drainage, the Salt River drainage, and several Ohio River tributaries from Breckinridige to Boone counties. In the Cumberland River drainage the species occurs upstream of Cumberland Falls and in the Rockcastle River drainage.

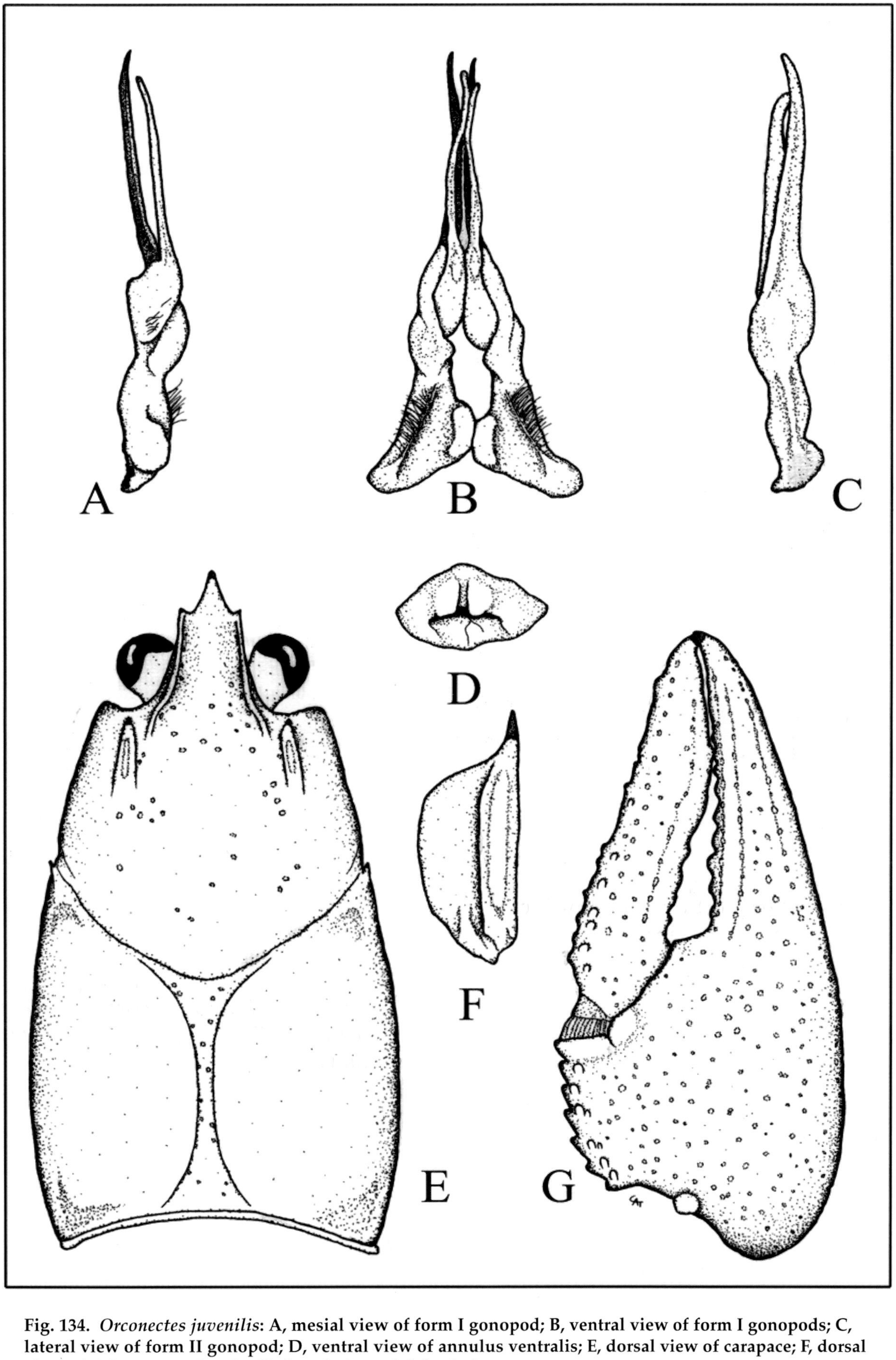

Fig. 134. *Orconectes juvenilis*: A, mesial view of form I gonopod; B, ventral view of form I gonopods; C, lateral view of form II gonopod; D, ventral view of annulus ventralis; E, dorsal view of carapace; F, dorsal view of right antennal scale; G, dorsal view of right chela.

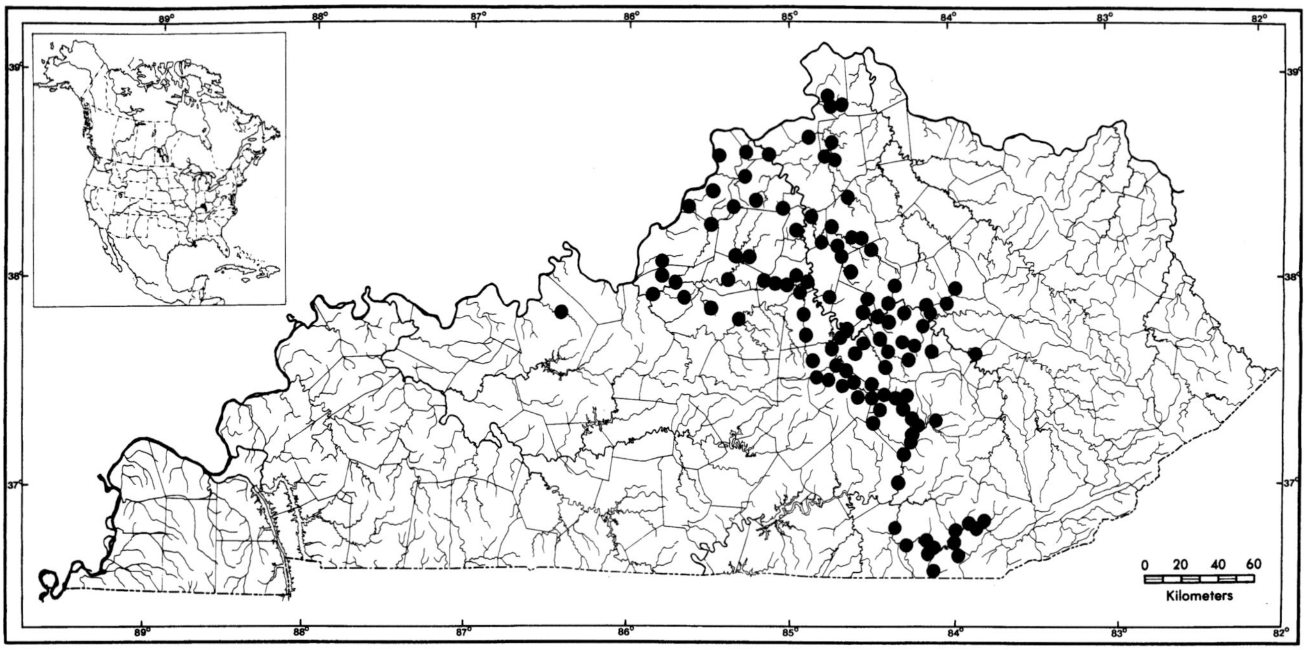

Fig. 135. Distribution of *Orconectes juvenilis* in Kentucky.

Orconectes juvenilis occurs commonly in almost every size of flowing upland creek or river within its range with substrates of gravel, cobble, boulders, or fractured bedrock. The species is usually encountered under cobble or large pieces of gravel in riffles or runs, but can also be found in woody debris or in undercut creek banks.

LIFE HISTORY – Little is known about the life history of *Orconectes juvenilis* despite the fact that over 150 collections of the species have been made in Kentucky. Within those collections, form I males occur in all months but February and are common in most months with the exception of July. Ovigerous females have been collected only in April. The absence of ovigerous females in other collections suggests that egg laying occurs only in the spring. The presence of form I males year-round suggests that copulation and gamete transfer may occur at any time. A 60.2-mm total length female collected in April carried 141 eggs that averaged 2.1 mm in diameter.

CRAYFISH ASSOCIATES – The following species have been collected from habitats containing *Orconectes juvenilis*: *Cambarus cumberlandensis, C. rusticiformis, C. tenebrosus,* and *O. cristavarius.*

TAXONOMIC NOTES – Until the publication of Taylor (2000), *O. juvenilis* had been considered a synonym of *O. rusticus* by most authors, including Hobbs (1989).

6/11/05 Little Ky. River Trimble Co., Ky.
7/5/05 Wilson Creek Bullitt Co. Ky.
7/10/05 Sinking Creek W. of Webster, Breckinridge Co.

Orconectes kentuckiensis Rhoades
Kentucky Crayfish

Fig. 136

LIFE COLORS – (Fig. 136) Base color of dorsal and lateral surfaces of chelae, carapace, and abdomen light brown to tan. Dorsum of abdomen, tail fan, and cephalothorax irregularly mottled with darker brown flecks and/or patches. Caudal margin of carapace with brown saddle that extends forward to just posterior of midpoint of lateral surfaces of the carapace. Saddle extends posteriorly to dorsum of first abdominal segment. Dorsum of chelae sometimes with dark brown flecks; fingers with orange tips bordered proximally by wide black or dark brown bands. Venter cream to whitish.

CHARACTERS – (Fig. 137) Rostrum deeply excavated; distally margins straight and slightly converging, proximally margins slightly convex, margins terminating in spines at base of acumen. Acumen equal to or slightly longer than width of rostrum at marginal spines. Carapace dorsoventrally flattened, more so in larger individuals. Areola of moderate width, 6.1 to 10.7 times longer than wide at its narrowest point, with room for three to five punctations. Antennal scale widest just distal to midpoint. Chelae large with moderately long fingers, dorsal surfaces of palm and fingers usually covered with short setae. Dorsal surfaces of fingers of chelae with well-developed longitudinal ridges. Mesial margin of palm of chela and dactyl with strong tubercles, palm with two or three rows of tubercles, 7 to 10 tubercles in mesialmost row. Males with significantly longer chelae than females of the same carapace length.

Gonopod of form I male terminating in two short caudodistally recurved terminal elements: a laterally flattened, corneous central projection with truncated distal tip and a shorter, noncorneous mesial process. Annulus ventralis of female subrhomboidal; with deeply recessed cephalic margin; poorly developed or absent fossa; sinus running through elevated central portion of the caudal margin.

MAXIMUM SIZE – The maximum known size for the species is 87 mm total length.

MOST LIKE – The form I male gonopod shape found on *Orconectes kentuckiensis* is rare among Kentucky crayfishes. Only one other species, *Cambarus bouchardi*, found in the state possesses a gonopod with two short, caudodistally recurved terminal elements and a central projection that is truncated at its distal tip. *Orconectes kentuckiensis* differs from *C. bouchardi* in having straight rostral margins that terminate in spines; the rostrum of *C. bouchardi* has rounded margins lacking spines. *Cambarus bouchardi* also has a much wider areola.

DISTRIBUTION AND HABITAT – (Fig. 138) *Orconectes kentuckiensis* has a limited distribution in the lower Ohio River drainage of western Kentucky and southern Illinois (Hobbs 1989). In Kentucky the species occurs throughout the Tradewater River drainage and in smaller Ohio River tributaries in Crittenden and Livingston counties. The species also occurs in the Livingston Creek and Little River drainages, both of which are tributaries of the lower Cumberland River. In Illinois the species is restricted to three creeks in Hardin County (Page 1985).

Orconectes kentuckiensis is most common in small to large creeks ranging in width from two to eight m with bottom substrates of cobble or large gravel. In these creeks the species can usually be found under the cobble or gravel. While Page (1985) reported the species almost exclusively from rocky pools, we found it to occur most commonly in flowing riffles averaging 0.4 m in depth. Like Rhoades (1944), we have occasionally collected the species from woody debris piles in mud-bottom creeks.

LIFE HISTORY – Form I males of *Orconectes kentuckiensis* have been collected in Kentucky

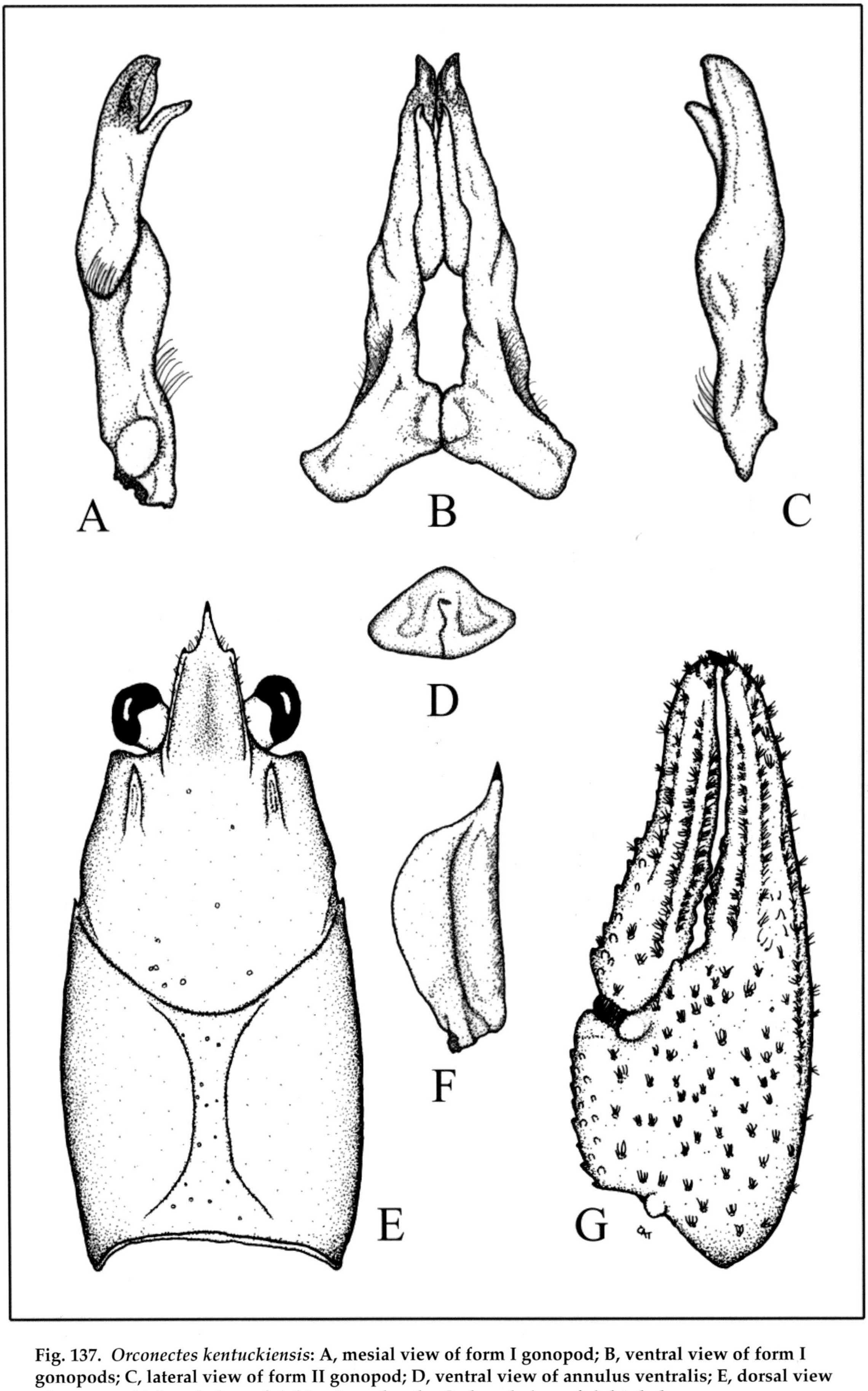

Fig. 137. *Orconectes kentuckiensis*: A, mesial view of form I gonopod; B, ventral view of form I gonopods; C, lateral view of form II gonopod; D, ventral view of annulus ventralis; E, dorsal view of carapace; F, dorsal view of right antennal scale; G, dorsal view of right chela.

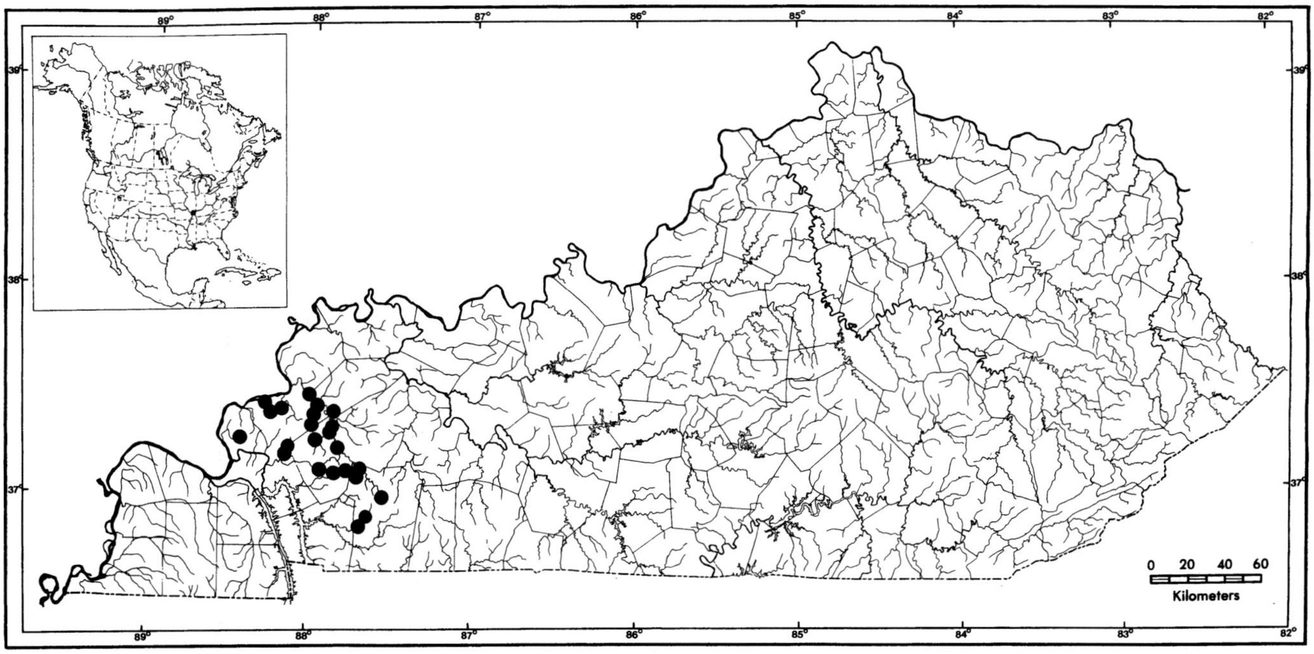

Fig. 138. Distribution of *Orconectes kentuckiensis* in Kentucky.

from August to May. Ovigerous females have been collected in April and May.

A detailed life history study of this species was conducted by Boyd and Page (1978) in southern Illinois. Given the geographic proximity of their study site, we believe that life history parameters reported by them are applicable to Kentucky populations. The following information is taken from that publication. Form I males occurred from July through April. Form I males reverted to form II in April and May and molted back to form I in July. Ovigerous females were collected in March and April; the number of eggs carried by females increased significantly with an increase in carapace length and egg diameter averaged 1.9 mm. Mating in *O. kentuckiensis* was observed in October and November. Most individuals lived for two years with a small percentage of the population surviving for a third year; reproductively mature first-year males and females were encountered in the months of September and October. The species had a 1:1 sex ratio and fed on vascular plants, amphipods, isopods, crayfishes, caddisflies, and midges. Growth of individuals occurred mostly in the spring and summer months.

CRAYFISH ASSOCIATES – The following species have been collected from habitats containing *Orconectes kentuckiensis*: *Cambarus diogenes, C. tenebrosus, Fallicambarus fodiens, O. immunis,* and *O. placidus*.

CONSERVATION STATUS – AFS Threatened.

Orconectes lancifer (Hagen)
Shrimp Crayfish

Fig. 139

LIFE COLORS – (Fig. 139) Base color of dorsal and lateral surfaces of chelae, carapace, and abdomen reddish brown to light gray and densely mottled with small light to dark brown flecks of varying size. Carapace usually with darker base color than chelae and abdomen. Tips of fingers of chelae lacking orange and/or black bands. Venter cream to whitish.

CHARACTERS – (Fig. 140) Rostrum deeply excavated; margins straight or slightly convex, terminating in small spines at the base of a very long acumen. Length of acumen equal to or slightly greater than twice the width of rostrum at marginal spines. Carapace neither laterally nor dorsoventrally flattened, almost subcircular in cross section. Strong cervical spines present. Areola obliterated. Antennal scale long and narrow, widest proximal to midpoint. Chelae long and thin, with short fingers. Dactyl of chela shorter than mesial margin of palm. Fingers and palm with rounded mesial and lateral margins, single setae covering dorsal surfaces. Chelae lacking longitudinal ridges on fingers and rows of tubercles on mesial margin of palm.

Gonopod of form I male terminating in two short, spatulate terminal elements: a corneous central projection and a pointed, noncorneous mesial process. Mesial process of equal length or slightly longer than central projection. Annulus ventralis of female subrhomboidal, rather flat; sinus extending from cephalic to caudal margin; lacking fossa.

MAXIMUM SIZE – The maximum known size for the species is 76.0 mm total length.

MOST LIKE – The extremely long acumen and form I gonopod with two short truncated terminal elements are characters found on no other Kentucky crayfish species.

DISTRIBUTION AND HABITAT – (Fig. 141) *Orconectes lancifer* occurs in the central Gulf Coastal Plain on both sides of the Mississippi River from extreme southern Illinois south to the coasts of eastern Texas and Louisiana (Hobbs 1989). In Kentucky the species has a limited distribution in the Ohio and Mississippi River floodplains in the extreme western portion of the state. It is known from a total of seven sites in Ballard, Carlisle, Fulton, and McCracken counties.

In Kentucky *Orconectes lancifer* is found in cypress swamps and floodplain and oxbow lakes with mud or sand substrates. In oxbow lakes the species frequents deep (> 0.5 m) open water. Pflieger (1996) reported the species from habitats that dried completely during summer and early fall months. We have noticed a similar habitat association, finding the species under dead and rotten tree branches up to 10 m away from standing water.

LIFE HISTORY – Given the small number of collections (< 8) of *Orconectes lancifer* from Kentucky, very little is known about its life history in the state. No form I males or ovigerous females have been collected in the state. Most of what is known about the species comes from data collected in other states, mostly Louisiana (Black 1972), Illinois (Page 1985), and Missouri (Pflieger 1996). The geographic proximity of southern Illinois and southeastern Missouri populations to Kentucky populations precludes significant variation in life history parameters among those populations. In Illinois form I's were collected in September and October while in Missouri they were collected only in September. No ovigerous females have been collected from either state. In Louisiana form I males have been collected from August to November, females with sperm plugs (indicating recent copulation) were collected in November, and ovigerous females were collected only in February. These data suggest a fall breed-

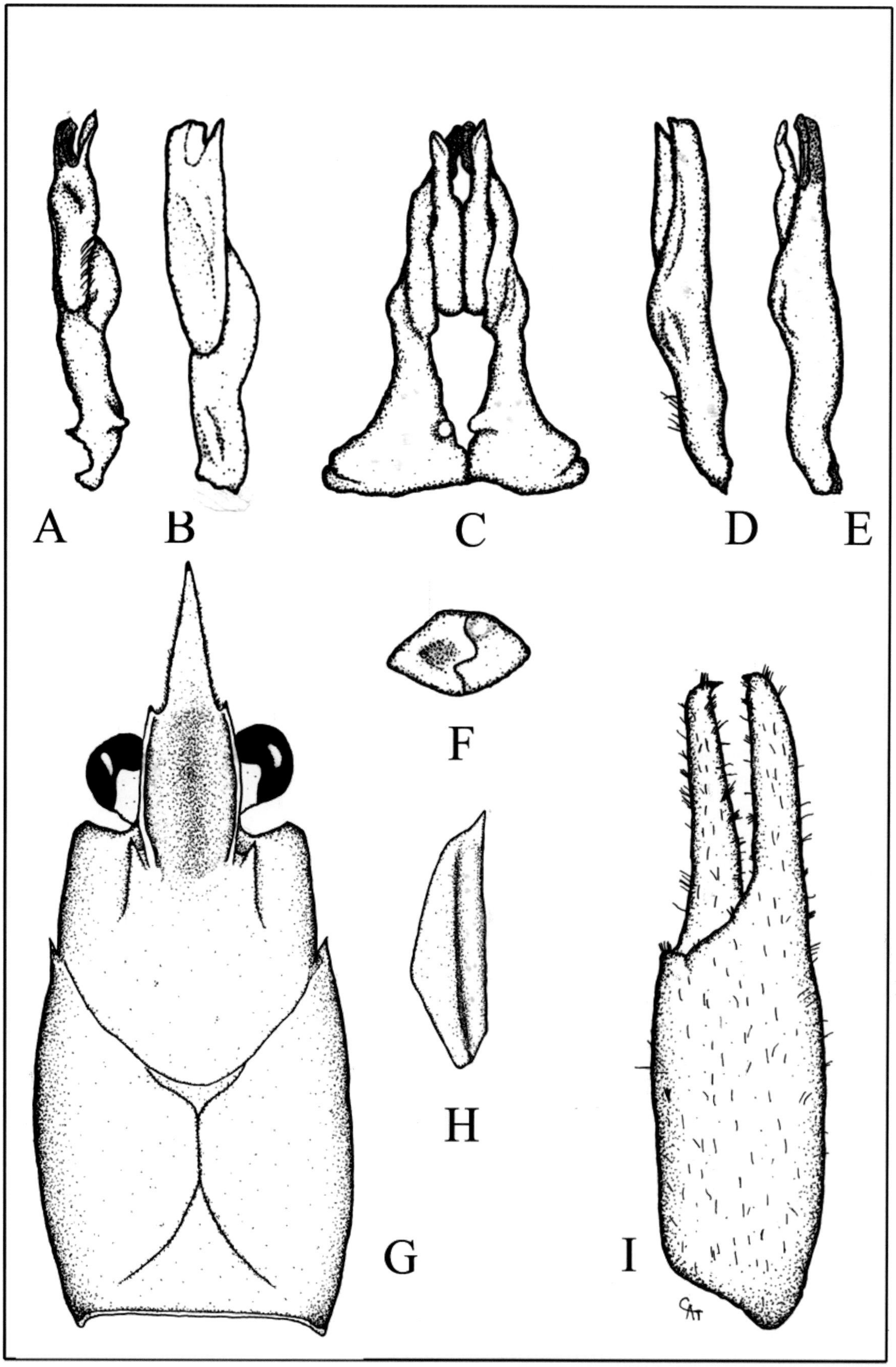

Fig. 140. *Orconectes lancifer*: A, mesial view of form I gonopod; B, mesial view of form II gonopod; C, ventral view of form I gonopods; D, lateral view of form II gonopod; E, lateral view of form I gonopod; F, ventral view of annulus ventralis; G, dorsal view of carapace; H, dorsal view of right antennal scale; I, dorsal view of right chela.

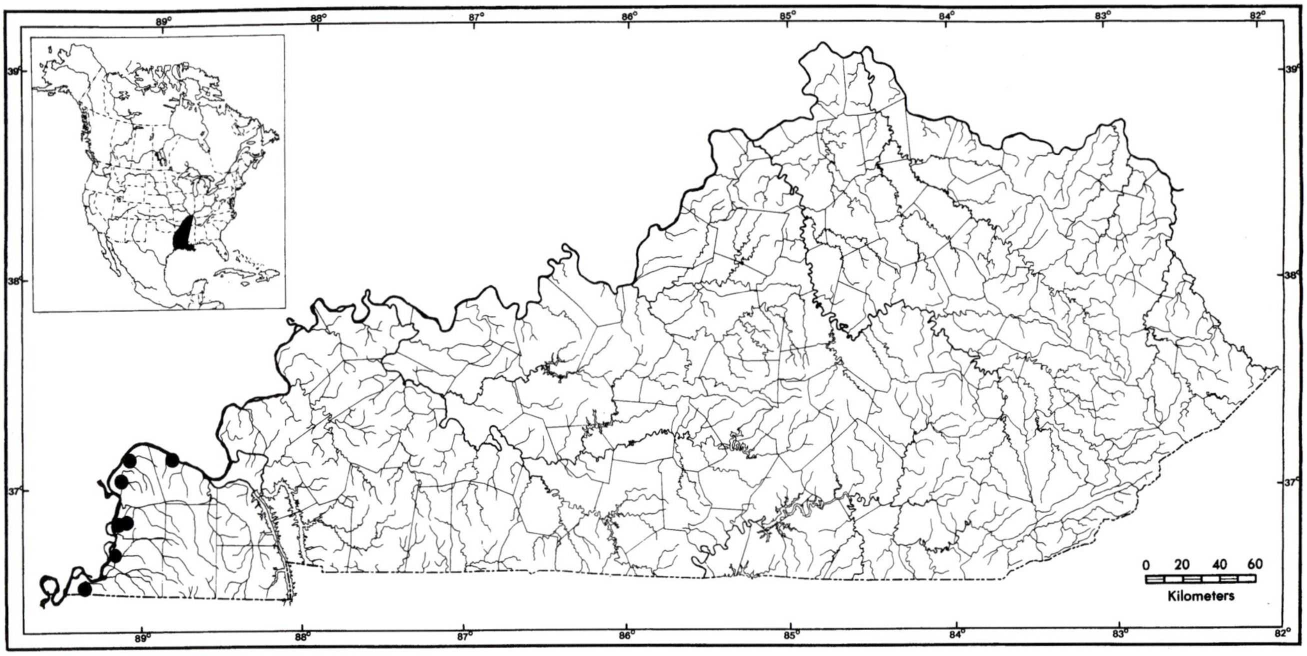

Fig. 141. Distribution of *Orconectes lancifer* in Kentucky.

ing season and winter egg-laying in southern populations of the species. Page (1985) counted 570 eggs averaging 1.5 mm in diameter from a female with a total length of 78.5 mm collected from northern Louisiana in February.

CRAYFISH ASSOCIATES – The following species have been collected from habitats containing *Orconectes lancifer*: *Cambarellus shufeldtii, Cambarus diogenes,* and *C. ludovicianus.*

CONSERVATION STATUS – KSNPC Endangered.

Orconectes margorectus Taylor
Livingston Crayfish

Fig. 142

LIFE COLORS – (Fig. 142) Base color of dorsal and lateral surfaces of chelae, carapace, and abdomen light brown to olive green. Dark dorsal patches just anterior of cervical groove and at caudal margin of carapace. Dorsum of abdominal segments darker than base color and usually flanked laterally by tan to light orange patches. Fingers of chelae with orange tips bordered proximally by wide subdistal black bands. Dorsum of chelae with numerous dark flecks; lateral margins of chelae with narrow dark stripe. Large knob at base of dactyl, spines on carapace, and tubercle at chela/carpus joint orange. Venter cream to whitish.

CHARACTERS – (Fig. 143) Rostrum moderately excavated; distally margins straight and converging anteriorly, proximally margins straight or slightly convex, margins terminating in spines at base of long acumen. Acumen longer than width of rostrum at marginal spines. Median carina absent. Carapace slightly flattened dorsoventrally, more so in larger individuals. Areola of moderate width, 4.9 to 10.5 times longer than wide at its narrowest point, with room for three to

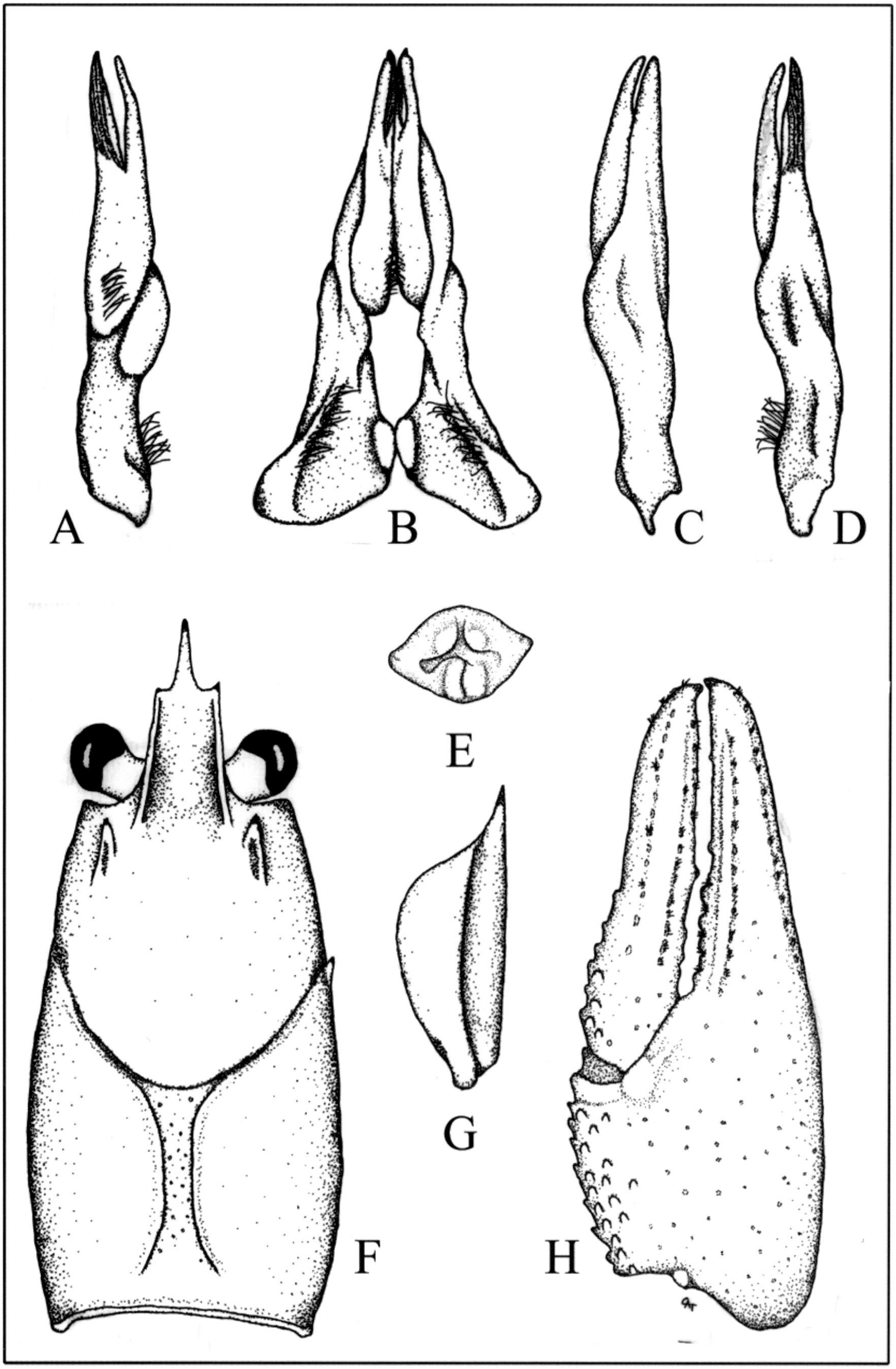

Fig. 143. *Orconectes margorectus*: A, mesial view of form I gonopod; B, ventral view of form I gonopods; C, lateral view of form II gonopod; D, lateral view of form I gonopod; E, ventral view of annulus ventralis; F, dorsal view of carapace; G, dorsal view of right antennal scale; H, dorsal view of right chela.

five punctations. Cervical spines present. Antennal scale widest at midpoint. Incisor region of mandible with straight edge. Chelae large with moderately long fingers, dorsal surfaces of palms usually covered with shallow punctations and short setae. Dorsal surfaces of fingers of chelae with weakly developed longitudinal ridges. Mesial margin of palm of chela and dactyl with rows of weak tubercles, palm with two or three rows of tubercles, third (lateral-most) row may take the form of several scattered tubercles; eight or nine tubercles in mesialmost row. Males with longer chelae than females of same carapace length.

Gonopod of form I male terminating in two short terminal elements: a corneous central with a sharply pointed tip and a pointed, noncorneous mesial process of equal or slightly shorter length. Mesial process recurved cephalodistally. Annulus ventralis of female subrhomboidal; with wide median trough running through cephalic half, two mesiocaudally directed proturberances overhanging shallow central fossa; sinus running from center of fossa to the caudal margin of annulus.

MAXIMUM SIZE – The maximum known size for the species is 83.0 mm total length.

MOST LIKE – *Orconectes margorectus* is most similar to other members of the genus *Orconectes* that possess short, more or less straight terminal elements on the form I gonopod. Those species are *O. bisectus*, *O. jeffersoni*, *O. rafinesquei*, *O. sanbornii*, and *O. tricuspis*. *Orconectes margorectus* differs from all of those species by possessing a mandible with a straight-edged incisor region. While an occasional specimen of *O. jeffersoni* may be encountered with a straight incisor region, *O. margorectus* has a form I gonopod with a central projection that is less than 25% of the total length of the gonopod. In *O. jeffersoni* the central projection is longer, measuring greater than 25% of the total length of the gonopod.

DISTRIBUTION AND HABITAT – (Fig. 144) *Orconectes margorectus* is endemic to the state of Kentucky. Within the state, it has an extremely limited distribution in the lower Ohio and Cumberland River drainages. The species occurs only in Ohio River tributaries in western Crittenden County and in extreme lower Cumberland River tributaries in Crittenden and Livingston counties.

Orconectes margorectus occurs in small to medium-sized creeks ranging in width from 2 to 10 m with bottom substrates of gravel, cobble, and

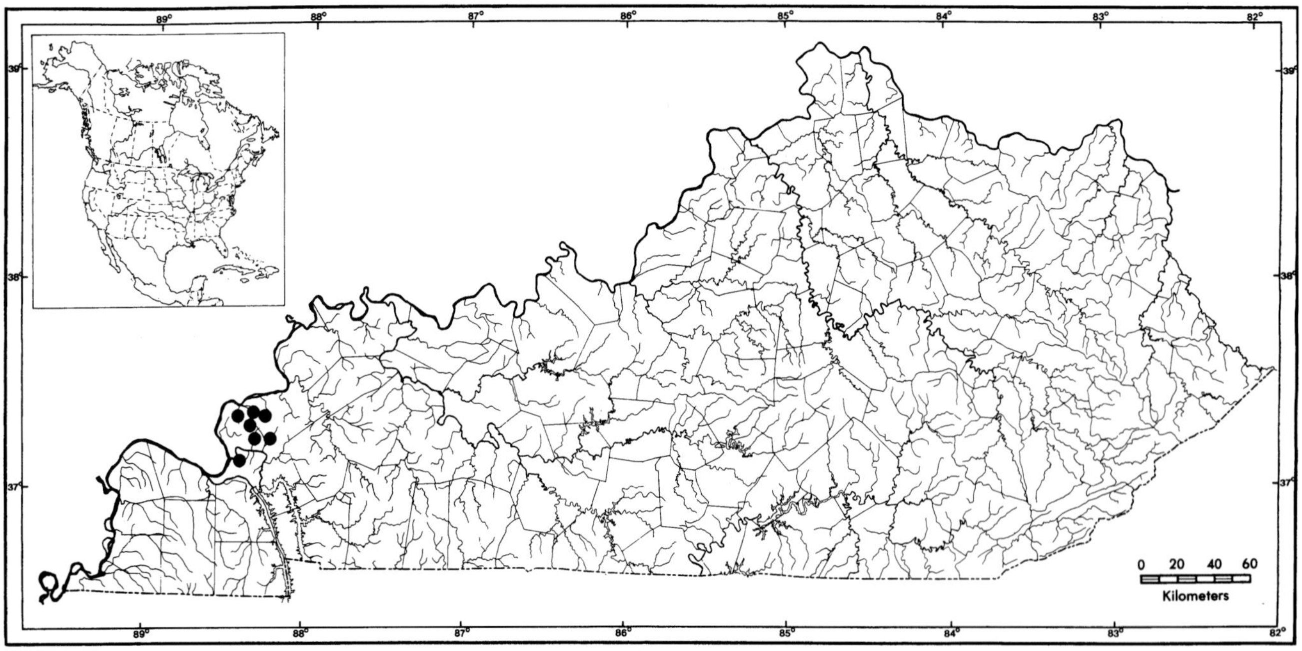

Fig. 144. Distribution of *Orconectes margorectus* in Kentucky.

some mud. The species is usually encountered under cobble in areas with flow or among woody debris or root masses along creek edges.

LIFE HISTORY – As a narrowly endemic species that was only recently described, very little is known about the life history of *Orconectes margorectus*. One site within the range of *O. margorectus* (Ferguson Creek, Livingston Co.) was sampled monthly from November 1972 to June 1973; those collections recorded form I males in the months of January through April and November. Females with sperm plugs were observed in January and February. Form I males have not been collected in other months from other sites. Ovigerous females have been collected in March and April. These data suggest that *O. margorectus* has a reproductive cycle similar to most other species of *Orconectes* found in the Midwest (e.g., late fall or early winter breeding, midspring egg laying). Two ovigerous females collected in April measured 77.0 mm and 63.3 mm in total length and carried 400 and 156 eggs, respectively. Average egg diameter for both females was 1.9 mm.

CRAYFISH ASSOCIATES – The following species have been collected from habitats containing *Orconectes margorectus*: *Cambarus diogenes*, *C. tenebrosus*, and *O. immunis*.

CONSERVATION STATUS – Not currently listed by AFS or KSNPC due to its recent description. Given its narrow range, *O. margorectus* should receive conservation status, most likely as Threatened.

Orconectes palmeri palmeri (Faxon) Gray–Speckled Crayfish

Fig. 145

LIFE COLORS – (Fig. 145) Base color of dorsal and lateral surfaces of chelae, carapace, and abdomen light brown to light gray (in freshly molted individuals) and densely mottled with small dark brown flecks of varying size. Dark flecks on chelae larger than flecks found on carapace and abdomen. Dorsum of carapace with pair of dark brown blotches anterior to cervical groove and another pair extending across junction of carapace and abdomen. Tips of fingers of chelae yellowish, bordered proximally by light brown or black patches. Venter cream to whitish.

CHARACTERS – (Fig. 146) Rostrum shallowly excavated; distally margins straight and converging, proximally margins slightly convex, margins terminating in spines at base of long acumen. Length of acumen slightly greater than width of rostrum at marginal spines. Carapace neither laterally nor dorsoventrally flattened, subcircular in cross section. Strong cervical spines present. Areola obliterated. Antennal scale widest at midpoint. Chelae large, with relatively long fingers; dorsal surfaces of fingers and palm covered with setae, setae usually dense near lateral margin of propodus. Chelae with weak longitudinal ridges on fingers. Mesial margin of palm of chela and dactyl with row of pointed, anteriorly directed tubercles. Mesial margin of palm also with second row of strong tubercles lateral to mesial row; mesialmost row with eight or nine tubercles.

Gonopod of form I male terminating in two long, thin, caudodistally recurved elements: a sharply pointed, corneous central projection recurved about at 90° to shaft and a noncorneous mesial process of about equal length. Annulus ventralis of female subrhomboidal; with elevated central region and deep fossa located in cephalic half; sinus extending from lateral margin of fossa to caudal margin of annulus.

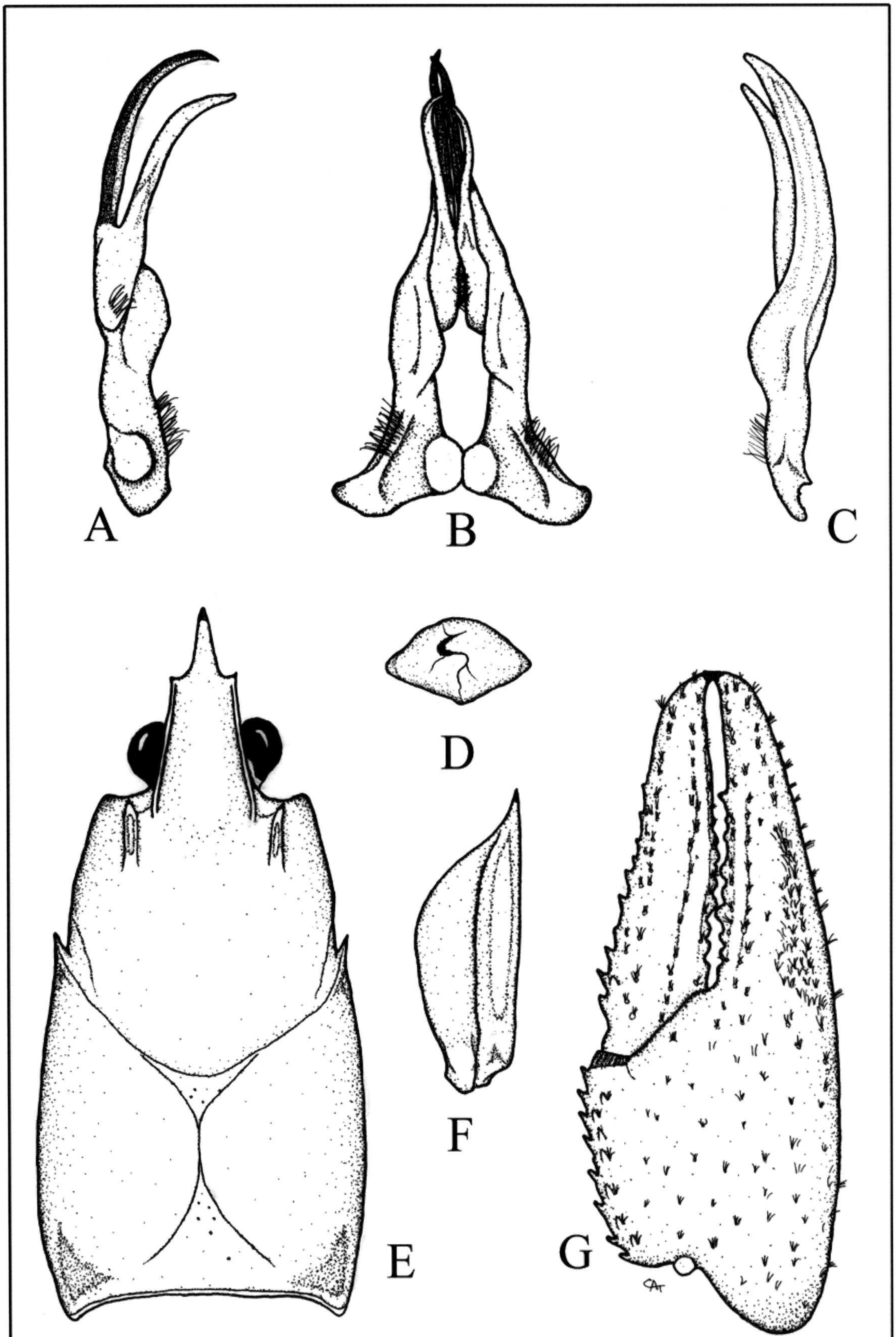

Fig. 146. *Orconectes palmeri palmeri*: A, mesial view of form I gonopod; B, ventral view of form I gonopods; C, lateral view of form II gonopod; D, ventral view of annulus ventralis; E, dorsal view of carapace; F, dorsal view of right antennal scale; G, dorsal view of right chela.

MAXIMUM SIZE – The maximum known size for the species is 87.0 mm total length.

MOST LIKE – With two long, strongly recurved elements, the form I gonopod of *Orconectes palmeri palmeri* is most similar to two other Kentucky species, *O. compressus* and *O. immunis*. *Orconectes palmeri palmeri* differs from both of those species in possessing an areola that is obliterated. *Orconectes palmeri palmeri* also differs from *O. compressus* in not being laterally compressed.

DISTRIBUTION AND HABITAT – (Fig. 147) *Orconectes palmeri palmeri* occurs in the central Gulf Coastal Plain on both sides of the Mississippi River from extreme southeastern Missouri and western Kentucky south to Louisiana (Hobbs 1989). In Louisiana, the subspecies only occurs east of the Mississippi River. In Kentucky, the species has a limited distribution in the Mississippi River drainage in the extreme western portion of the state. It is currently known from a total of four sites in the Mayfield Creek, Obion Creek, and Bayou du Chien drainages.

In Kentucky *O. palmeri palmeri* is found in flowing lowland creeks and ditches with mud or sand substrates. Within these habitats it is usually found within woody debris or among submerged tree roots or rooted vegetation.

LIFE HISTORY – Given the small number of collections (< 7) of *Orconectes palmeri palmeri* from Kentucky, very little is known about the species' life history in the state. Form I males have been collected in October and November, but no ovigerous females have been collected from within the state. Pflieger (1996) reported that form I males were commonly collected in southeastern Missouri from October through February and that four ovigerous females were collected in March. Those females ranged from 43 to 58.4 mm total length and carried 1 to 354 eggs. A detailed life history study of *O. palmeri palmeri* was conducted in western Tennessee by Payne and Price (1983). The information that follows is from that study. Form I males were abundant in collections made from October through March. Most males molted to form II in April and remained as form II until molting back to form I in September and October. Ovigerous females were collected in late February through April and were collected from shallow burrows under stones or accumulations of debris. Longevity in the subspecies was estimated to be three or four years; however, most individuals reproduced once and died before their second breeding season. The population showed a near 1:1 sex ratio regardless of time of year.

CRAYFISH ASSOCIATES – The following species have been collected from habitats containing *Orconectes palmeri palmeri*: *Cambarus diogenes, Fallicambarus fodiens, Procambarus acutus, P. clarkii,* and *P. viaeviridis*.

CONSERVATION STATUS – KSNPC Endangered.

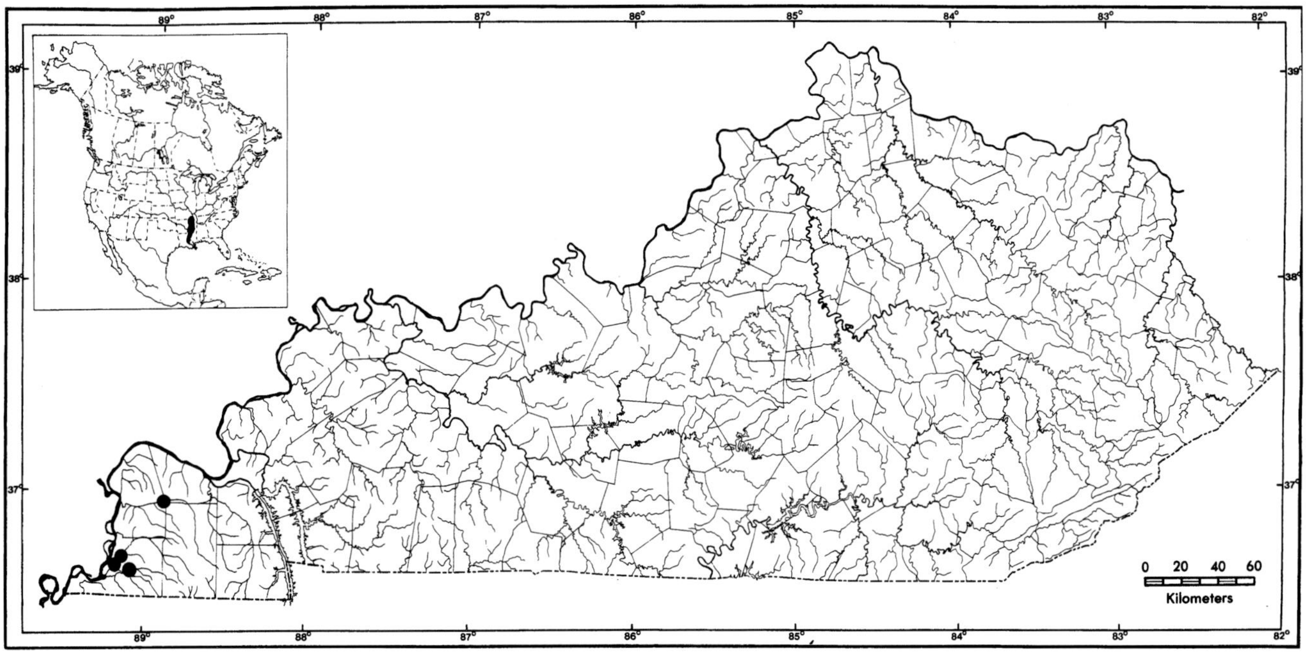

Fig. 147. Distribution of *Orconectes palmeri palmeri* in Kentucky.

Orconectes pellucidus (Tellkampf)
Mammoth Cave Crayfish

Fig. 148

LIFE COLORS – (Fig. 148) Albinistic; dorsal, ventral, and lateral surfaces of chelae, carapace, and abdomen white to cream to light yellow in color. Larger individuals may have several shades of cream and light yellow; yellows highlighting various grooves and joints of the individual.

CHARACTERS – (Fig. 149) Eyes greatly reduced and lacking pigment. Rostrum moderately excavated, distally margins straight and slightly converging, proximally margins straight or slightly convex, margins terminating in spines or tubercles at base of long acumen. Acumen longer than width of rostrum at marginal spines. Carapace dorsoventrally flattened. At least one pair of cervical spines present; if multiple cervical spines or tubercles present, size gradually decreases dorsally. Hepatic region highly variable, with 0–12 spines of various sizes on each side. Areola wide, 3.7 to 6.0 times longer than wide at its narrowest point, with room for three to seven punctations. Antennal scale widest at midpoint. Chelae long and thin, mesial margin of palm with 10 to 14 small tubercles, dorsal surfaces of palms with several irregular rows of tubercles. Fingers of chelae smooth or with weak longitudinal ridges. Opposable margin of propodus with enlarged third or fourth and eighth or ninth tubercles, opposable margin of dactyl usually with enlarged fifth tubercle.

Gonopod of form I male terminating in two short elements: a short, straight, and truncated corneous central projection and a much longer, noncorneous mesial process. Mesial process at least twice as long as central projection, broad basally, and constricting rapidly to form pointed tip; tip arched caudodistally. Annulus ventralis of female subcircular; lacking fossa; with centrally located and smoothly rounded longitudinal ridge; short sinus running to caudal edge of annulus.

MAXIMUM SIZE – The maximum known size for the species is 86.0 mm total length.

MOST LIKE – Three species of Kentucky crayfishes are albinistic and inhabit caves: *Orconectes australis packardi, O. inermis inermis,* and *O. pellucidus*. Males of *O. pellucidus* differ from males of *O. australis packardi* and *O. inermis inermis* in possessing a mesial process that is significantly longer than the central projection and by lacking an angular shoulder at the base of the central projection.

DISTRIBUTION AND HABITAT – (Fig. 150) *Orconectes pellucidus* occurs in a narrow arc of karst extending in Kentucky from Trigg County east to Hart County, dipping southward into Montgomery County, Tennessee (Hobbs and Barr 1972). Within Kentucky the species occurs in subterranean passages of the lower Cumberland River drainage in Christian, Logan, and Trigg counties and in subterranean passages of the Green River drainage, most notably Mammoth Cave and in Barren, Edmonson, Hart, Logan, and Warren counties.

Orconectes pellucidus occurs exclusively in subterranean creeks. Like the other cave-dwelling crayfish species found in Kentucky, the species is most frequently encountered in shallow pools with mud and silt substrates where it can be usually be seen out in the open. As mentioned for *O. australis parkardi* and *O. inermis inmeris*, the lack of large predators in subterranean creeks may also account for *O. pellucidus'* habitat choices.

LIFE HISTORY – Very little is known of the life history of *Orconectes pellucidus*, with brief notes included in two publications. Hobbs and Barr (1972) recorded a female with eggs in September 1965 from Hart County, Kentucky, while Eigenmann (1899) collected a female with young in November 1898 from Mammoth Cave. Neither paper provided egg size or clutch size information. In Kentucky form I males have been col-

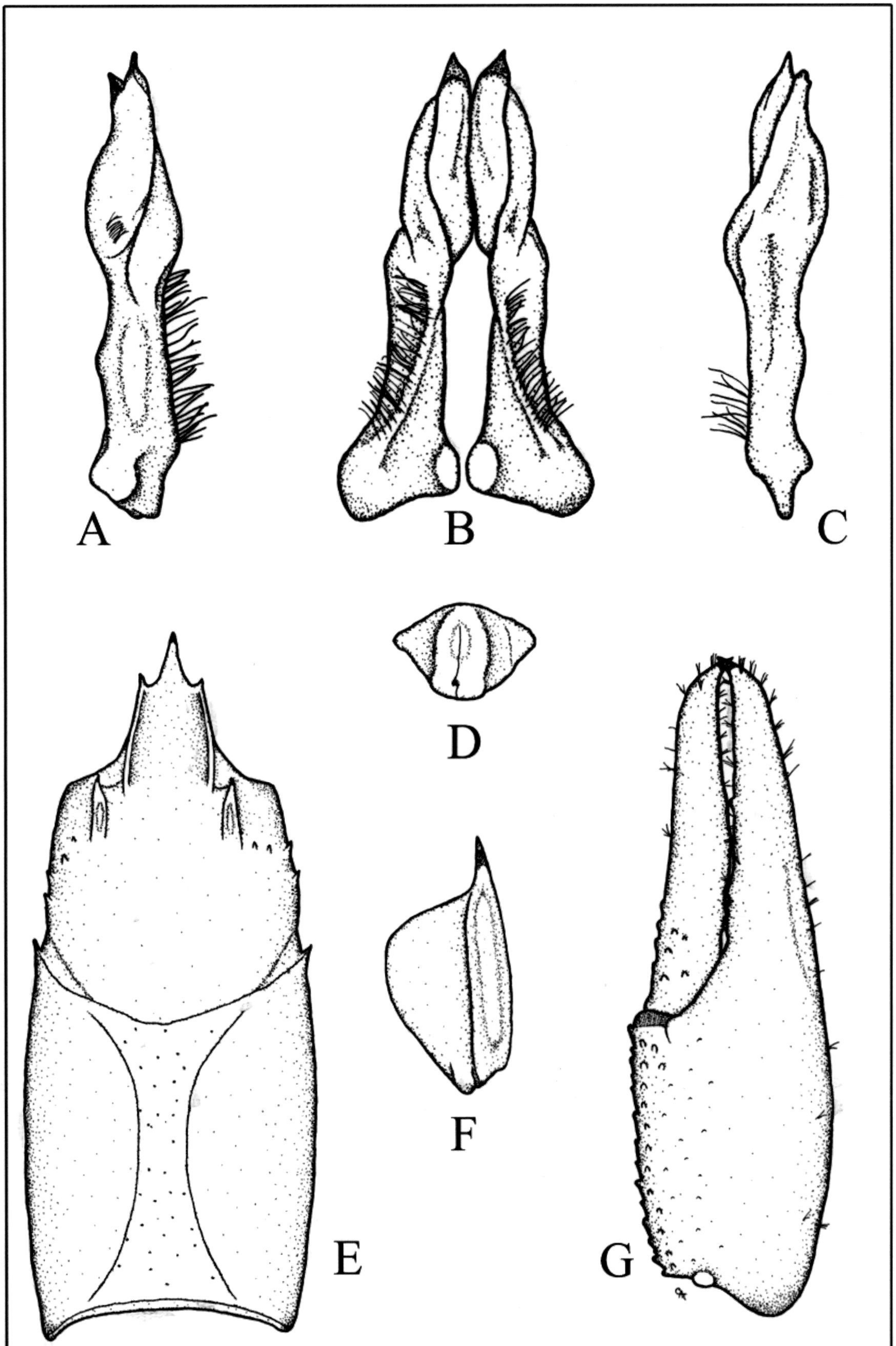

Fig. 149. *Orconectes pellucidus*: A, mesial view of form I gonopod; B, ventral view of form I gonopods; C, lateral view of form II gonopod; D, ventral view of annulus ventralis; E, dorsal view of carapace; F, dorsal view of right antennal scale; G, dorsal view of right chela.

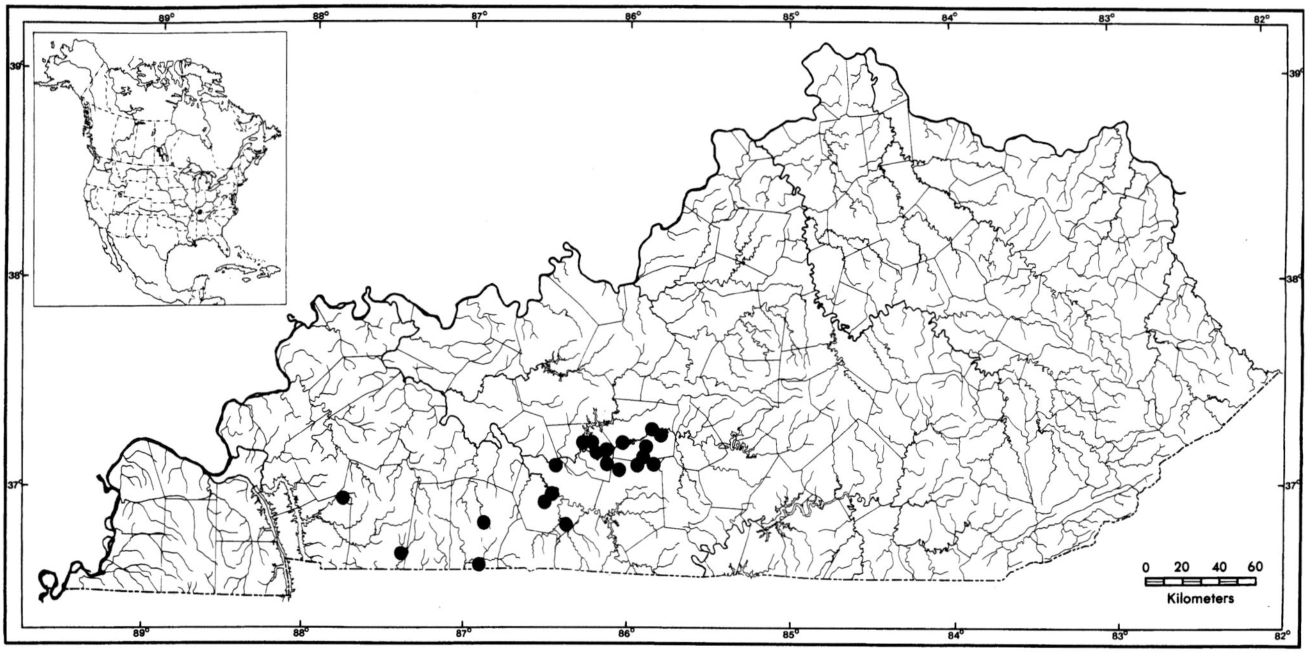

Fig. 150. Distribution of *Orconectes pellucidus* in Kentucky.

lected in all months but January, September, and December.

CRAYFISH ASSOCIATES – No other crayfish species have been collected from habitats containing *Orconectes pellucidus*.

CONSERVATION STATUS – KSNPC Special Concern.

Orconectes placidus (Hagen)
Bigclaw Crayfish

Figs. 151A: a form II male.

Figs. 151B: a large, mature form I male.

LIFE COLORS – (Figs. 151A, B) Base color of dorsal and lateral surfaces of chelae, carapace, and abdomen variable, tan to olive green. Carapace with two dark brown or black saddles, one immediately posterior to cervical groove and one at caudal margin. Saddle at caudal margin extending anterior to midlength of lateral surfaces of carapace. Dorsum of abdominal segments with dark anterior and posterior margins, lateral margins dark with light yellow patch on each side. Each abdominal segment with pair of thin, dark, laterally directed bars on dorsal surface. Fingers of chelae with orange tips bordered proximately by wide black bands. Dorsal surfaces of chelae with large, dark spot at base of dactyl, lateral edges of chelae with thin dark stripe. Venter cream to whitish.

CHARACTERS – (Fig. 152) Rostrum moderately excavated; distally margins straight or slightly concave and converging, proximally margins convex, margins terminating in spines at base of long acumen. Acumen slightly longer than width of rostrum at marginal spines. Weak median carina occasionally present. Carapace slightly compressed dorsoventrally. Areola of moderate width, 5.6 to 9.1 times longer than wide at it narrowest point, with room for three to five punctations. Cervical spines present. Antennal scale widest distal to midpoint. Chelae large with long fingers, dactyl greater than twice the length of mesial margin of palm. Gap between opposable margins of bases of fingers. Finger length and gap width greater in larger individuals. Dorsal surfaces of fingers of chela with weak longitudinal ridges and dorsal surfaces of palm usually covered with shallow punctations and short setae. Mesial margin of palm of chela with two or occasionally three rows of weak tubercles; 8 to 11 tubercles in mesialmost row. Males with slightly longer chelae than females of same carapace length.

Gonopod of form I male terminating in two long, thin elements: a corneous central projection with a caudodistally recurved tip and a shorter, straight noncorneous mesial process. Central projection weakly recurved caudodistally. Caudal surface of gonopod at base of central projection straight or slightly arched. Annulus ventralis of female subrhomboidal; with narrow median trough extending through cephalic half; laterally wide, centrally located fossa; sinus extending from center of fossa to caudal edge of annulus.

MAXIMUM SIZE – The maximum known size for the species is 101.5 mm total length.

MOST LIKE – Numerous species of crayfishes in Kentucky possess form I gonopods with two long, thin terminal elements. Of those species, three lack a strong angular shoulder at the caudal base of the central projection: *Orconectes barrenensis, O. durelli*, and *O. placidus*. *Orconectes placidus* differs from both *O. barrenensis* and *O. durelli* in possessing very long fingers on the chelae. The fingers of the chelae on *O. placidus* are more than twice the length of the mesial margin of the palm. *Orconectes placidus* also differs from *O. barrenensis* in possessing cervical spines.

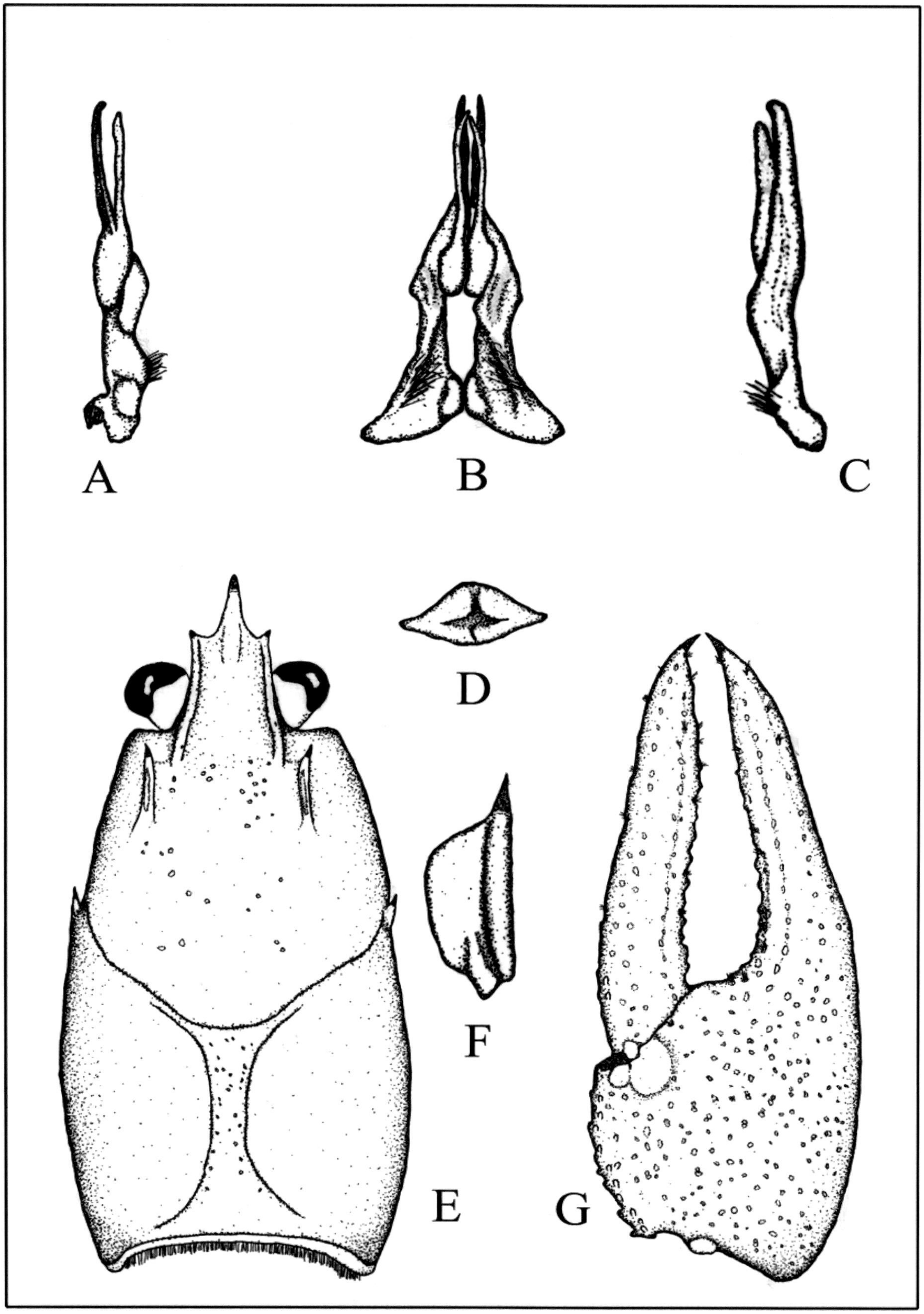

Fig. 152. *Orconectes placidus*: A, mesial view of form I gonopod; B, ventral view of form I gonopods; C, lateral view of form II gonopod; D, ventral view of annulus ventralis; E, dorsal view of carapace; F, dorsal view of right antennal scale; G, dorsal view of right chela.

DISTRIBUTION AND HABITAT – (Fig. 153). *Orconectes placidus* is a relatively wide-ranging species, occurring in upland creeks of the middle and lower Cumberland and Tennessee and middle Green River drainages of southern Kentucky, Tennessee, and northern Alabama (Hobbs 1989). The species also occurs in a limited stretch of the Ohio River near the mouths of the Tennessee and Cumberland rivers and has crossed the Ohio River to occur in the Big Creek drainage of southern Illinois (Page 1985). Within Kentucky the species is widespread and common in the Cumberland River drainage primarily below Cumberland Falls and in the Clarks River drainage in western Kentucky. It is also known from two widely disjunct sites in the Barren River drainage.

Orconectes placidus occurs commonly in almost every size creek or river. Substrates at sites where *O. placidus* occurs are predominately gravel, cobble, and boulders. Within the Clarks River drainage the species is found over substrates of fine gravel and sand. The species is usually encountered under cobble or large pieces of gravel in shallow riffles or runs but can also be found in woody debris or in undercut creek banks.

LIFE HISTORY – In Kentucky form I males of *Orconectes placidus* have been collected from March through October but were very uncommon in July and August collections. Ovigerous females have been collected only in April. The lack of ovigerous females in numerous other collections made throughout the year suggest that egg laying only occurs in the midspring. Two females collected in Kentucky in April measured 41.8 and 49.1 mm total length and carried 44 and 121 eggs, respectively. Egg diameter in both females averaged 1.9 mm. Page (1985) presented data for *O. placidus* from monthly collections made year-round in Big Creek, Illinois. In those collections, form I males were found from September through January and in March and ovigerous females were collected in March and April. Page (1985) reported 12, 88, and 134 eggs on females measuring 30.9, 40.4, and 53.6 mm total length, respectively. These females had an average egg diameter of 2.1 mm, which was slightly larger than those of Kentucky females. Using a length-frequency histogram, Page (1985) estimated longevity of the species to be three years with very few individuals living to their third year. The 30.9-mm ovigerous female collected in April with 12 eggs indicates that females can reproduce the first spring after they were born, albeit with very low fecundity. Using genetic analysis of eggs carried by *O. placidus* collected in Tennessee, Walker et al. (2002) determined that females can mate with and carry the young of multiple males.

CRAYFISH ASSOCIATES – The following species have been collected from habitats containing *Orconectes placidus*: *Cambarus friaufi*, *C. graysoni*, *C. rusticiformis*, *C. tenebrosus*, *O. compressus*, *O. durelli*, *O. kentuckiensis*, *O. putnami*, *O. tricuspis*, and *Procambarus acutus*.

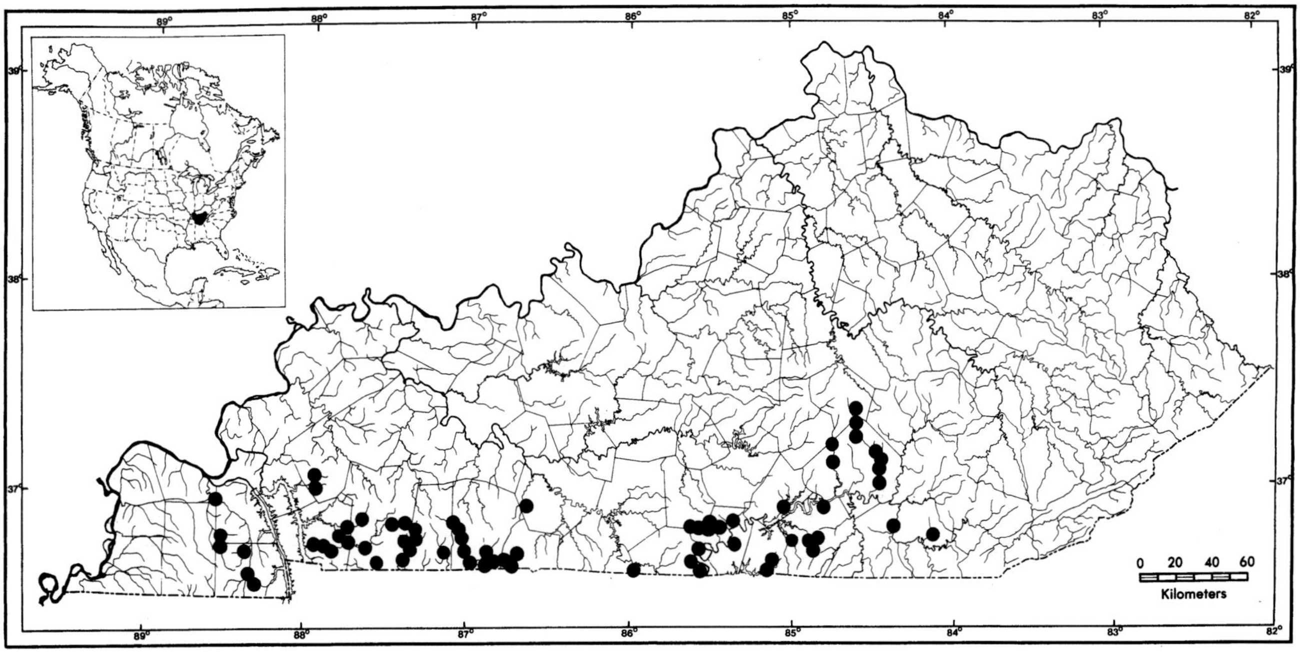

Fig. 153. Distribution of *Orconectes placidus* in Kentucky.

Orconectes putnami (Faxon)
Phallic Crayfish

Fig. 154

LIFE COLORS – (Fig. 154) Base color of dorsal and lateral surfaces of chelae, carapace, and abdomen variable, tan to light brown to olive green. Carapace usually with two dark brown saddles, one immediately anterior to cervical groove, other at caudal margin. Intensity of saddles highly variable among populations; saddle anterior of cervical groove occasionally absent. Abdomen with or without long, dark brown wedge extending down center of dorsum; posterior edges of abdominal segments always dark brown. Fingers of chelae with orange tips bordered proximately by black bands of variable width. Dorsal knob at base of dactyl and joints of pereiopods orange. Venter cream to whitish.

CHARACTERS – (Fig. 155) Rostrum moderately excavated; distally margins straight and slightly converging, proximally margins straight or slightly convex, margins terminating in spines at base of acumen. Acumen slightly longer than width of rostrum at marginal spines. Median carina absent. Carapace slightly compressed dorsoventrally. Areola of moderate width, 4.9 to 8.9 times longer than wide at its narrowest point, with room for two to six punctations. Cervical spines present. Antennal scale widest at midpoint. Chelae large with moderately long fingers, dorsal surfaces of palms usually covered with shallow punctations. Dorsal surfaces of fingers of chelae with longitudinal ridges. Mesial margin of palm of chela and dactyl lined with smoothly rounded tubercles, palm with two or occasionally three rows of tubercles; 7 to 10 tubercles in mesialmost row. Males with slightly longer chelae than females of same carapace length.

Gonopod of form I male terminating in two very long, thin elements that are cephalodistally angled at their bases: a sharply pointed corneous central projection, slightly recurved caudodistally and a shorter, noncorneous mesial process, slightly recurved cephalodistally at its distal end. Caudal surface of gonopod with strong angular shoulder at base of central projection. Annulus ventralis of female subrhomboidal; with wide median trough running through cephalic half; two caudally directed protuberances overhanging deep, centrally located, laterally wide fossa; sinus extending from center of fossa to caudal edge of annulus.

MAXIMUM SIZE – The maximum known size for the species is 90.7 mm total length.

MOST LIKE – Numerous species of crayfishes in Kentucky possess form I gonopods with two long, thin terminal elements. Of those species *Orconectes putnami* is most similar to *O. cristavarius, O. juvenilis, O. ronaldi,* and *O. rusticus* in that they all possess a strong angular shoulder at the base of the central projection. *Orconectes putnami* differs from those four species, and all other Kentucky species, by having extremely long terminal elements. The length of the central projection of *O. putnami* is 50% or greater than the total length of the gonopod; all other species have central projections less than 50% of total gonopod length.

DISTRIBUTION AND HABITAT – (Fig. 156). The range of *Orconectes putnami* was restricted by Taylor (2000) to include portions of the middle Ohio, Cumberland, and Tennessee and middle and upper Green River drainages of Kentucky and Tennessee. In Kentucky the species is widespread and common in the Green River drainage upstream of the Rough River. It also occurs sporadically in the upper Cumberland River drainage primarily below the Cumberland Falls and in several Ohio River tributaries in Breckinridge County.

Orconectes putnami occurs commonly in almost every size creek or river with substrates of gravel, cobble, and boulders. The species is usually encountered under cobble or large pieces of gravel in riffles or runs but can also be found in woody debris.

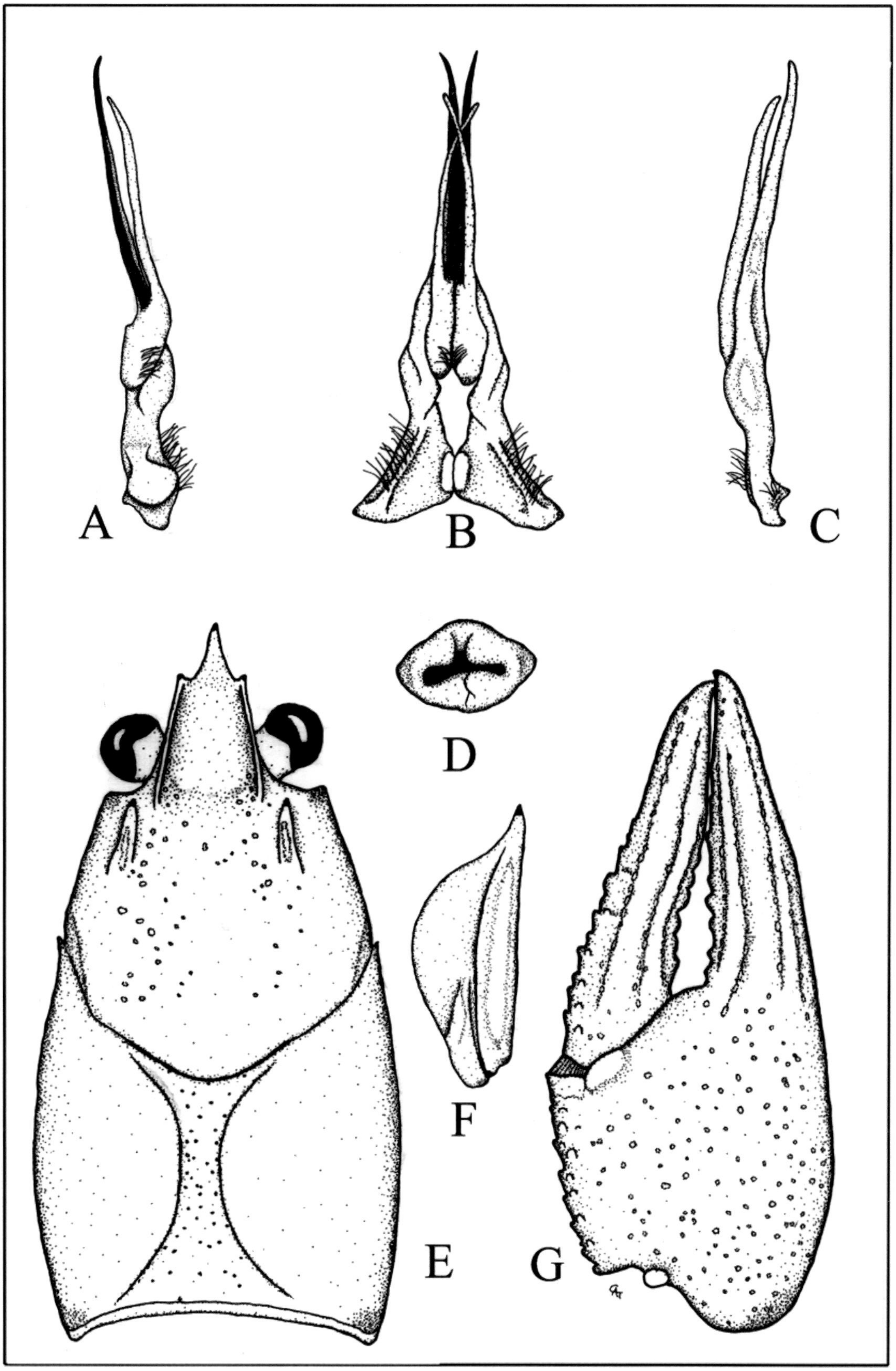

Fig. 155. *Orconectes putnami*: A, mesial view of form I gonopod; B, ventral view of form I gonopods; C, lateral view of form II gonopod; D, ventral view of annulus ventralis; E, dorsal view of carapace; F, dorsal view of right antennal scale; G, dorsal view of right chela.

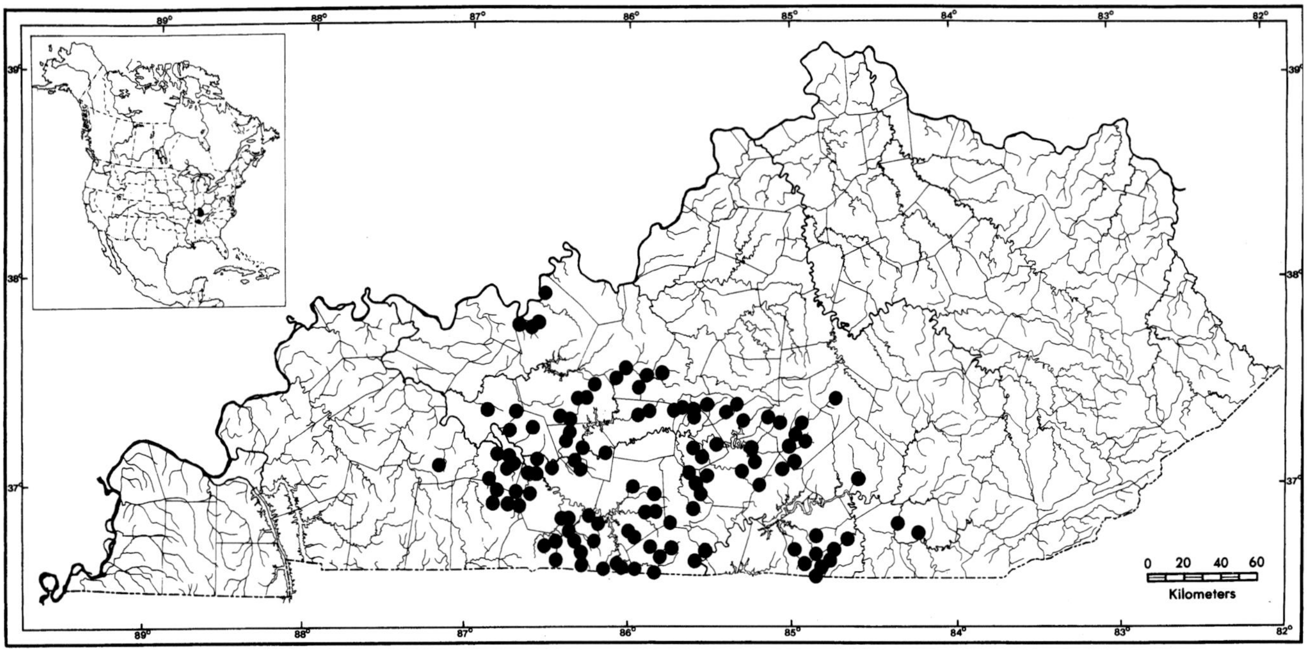

Fig. 156. Distribution of *Orconectes putnami* in Kentucky.

LIFE HISTORY – Numerous collections of *O. putnami* have been made in Kentucky from all months of the year. Form I males have been collected in every month except for December. Ovigerous females have been collected in March, April, and May. The absence of ovigerous females in numerous collections from other months suggests that egg–laying occurs only in the spring. Two ovigerous females collected in late March measured 55.6 and 52.4 mm total length and carried 175 and 134 eggs, respectively. Average egg diameter was 1.8 mm for eggs found on the larger female and 1.9 mm for those found on the smaller female. Nothing else is known about the life history of this species.

CRAYFISH ASSOCIATES – The following species have been collected from habitats containing *Orconectes putnami*: *Barbicambarus cornutus, Cambarus cumberlandensis, C. diogenes, C. graysoni, C. rusticiformis, C. tenebrosus, O. barrenensis, O. compressus, O. durelli, O. immunis, O. placidus, O. rafinesquei,* and *O. rusticus.*

Orconectes rafinesquei Rhoades
Rough River Crayfish

Fig. 157

LIFE COLORS – (Fig. 157) Base color of dorsal and lateral surfaces of chelae, carapace, and abdomen tan to light brown. Dorsum of carapace anterior to cervical groove with laterally wide dark brown patch. Carapace with wide U–shaped dark saddle centered at dorsal surface of caudal margin and extending forward to midpoint of lateral surfaces. Dorsum of abdominal segments with either pair of small, longitudinal dark brown patches or single dark brown patch with lighter brown patches at center. If latter color pattern present on each abdominal segment, individual appears to have a dark brown wedge on the dorsum of abdomen. Fingers of chelae with orange tips bordered proximally by black bands. Large knob at base of dactyl orange. Venter cream to whitish.

CHARACTERS – (Fig. 158) Rostrum moderately excavated; distally margins straight and converging; proximally margins convex, margins terminating in small spines at base of long acumen. Acumen longer than width of rostrum at marginal spines. Median carina absent. Carapace dorsoventrally flattened, more so in larger individuals. Areola of moderate width, 5.4 to 9.5 times longer than wide at its narrowest point, with room for three to five punctations. Cervical spines present. Antennal scale widest at midpoint. Chelae large with moderately long fingers, dorsal surfaces of palms usually covered with shallow punctations and short setae. Dorsal surfaces of fingers of chelae with well-developed longitudinal ridges, more strongly developed on larger individuals. Mesial margin of palm of chela and dactyl with tubercles, palm with two rows and several additional scattered tubercles lateral to second row; 8 to 10 tubercles in mesialmost row. Males with longer chelae than females of same carapace length.

Gonopod of form I male terminating in two short terminal elements: a corneous central projection with a sharply pointed, caudodistally recurved tip and a slightly shorter or equal length noncorneous mesial process. Mesial processes usually slightly recurved cephalodistally. Annulus ventralis of female subrhomboidal; wide median trough extending to cephalic margin; centrally located fossa; sinus extending from the center of fossa to caudal margin of annulus.

MAXIMUM SIZE – The maximum known size for the species is 73.0 mm total length.

MOST LIKE – *Orconectes rafinesquei* is most similar to other members of the genus *Orconectes* that possess short, more or less straight terminal elements on the form I gonopod. Those species are *O. bisectus, O. jeffersoni, O. margorectus, O. sanbornii*, and *O. tricuspis*. Of those five species, *O. rafinesquei* is most similar to *O. bisectus* in that both possess a mesial process that is not dorsoventrally flattened and a central projection that is less than 25% of the total length of the gonopod. *Orconectes rafinesquei* differs from *O. bisectus* by possessing a central projection with a pointed tip and mesial processes that do not diverge laterally from one another in a ventral view.

DISTRIBUTION AND HABITAT – (Fig. 159) *Orconectes rafinesquei* is endemic to the state of Kentucky. Within the state the species has a limited distribution in the lower Ohio and Green River drainages. It occurs throughout the Rough River drainage and sporadically in Green River tributaries downstream of the Mud River in Muhlenberg County. *Orconectes rafinesquei* also occurs in some Ohio River tributaries from Breckinridge to Henderson counties.

Orconectes rafinesquei is most common in small creeks and rivers ranging in width from 2 to 20 m with bottom substrates of gravel, cobble, and mud. The species is usually encountered under cobble in areas with moderate flow or among woody debris in creeks with mud substrates.

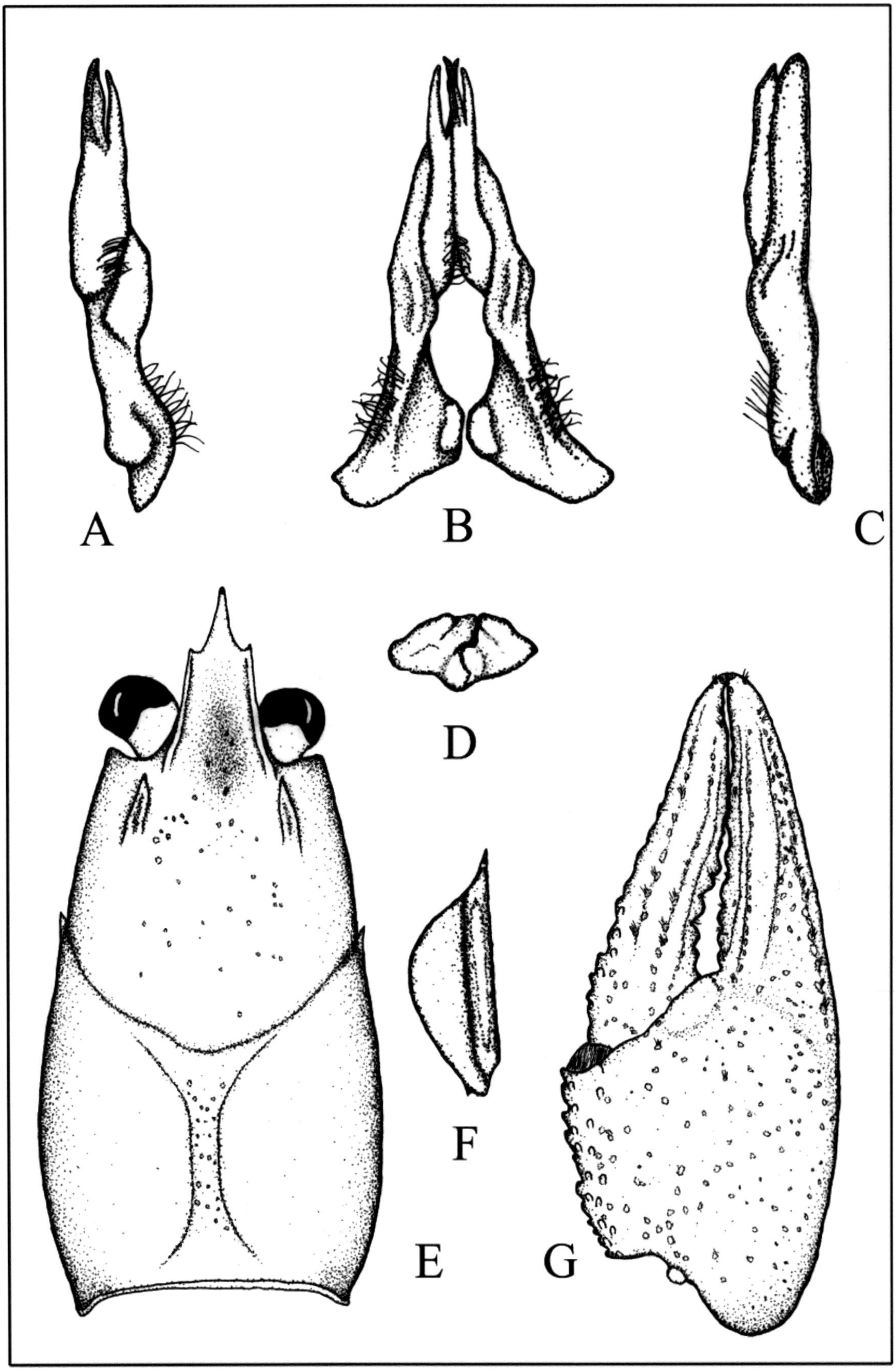

Fig. 158. *Orconectes rafinesquei*: A, mesial view of form I gonopod; B, ventral view of form I gonopods; C, lateral view of form II gonopod; D, ventral view of annulus ventralis; E, dorsal view of carapace; F, dorsal view of right antennal scale; G, dorsal view of right chela.

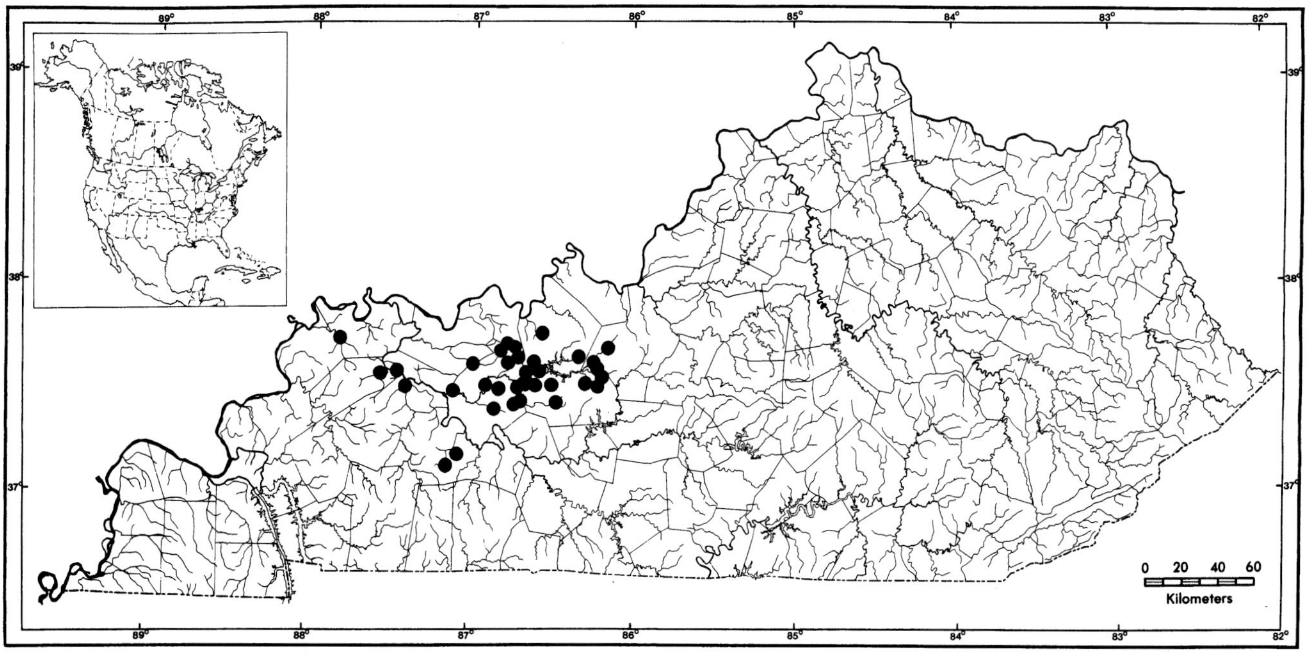

Fig. 159. Distribution of *Orconectes rafinesquei* in Kentucky.

LIFE HISTORY – Little is known about the life history of *Orconectes rafinesquei*. Form I males have been collected in every month but February, June, and December. The lack of form I males in February and December is most likely due to an absence of field collections made in those months. Only two ovigerous females have been collected, one in March and one in April. These data seem to indicate that the timing of reproductive activities in *O. rafinesquei* is similar to most species of *Orconectes* found in Kentucky. The single ovigerous female collected in late March measured 68.2 mm total length and carried 312 eggs. Average egg diameter was 1.9 mm. Rhoades (1944) reported that the species will build shallow burrows if "cover is absent and the creek bed is muddy."

CRAYFISH ASSOCIATES – The following species have been collected from habitats containing *Orconectes rafinesquei*: *Cambarus diogenes*, *C. graysoni*, *C. rusticiformis*, *C. tenebrosus*, *C.* sp. A, *Fallicambarus fodiens*, *O. barrenensis*, *O. immunis*, *O. putnami*, and *O. rusticus*.

CONSERVATION STATUS – AFS Special Concern.

Orconectes ronaldi Taylor
Mud River Crayfish

Fig. 160

LIFE COLORS – (Fig. 160) Base color of dorsal and lateral surfaces of chelae and carapace light brown to olive green. Dorsum of carapace with laterally elongate dark brown patch immediately anterior to cervical groove and dark brown U–shaped saddle connected at caudal margin, extending forward to just posterior of midpoint of lateral surfaces. Abdomen with a dark brown wedge extending down center of dorsum. Fingers of chelae with orange tips bordered proximally by wide black bands. Large knob at base of dactyl orange. Venter cream to whitish.

CHARACTERS – (Fig. 161) Rostrum moderately excavated; distally margins straight and slightly converging, proximally margins straight or slightly convex, margins terminating in spines at base of acumen. Acumen equal to or slightly longer than width of rostrum at marginal spines. Median carina absent. Carapace slightly compressed dorsoventrally. Areola of moderate width, 4.6 to 9.5 times longer than wide at its narrowest point, with room for three to five punctations. Cervical spines present. Antennal scale widest at midpoint. Chelae large and stocky, surfaces of palms usually covered with shallow punctations; fingers slightly longer than length of mesial margin of palm, dorsal surfaces of fingers with strong longitudinal ridges. Mesial margin of palm of chela and dactyl covered with tubercles, palm with two discreet rows and several additional scattered tubercles lateral to those rows; seven to nine tubercles in mesialmost row. Slight gap between basal opposable margins of fingers. Males with larger chelae than females of same carapace length.

Gonopod of form I male terminating in two long, thin elements: a sharply pointed, straight corneous central projection and a shorter, straight noncorneous mesial process. Both elements slightly angled cephalodistally at their bases. Caudal surface of gonopod with strong angular shoulder at base of central projection. Annulus ventralis of female subcircular; with median trough extending through cephalic half; two caudally directed protuberances overhanging centrally located fossa; fossa laterally wide; sinus running from center of fossa to caudal edge of annulus.

MAXIMUM SIZE – The maximum known size for the species is 74.0 mm total length.

MOST LIKE – Numerous species of crayfishes in Kentucky possess form I gonopods with two long, thin terminal elements. Of those species, *Orconectes ronaldi* is most similar to *O. cristavarius, O. juvenilis, O. rusticus,* and *O. putnami* in that all five possess a strong angular shoulder at the base of the central projection. *Orconectes ronaldi* differs from those species in possessing terminal elements of the gonopod that are straight and a mandible with a dentate incisor region.

DISTRIBUTION AND HABITAT – (Fig. 162). Initially thought to be endemic to the Green River drainage of Kentucky, new records indicate that disjunct populations of *Orconectes ronaldi* may occur in several Ohio River tributaries in south-central Indiana. In Kentucky *O. ronaldi* is widespread and common in the Mud River and Muddy Creek drainages in Butler, Logan, and Muhlenberg counties.

Orconectes ronaldi occurs in creeks and small rivers with substrates of gravel, cobble, and mud. The species is usually encountered under cobble in shallow riffles or among woody debris in more slowly flowing runs.

LIFE HISTORY – Very few collections of *Orconectes ronaldi* have been made given its recent discovery and limited range (Taylor 2000). As such, little is known about its life history. Form I males have been collected in March, April, September, and October and two ovigerous females have been collected in late March. Those females

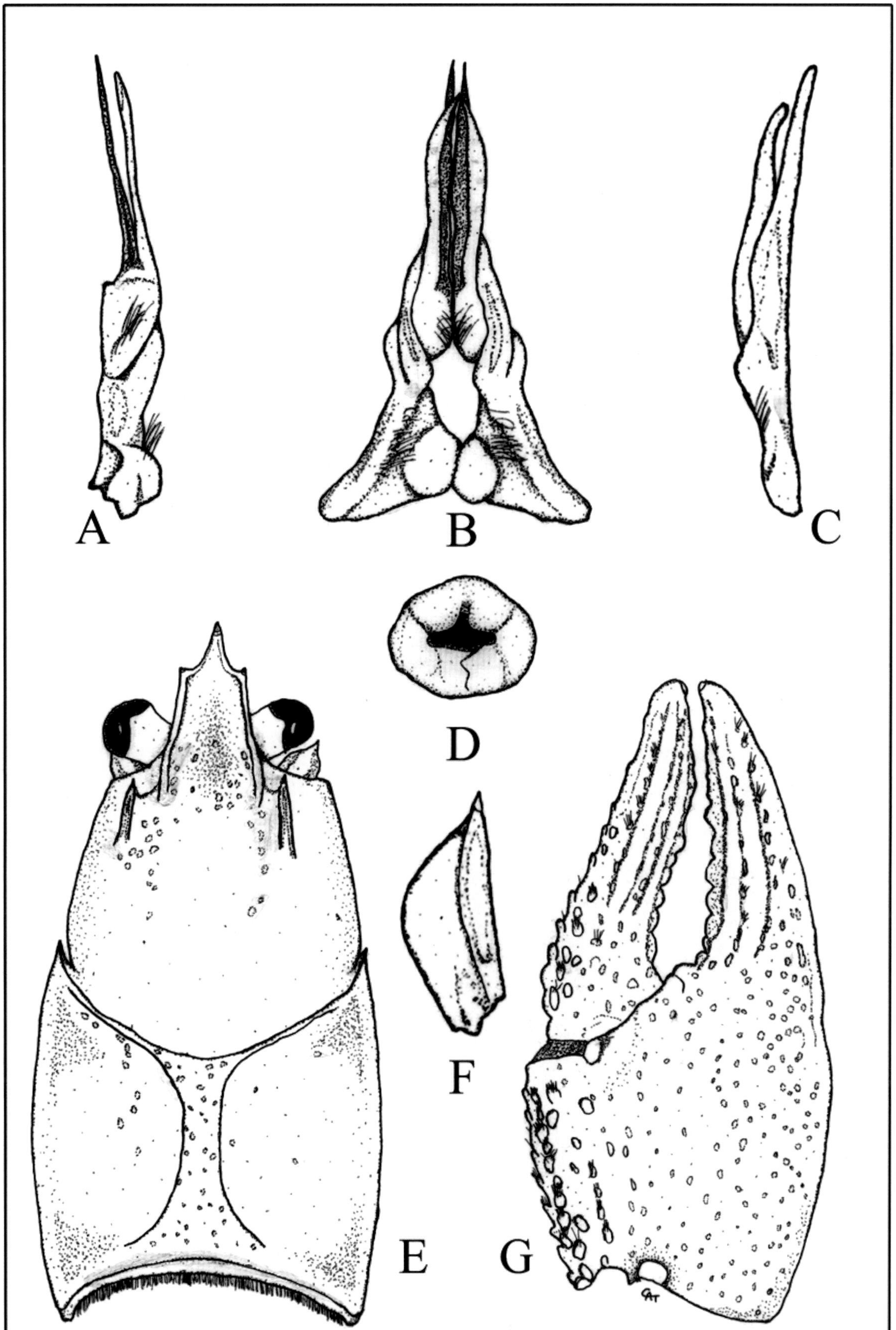

Fig. 161. *Orconectes ronaldi*: A, mesial view of form I gonopod; B, ventral view of form I gonopods; C, lateral view of form II gonopod; D, ventral view of annulus ventralis; E, dorsal view of carapace; F, dorsal view of right antennal scale; G, dorsal view of right chela.

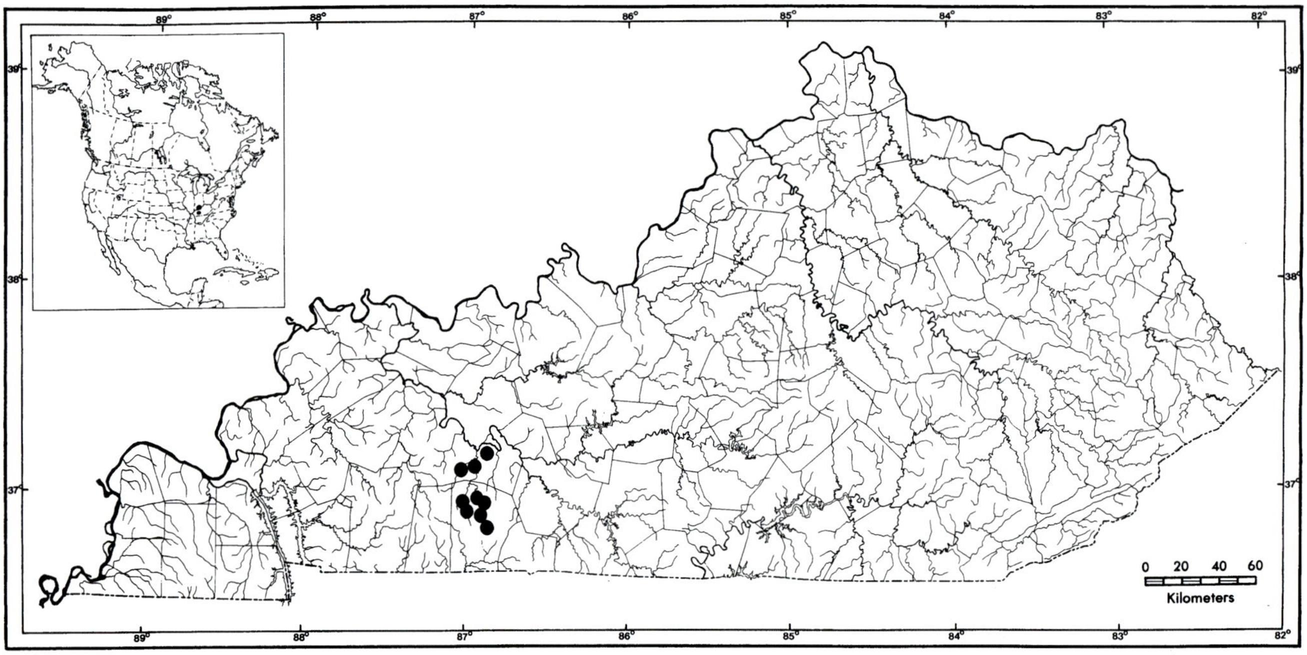

Fig. 162. Distribution of *Orconectes ronaldi* in Kentucky.

measured 53.7 and 69.7 mm in total length and carried 142 and 256 eggs, respectively. Average egg diameter for the smaller female was 1.9 mm while for the larger female it was 2.1 mm.

CRAYFISH ASSOCIATES – The following species have been collected from habitats containing *Orconectes ronaldi*: *Cambarus diogenes, C. striatus, C. tenebrosus,* and *O. immunis.*

Orconectes rusticus (Girard)
Rusty Crayfish

Fig. 163

LIFE COLORS – (Fig. 163) Base color of dorsal and lateral surfaces of chelae and carapace light brown to olive green; dorsum of chelae variable, usually having grayish or bluish tint. Coloration of dorsal surface of carapace variable; occasionally having laterally elongate dark brown patch immediately anterior to cervical groove or dark brown areolar region. Dorsum of carapace with dark brown U-shaped saddle connected at caudal margin and extending somewhat anteriorly on lateral surfaces, forward extensions may be discontinuous (see paragraph below). Abdominal segments with dark brown W-shaped patches, patches give appearance of wedge extending down center of dorsum of abdomen. Fingers of chelae with orange tips bordered proximally by wide black bands. Large knob at base of dactyl orange. Venter cream to whitish.

The common name of *Orconectes rusticus* is most likely derived from the presence of a large rust or brown colored spot located on each lateral surface at the posterior end of the carapace. These spots are much more pronounced on specimens found in northern portions of the species' range (e.g., Ohio, Indiana, and on specimens that have been introduced across the northern United States). The spots seem to be formed from the fading of the dorsal junction of the dark saddle found at the caudal edge of the carapace. The intensity of the lateral spots, and the accom-

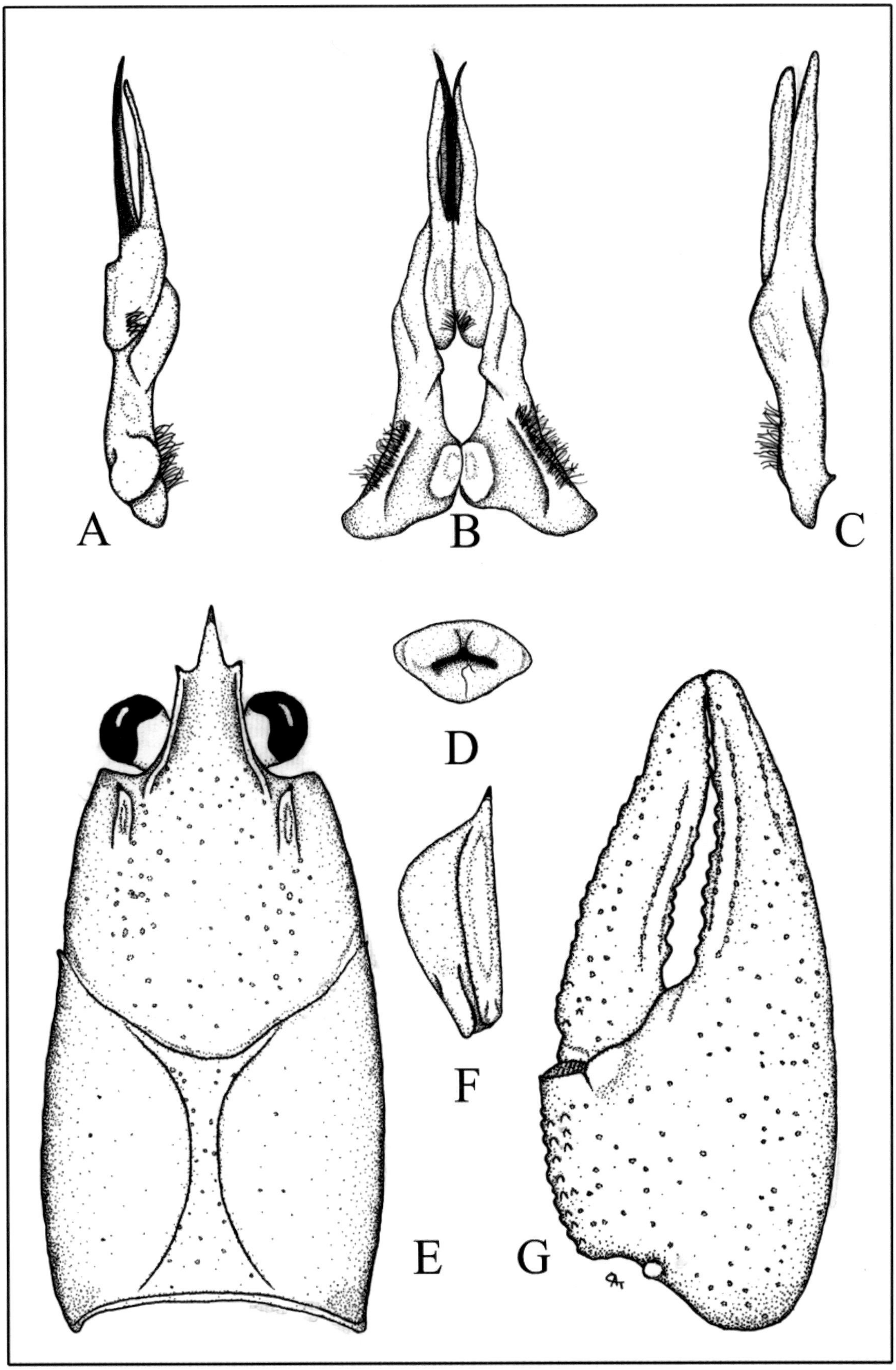

Fig. 164. *Orconectes rusticus*: A, mesial view of form I gonopod; B, ventral view of form I gonopods; C, lateral view of form II gonopod; D, ventral view of annulus ventralis; E, dorsal view of carapace; F, dorsal view of right antennal scale; G, dorsal view of right chela.

panying lack of intensity of the dorsal junction of the saddle, is highly variable among Kentucky populations.

CHARACTERS – (Fig. 164) Rostrum deeply excavated; distally margins straight or slightly concave, proximally margins convex, margins terminating in small spines at base of acumen. Acumen equal to or slightly longer than width of rostrum at marginal spines. Median carina absent. Carapace slightly compressed dorsoventrally. Areola of moderate width, 5.8 to 10.8 times longer than wide at its narrowest point, with room for two to five punctations. Cervical spines present. Antennal scale widest at midpoint. Incisor region of mandible with straight edge. Chelae large and with moderately long fingers; surfaces of palms smooth and usually covered with shallow punctations; dorsal surfaces of fingers with weak longitudinal ridges. Mesial margin of palm of chela and dactyl covered with weak, rounded tubercles, palm with two discreet rows and several additional scattered tubercles lateral to those rows; seven to nine tubercles in mesialmost row. Slight gap between basal opposable margins of fingers. Males with larger chelae than females of same carapace length.

Gonopod of form I male terminating in two long, thin elements: a sharply pointed, slightly caudodistally recurved corneous central projection and a shorter, noncorneous mesial process. Mesial process variable, either straight, slightly recurved cephalodistally, or slightly recurved caudodistally. Caudal surface of gonopod with strong angular shoulder at base of central projection. Annulus ventralis of female subrhomboidal; with median trough extending through cephalic half; two caudally directed protuberances overhanging centrally located fossa; fossa laterally wide; sinus extending from center of fossa to caudal edge of annulus.

MAXIMUM SIZE – The maximum known size for the species is approximately 106.0 mm total length.

MOST LIKE – Numerous species of crayfishes in Kentucky possess form I gonopods with two long, thin terminal elements. Of those species, *Orconectes rusticus* is most similar to *O. cristavarius* and *O. juvenilis* in that all three possess a strong angular shoulder, cervical spines, and a gonopod with an arched central projection less than 50% of the total length of gonopod. *Orconectes rusticus* differs from *O. cristavarius* in possessing a mandible with a straight-edged incisor region. While *O. juvenilis* also has a straight edged incisor region, *O. rusticus* differs from that species in possessing shorter terminal elements of the gonopod. The length of the central projection of *O. rusticus* is less than 44% of the total length of the gonopod, while the length of the central projection of *O. juvenilis* is at least 44% of the total length of its gonopod, and usually much greater.

DISTRIBUTION AND HABITAT – (Fig. 165) The native range of *Orconectes rusticus* was described by Taylor (2000) to include the lower middle Ohio River drainage of central Kentucky, western Ohio, and eastern and central Indiana and the western Lake Erie drainage in southeastern Michigan and northwestern Ohio. Mostly because of its use as fishing bait, *O. rusticus* has been introduced across the United States. Large populations of the species now occur in Connecticut, Illinois, Iowa, Maine, Massachusetts, Minnesota, New Hampshire, New Jersey, New Mexico, New York, Pennsylvannia, Tennessee, Vermont, West Virginia, and Wisconsin. In Kentucky *O. rusticus* is native, widespread, and common in the lower Licking, Salt, and middle and upper Green River drainages. The species also occurs at a few sites in the middle Kentucky and middle Ohio River drainages. The single upper Little Laurel River (Cumberland River drainage) site in Laurel County was hypothesized by Taylor (2000) to be an introduced population. Additional collecting efforts in the mainstem of the Ohio River between the mouths of the Licking and Green rivers will likely find *O. rusticus* to be more common than indicated on our range map.

Orconectes rusticus occurs in reservoirs, creeks, and rivers of all sizes with a wide variety of substrates. The species is usually encountered under cobble or among woody debris in lotic habitats or under riprap or among rooted vegetation or submersed logs in lentic habitats. The species also has a wide range of water depth tolerances, occurring in creek riffles 0.1 m deep to lake bottoms up to 15 m deep.

LIFE HISTORY – Within the genus *Orconectes* more is known about the ecology and life history of *O. rusticus* than perhaps any other species.

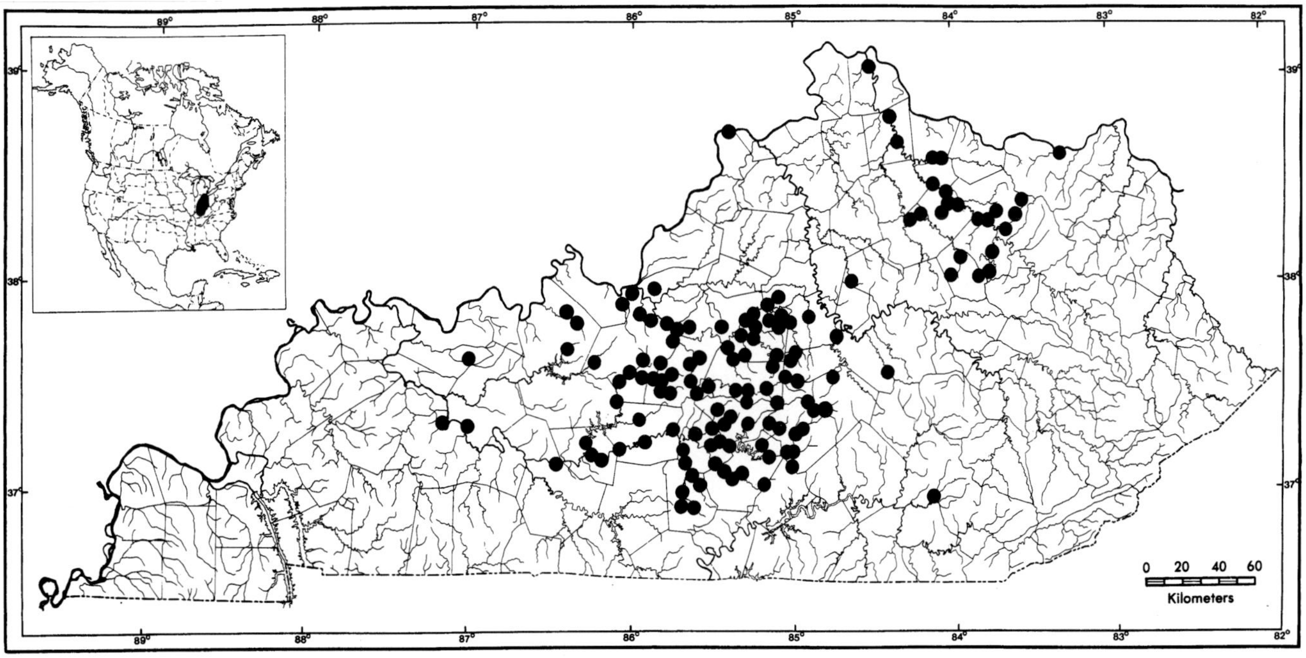

Fig. 165. Distribution of *Orconectes rusticus* in Kentucky.

This is partially due to its widespread introduction and destructive effects on native species and habitats where it has been introduced. Prins (1968) conducted a comparative ecological study involving *O. rusticus* in Doe Run, Meade County, Kentucky. He reported that the species became reproductively mature at a size of approximately 34 mm total length and mated in the fall and spring. Females were ovigerous from March to May but were most abundant in April. Eggs hatched in May after four to six weeks of development and juveniles remained attached to their mothers for another three weeks. After their first summer, young attained a size of 15 to 30 mm total length. Some males were able to attain form I status by their first fall and were able to mate. Little growth occurred over the winter months and at one year of age individuals ranged in size from 23 to 36 mm. By their second fall, individuals ranged in size from 36 to 51 mm total length. Fecundity in the species was highly correlated with female body size. Prins (1968) recorded egg counts ranging from 54 on a 34-mm individual to 357 on a 70-mm individual. Prins estimated average longevity of the species to be 2.5 years. Detritus and vascular plants were the dominant food items for *O. rusticus*. Our statewide collections of *O. rusticus* found form I males in every month but May.

CRAYFISH ASSOCIATES – The following species have been collected from habitats containing *Orconectes rusticus*: *Barbicambarus cornutus, Cambarus cumberlandensis, C. diogenes, C. graysoni, C. rusticiformis, C. tenebrosus, O. barrenensis, O. immunis, O. putnami, O. rafinesquei,* and *O. sanbornii.*

CONSERVATION NOTES – As mentioned above, *Orconectes rusticus* has been widely introduced across the United States. At many of these locations the species has rapidly expanded its range and displaced populations of native crayfishes through a variety of mechanisms such as direct competition or reproductive interference (Lodge et al. 2000; Perry et al. 2001). At some sites *O. rusticus* has driven local native populations extinct. While *O. rusticus* is native to central Kentucky, it is imperative that the species, or any other crayfish species, not be transported across drainage basin divides. Kentucky is home to several narrowly endemic species. These species could be rapidly extirpated if *O. rusticus* were introduced into their native watersheds.

6-11-05 ; Corn Creek, Trimble Co. Ky.

Orconectes sanbornii (Faxon)
Sanborn's Crayfish

Fig. 166

LIFE COLORS – (Fig. 166) Base color of dorsal and lateral surfaces of chelae and carapace light brown to tan. Dorsum of carapace with a laterally elongate dark brown patch immediately anterior to cervical groove and dark brown U-shaped saddle connected at caudal margin extending anterior to just posterior of midpoint of lateral surfaces. Intensity of saddle variable abdomen with unevenly colored dark brown wedge extending down center of dorsum, anterior edges of abdominal segments darker than posterior dorsal surfaces. Fingers of chelae with orange tips bordered proximally by narrow black bands. Large knob at base of dactyl orange. Venter cream to whitish.

CHARACTERS – (Fig. 167) Rostrum moderately excavated, distally margins straight and slightly converging; proximally margins convex, margins terminating in small spines at base of acumen. Acumen equal to or slightly longer than width of rostrum at marginal spines. Median carina absent. Carapace dorsoventrally flattened, more so in larger individuals. Areola of moderate width, 5.7 to 8.3 times longer than wide at its narrowest point, with room for two to five punctations. Cervical spines present. Antennal scale widest at midpoint. Chelae large with moderately long fingers, dorsal surfaces of palms and fingers usually covered with shallow punctations and short setae. Dorsal surfaces of fingers of chelae with well-developed longitudinal ridges, ridges more strongly developed on larger individuals. Mesial margin of palm of chela and dactyl with tubercles, palm with two rows, 7 to 10 tubercles in mesialmost row. Males with longer chelae than females of same carapace length.

Gonopod of form I male terminating in two short terminal elements: a corneous central projection with a sharply pointed and caudodistally recurved tip and a slightly shorter or equal length noncorneous mesial process. Mesial process dorsoventrally flattened. Annulus ventralis of female subrhomboidal; lacking median trough through its cephalic margin; centrally located fossa; sinus running from center of fossa to the caudal margin of annulus.

MAXIMUM SIZE – The maximum known size for the species is 85.0 mm total length.

MOST LIKE – *Orconectes sanbornii* is most similar to other members of the genus that possess short, more or less straight terminal elements on the form I gonopod. Those species are: *O. bisectus, O. jeffersoni, O. margorectus, O. rafinesquei,* and *O. tricuspis*. Of those five species, *O. sanbornii* is most similar to *O. margorectus* and *O. tricuspis* in that all three possess a mesial process that is dorsoventrally flattened and a central projection that is less than 25% of the total length of the gonopod. *Orconectes sanbornii* differs from *O. margorectus* by possessing a dentate incisor region on the mandible and differs from *O. tricuspis* by possessing a mesial process that is not recurved cephalodistally and by lacking a median carina.

DISTRIBUTION AND HABITAT – (Fig. 168) *Orconectes sanbornii* occurs in the middle Ohio River drainage from northeastern Kentucky upstream to the Fishing Creek drainage of northern West Virginia (Jezerinac et al. 1995). In Ohio the species also occurs in Lake Erie tributaries in the north–central portion of the state (Thoma and Jezerinac 2000). In Kentucky *O. sanbornii* has a limited distribution. It occurs commonly in the Tygarts and Grassy Fork Creek drainages and the Ohio River in Carter, Lewis, and Greenup counties in the extreme northeastern portion of the state. A single specimen was also collected from the Ohio River in Trimble County in 1998. Additional collecting in the mainstem of the Ohio River is needed to determine if the Trimble County specimen is a waif, an introduction, or if *O. sanbornii* occurs in a much larger stretch of that river than is currently known.

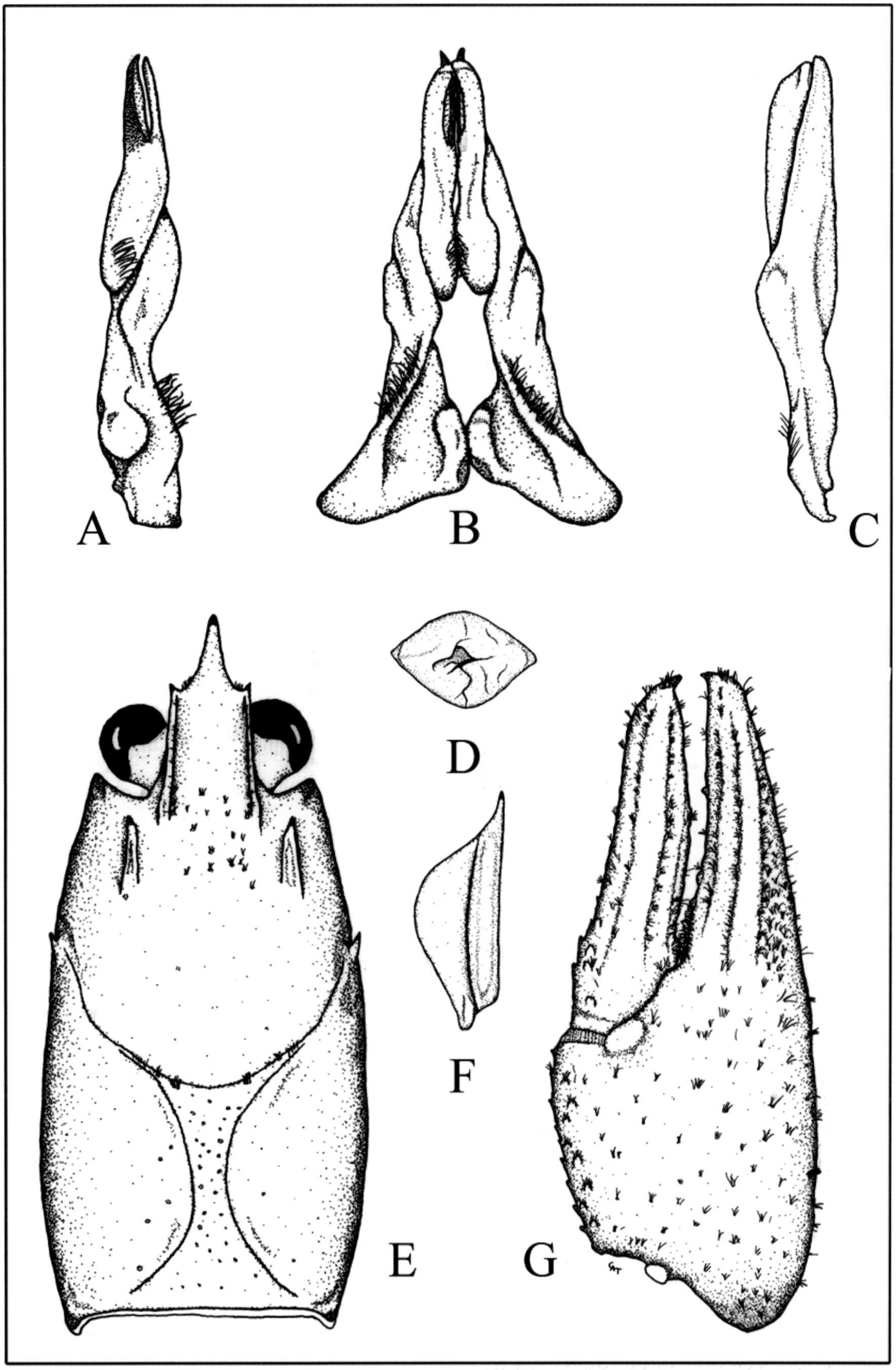

Fig. 167. *Orconectes sanbornii*: A, mesial view of form I gonopod; B, ventral view of form I gonopods; C, lateral view of form II gonopod; D, ventral view of annulus ventralis; E, dorsal view of carapace; F, dorsal view of right antennal scale; G, dorsal view of right chela.

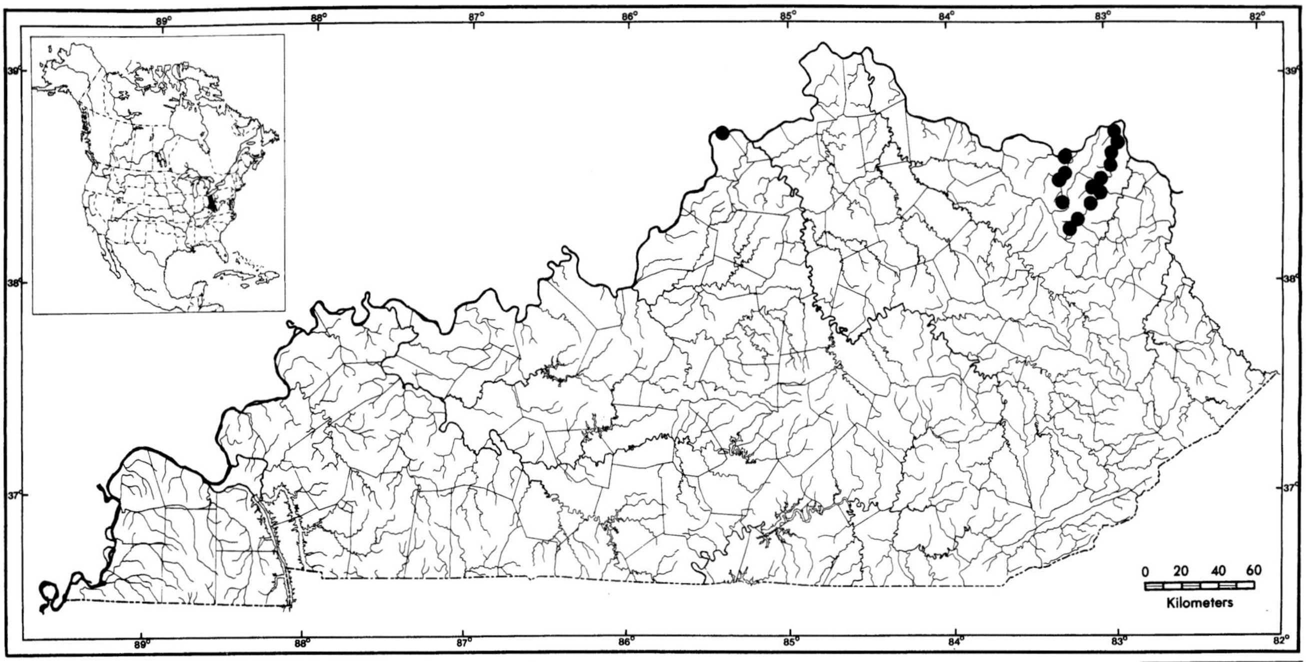

Fig. 168. Distribution of *Orconectes sanbornii* in Kentucky.

A record for *Orconectes sanbornii* from Two Mile Creek, Breathitt County, exists in the EKU Crayfish Collection. These two specimens, collected in May of 1968, represent the only known record for the species from the Kentucky River drainage. Return visits to the site, and to many other nearby locations, have failed to recollect the species. Unless additional specimens can be acquired from the Kentucky River drainage, we consider the record to be dubius.

Orconectes sanbornii is most common in small to moderately-wide creeks up to 20 m in width with substrates of fine gravel and cobble. The species is usually encountered under cobble in areas with moderate to no flow. The species has been collected under cobble along the edges of the Ohio River or in woody debris and undercut banks in smaller creeks. Jezerinac et al. (1995) reported that during a drought year, the species was found in small depressions under cobble in dry creek reaches.

LIFE HISTORY – Little is known about the life history of *Orconectes sanbornii.* Reproductive information for Ohio populations of the species was reported by Fielder (1972) and West Virginia populations by Jezerinac et al. (1995). Those authors collected form I males from early April through October and ovigerous females in April and May. Females with young were collected in April, June, and September. Mating was observed in July, August, and October. The collection of females with young in September is unusual for creek-dwelling *Orconectes* and indicates that egg–laying may take place twice during a single year or that it may occur later in some individuals. More data are needed to determine if two spawns per season are occurring. Fielder (1972) also reported that egg diameter for ovigerous females ranged from 0.01 to 2.26 mm. Juveniles were independent of females and had reached 12.0 to 16.0 mm total length by early June. Jezerinac et al. (1995) reported males were capable of becoming form I at approximately 33 mm total length and suggested that West Virginia individuals lived on average to three years of age. Both Fielder (1972) and St. John (1976) suggested that four age classes were present in Ohio populations. Given the proximity of the Ohio and West Virginia populations, we believe that the above reproductive data are applicable to Kentucky populations. In Kentucky form I males were collected in April, May, and July through November. No ovigerous females have been collected in the state.

CRAYFISH ASSOCIATES – The following species have been collected from habitats containing *Orconectes sanbornii*: *Cambarus bartonii cavatus, C. sciotensis, O. cristavarius,* and *O. rusticus.*

Orconectes tricuspis Rhoades
Western Highland Crayfish

Fig. 169

LIFE COLORS – (Fig. 169) Base color of dorsal and lateral surfaces of chelae, carapace, and abdomen light to dark brown. Dorsum of carapace anterior to cervical groove and caudal margin of carapace with dark brown patches. Dorsum of abdominal segments usually with pair of small dark brown patches or bars. Fingers of chelae with orange tips bordered proximally by black bands. Large knobs at base of dactyl orange. Venter cream to whitish.

CHARACTERS – (Fig. 170) Rostrum moderately excavated; distally margins straight and slightly converging, proximally margins slightly convex, margins terminating in spines at base of acumen. Acumen equal to or slightly longer than width of rostrum at marginal spines. Weak median carina present. Carapace dorsoventrally flattened, more so in larger individuals. Areola wide, 5.5 to 8.0 times longer than wide at its narrowest point, with room for three to five punctations. Antennal scale widest at midpoint. Chelae large with moderately long fingers; dorsal surfaces of palms covered with shallow punctations and short setae. Dorsal surfaces of fingers of chelae with well-developed longitudinal ridges and one row of deep punctuations that coalesce; ridges more strongly developed on larger individuals. Mesial margin of palm of chela and dactyl with tubercles, palm with two or three rows of tubercles, 7 to 10 tubercles in mesialmost row. Two or three tubercles usually present at base of movable finger lateral to mesial margin. Males with longer chelae than females of same carapace length.

Gonopod of form I male terminating in two short terminal elements: a laterally flattened, corneous central projection, distal tip of which is pointed and slightly recurved caudodistally and a noncorneous mesial process of equal length or slightly longer. Mesial process dorsoventrally flattened and slightly recurved cephalodistally. Distal tip of mesial process occasionally corneous. Annulus ventralis of female subrhomboidal; with narrow trough running between two strongly elevated protuberances at center of cephalic margin; centrally located, poorly developed fossa; sinus running to the caudal margin of annulus.

MAXIMUM SIZE – The maximum known size for the species is 83.0 mm total length.

MOST LIKE – *Orconectes tricuspis* will most likely be confused with other members of the genus that possess short, more or less straight terminal elements on the form I gonopod. Those species are *O. bisectus, O. jeffersoni, O. margorectus, O. rafinesquei*, and *O. sanbornii*. *Orconectes tricuspis* differs from these species in possessing three characters: a gonopod with a central projection less than 25% of the total length of the gonopod, a dorsoventrally flattened mesial process that is slightly recurved cephalodistally, and a weak median carina on the rostrum. *Orconectes tricuspis* also differs from *O. margorectus* by possessing an incisor region with a straight edge.

DISTRIBUTION AND HABITAT – (Fig. 171) *Orconectes tricuspis* is endemic to the state of Kentucky. Within the state it occurs in the lower Tennessee, lower Cumberland, and Pond (Green River drainage) River drainages. In the Christian and Todd County region, the species occurs in the headwaters of three different drainages; the Little River (Cumberland River drainage), the Pond River, and Whipperwill Creek (Red River - Cumberland River drainage).

Orconectes tricuspis is most common in small to large creeks ranging in width from two to eight m with bottom substrates of bedrock, cobble, or large gravel. In larger creeks the species can occur in slower flowing runs with mud and scattered cobble substrates. In all creeks the species can usually be found under the cobble or gravel.

LIFE HISTORY – Little is known about the life history of *Orconectes tricuspis*. Form I males have been collected in March, April, June, July, August,

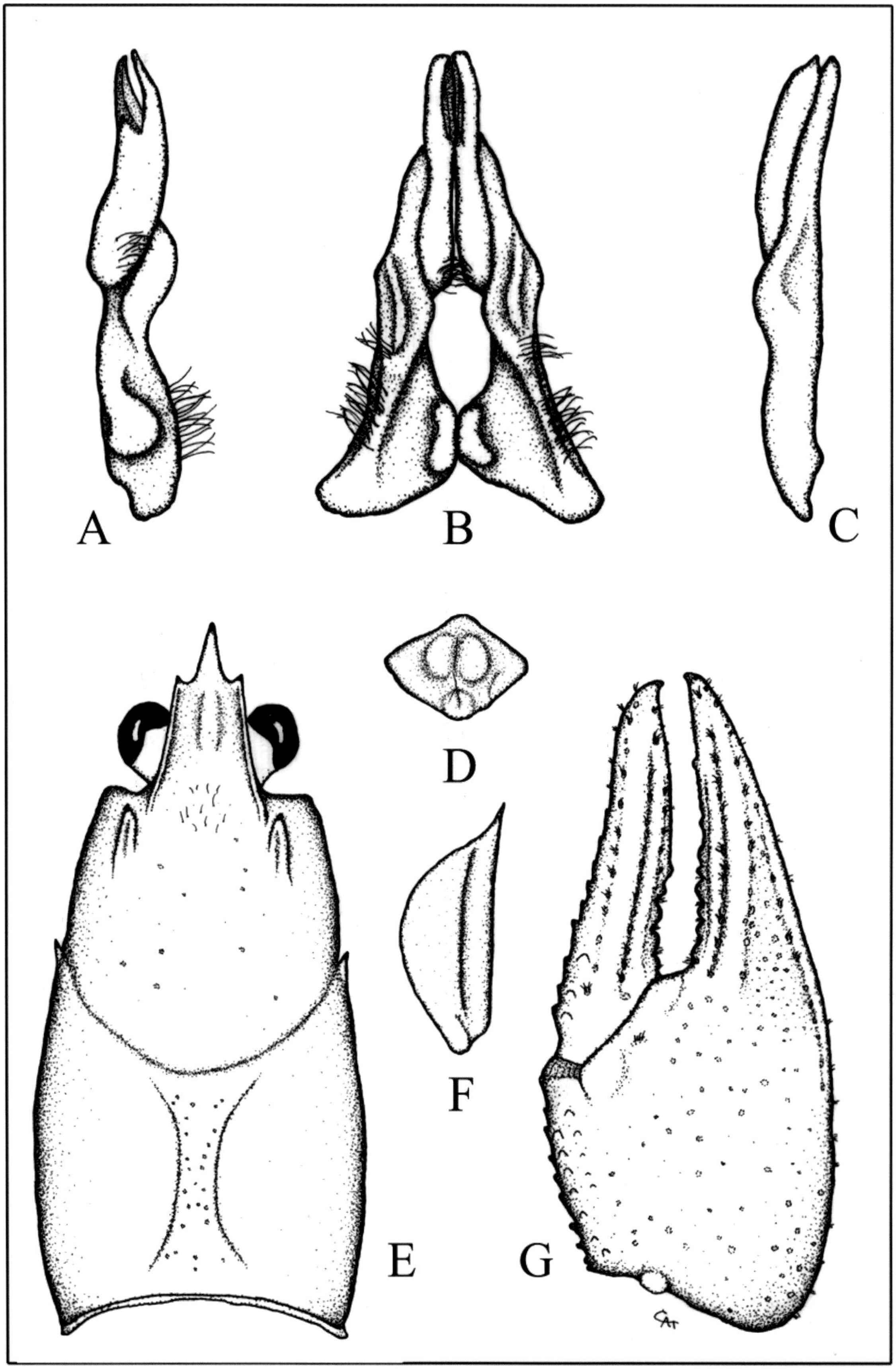

Fig. 170. *Orconectes tricuspis*: A, mesial view of form I gonopod; B, ventral view of form I gonopods; C, lateral view of form II gonopod; D, ventral view of annulus ventralis; E, dorsal view of carapace; F, dorsal view of right antennal scale; G, dorsal view of right chela.

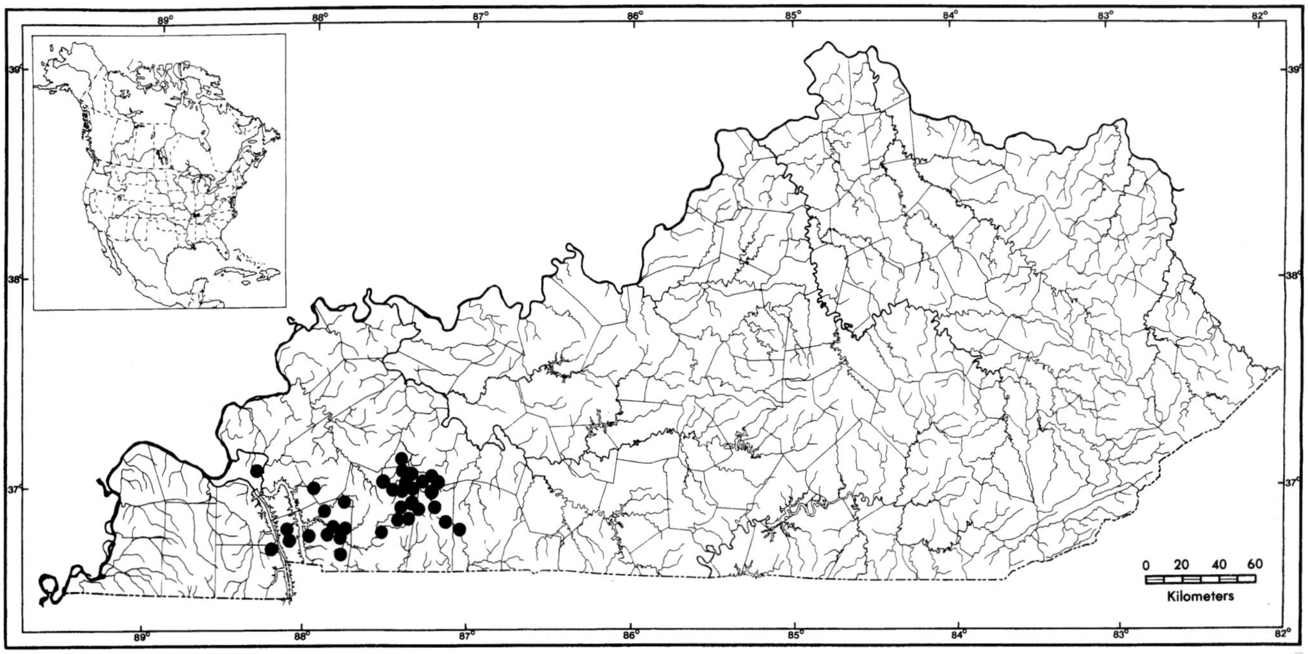

Fig. 171. Distribution of *Orconectes tricuspis* in Kentucky.

September, November, and December. Ovigerous females have been collected only in April. These limited data suggest that *O. tricuspis* has life history traits similar to most temperate North American crayfishes (i.e., copulation activities occurring in fall and/or spring months, egg laying occurring in the spring). Two ovigerous females collected in early April measured 55.8 mm and 50.0 mm total length and carried 128 and 70 eggs, respectively. Average egg diameter was 1.9 mm for the larger female and 1.8 mm for the smaller female. While true burrowing by the species has not been reported in the literature or observed by us, Rhoades (1944) reported that at one site the species had "moved a considerable quantity of gravel from under a concrete wall to form roomy burrows."

CRAYFISH ASSOCIATES – The following species have been collected from habitats containing *Orconectes tricuspis*: *Cambarus diogenes, C. friaufi, C. tenebrosus, O. compressus, O. immunis, O. placidus,* and *Procambarus acutus.*

Procambarus clarkii **Photo by C. Lukhaup**

Procambarus Ortmann

Procambarus is the largest genus of crayfishes in the world with over 160 species and subspecies. Its members are found natively across the central and eastern United States, Mexico, parts of Latin America, and Cuba. Only a handful of species occur north of 40° latitude. Members of the genus *Procambarus* occur in almost all habitat types including subterranean creeks, lowland swamps and ditches, and creeks; however, most species show a preference for more lentic, well–vegetated habitats in lowland areas. Many members of the genus are proficient burrowers. Males have two to four terminal elements on the gonopod and hooks on the ischia of the third, and occasionally third and fourth, pair of pereiopods. Kentucky is home to three species of *Procambarus*: *P. acutus, P. clarkii*, and *P. viaeviridis*.

Procambarus acutus (Girard)
White River Crawfish

Fig. 172A: a large mature form I male.

Fig. 172B: a smaller form II adult male.

LIFE COLORS – (Figs. 172A, B) Base color of dorsal and lateral surfaces of chelae, carapace, and abdomen of large, mature adults dark red. Abdomen with long, dark black wedge extending down center of dorsum. Tubercles on carapace and chelae whitish or cream colored. Larger form II males and females can also be light brown in overall color and mottled with dark spots. Dorsal and lateral surfaces of smaller individuals grayish or light brown and mottled with black and white spots. Venter of all individuals cream to whitish.

CHARACTERS – (Fig. 173) Rostrum moderately excavated, margins converging anteriorly and terminating in small spines at base of short acumen. Acumen equal to or shorter than width of rostrum at marginal spines. Carapace laterally compressed and moderately arched dorsoventrally, dorsal and lateral surfaces of larger individuals covered with tubercles. Cervical spines present. Areola variable, from narrow to very narrow, 8.5 to 18.0 times longer than wide at its narrowest point, with room for one to three punctations. Antennal scale widest proximal to midpoint. Palm and fingers of chelae relatively narrow, dorsal surface of palm covered with tubercles. Basal third of dactyl with scattered tubercles on dorsal surface. Mesial margin of propodus with greatly enlarged third or fourth tubercle and with enlarged and pointed terminal tubercle. Mesial margin of palm of chela with single row of eight or nine strong, anteriorly directed tubercles. Lateral margin of dactyl with four to six tubercles. Males with longer chelae than females of same carapace length.

Gonopod of form I male terminating in four short elements all of which are closely aligned, recurved laterally and caudodistally, and usually obscured by long setae originating from cephalic process. Elements consist of the following: a corneous pointed central projection; a shorter corneous, pointed caudal process; an elongated and pointed mesial process; and an elongated and pointed, corneous cephalic process. Annulus ventralis of female with centrally located wide trough; trough not running to cephalic margin; fossa located within central trough. Divergent sinus running through slightly elevated posterior central region.

MAXIMUM SIZE – The maximum known size for the species is 130.0 mm total length.

MOST LIKE – In possessing four terminal elements on the form I male gonopod and a dark wedge on the dorsal surface of its abdomen, *Procambarus acutus* is most likely to be confused with the other two members of the genus *Procambarus* found in Kentucky, *P. clarkii* and *P. viaeviridis*. *Procambarus acutus* differs from *P. clarkii* by possessing an open areola and not possessing a strong, angular shoulder on the caudal surface to the male gonopod. *Procambarus acutus* differs from *P. viaeviridis* by possessing rostral margins that terminate in spines and by not possessing a strong, rounded knob on the caudal surface of the male gonopod.

DISTRIBUTION AND HABITAT – (Fig. 174) *Procambarus acutus* has a disjunct distribution across a large portion of the eastern United States. The species occurs along the Atlantic Slope from

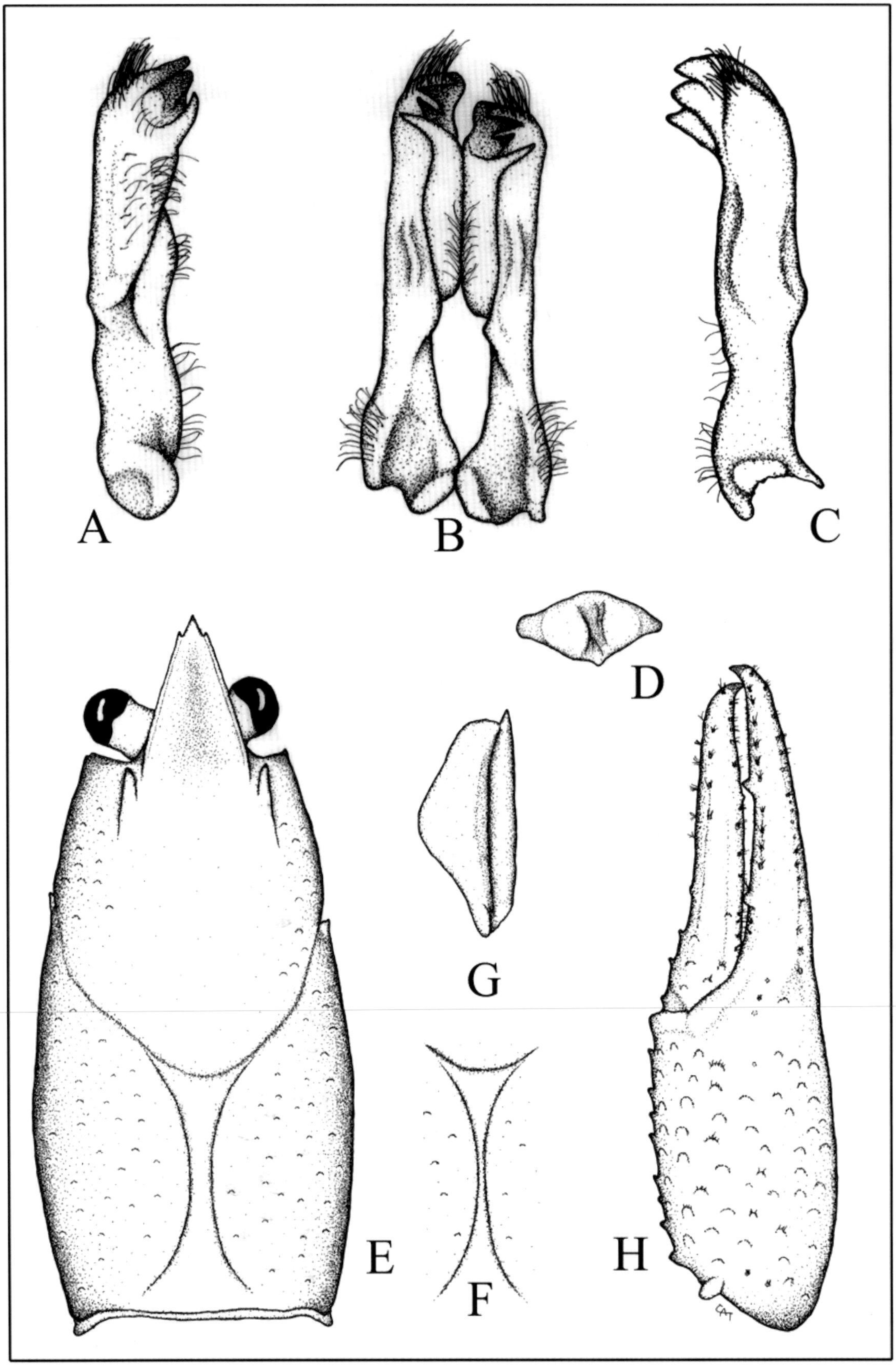

Fig. 173. *Procambarus acutus*: A, mesial view of form I gonopod; B, ventral view of form I gonopods; C, lateral view of form II gonopod; D, ventral view of annulus ventralis; E, dorsal view of carapace; F, dorsal view of areola; G, dorsal view of right antennal scale; H, dorsal view of right chela.

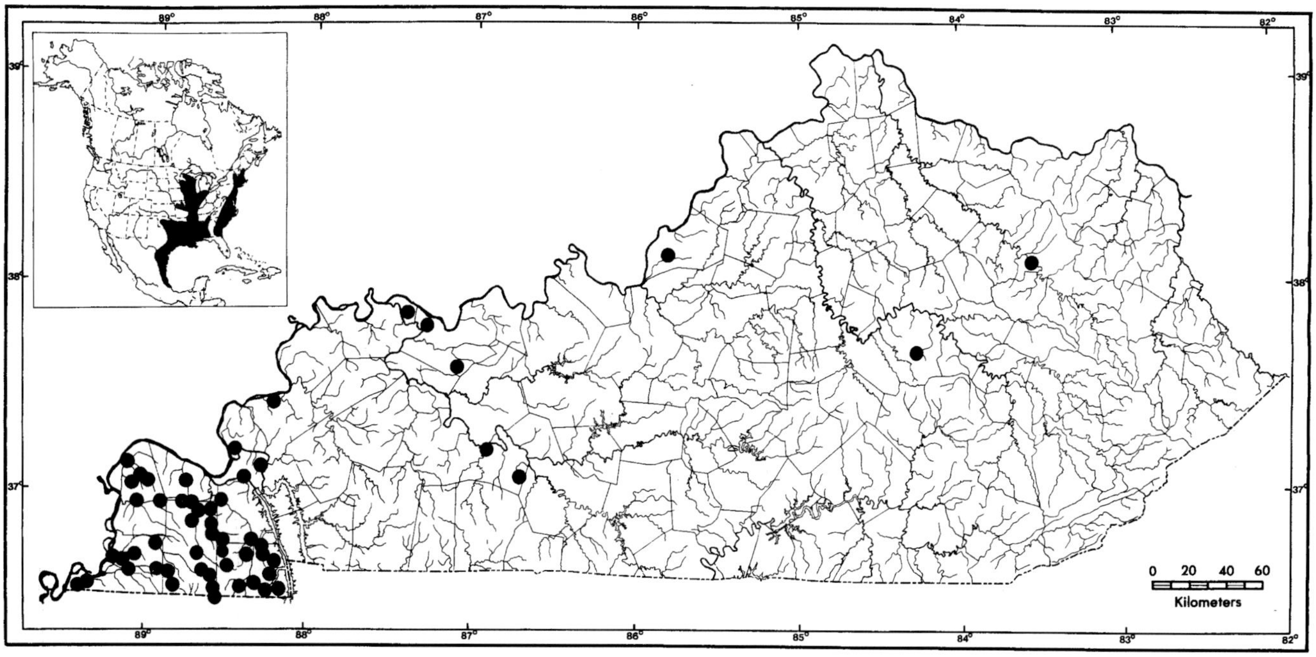

Fig. 174. Distribution of *Procambarus acutus* in Kentucky.

Maine to southern Georgia and also from southern Great Lakes drainages south to the Gulf of Mexico (Hobbs 1989). The species ranges west into the Red River drainage in Oklahoma and east to the Florida panhandle. The species has also been widely introduced in North America. In the United States established non–native populations now occur in California and several New England states (Hobbs 1989). In Kentucky, the species occurs commonly in most aquatic habitats west of the lower Cumberland River and sporadically in the middle and lower Green River drainage. The species is also known from several probably introduced populations, including the Pond Creek drainage in Jefferson County, from ponds in the Bluegrass Army Depot and Central Kentucky Wildlife Management Area in Madison County, and Minor Clark Fish State Hatchery in Rowan County. *Procambarus acutus* is widely used and sold as fishing bait in Kentucky and this, in combination with its presence in ponds in almost all hatcheries in the state, explains it's sporadic distribution in central and eastern Kentucky. It is highly likely that the species will become more widespread in the state.

In Kentucky *Procambarus acutus* is found in a wide variety of habitats including swamps, ponds, seasonally flooded ditches, creeks, and medium-sized rivers. Substrates in lentic habitats are usually soft or compacted mud, while in flowing creeks substrates range from compacted sand and mud to gravel. The species is frequently collected from under woody debris and amongst rooted vegetation in creeks. As water levels recede in seasonally flooded habitats such as ponds and ditches during summer and fall months, individuals can be found in simple vertical burrows.

LIFE HISTORY – *Procambarus acutus* is one of the most economically important species of crayfish in North America. It trails only *P. clarkii* and *P. zonangulus* in terms of annual harvest from culture ponds and wild populations. Like *P. clarkii* and *P. zonangulus*, it is commonly consumed by humans in the southern United States; however, it differs from *P. clarkii* in also being frequently used in the fishing bait industry. Compared to *P. clarkii*, little specific life history data exists for *P. acutus*. In Kentucky, form I males have been collected in January, February, March, May, June, September, and November while juveniles less than 25 mm in total length have been collected in April. These collections suggest that in Kentucky the species is capable of mating in fall or early spring with egg-laying occurring in early spring. However, data from neighboring states suggest that egg laying may occur throughout the year. Turner (1926) found ovigerous females in Ohio in March, July, and September while in Illinois ovigerous females have been collected in April, July, August, and December (Page 1985). No ovigerous females from Kentucky are present in museum collections nor were any collected by us.

Procambarus acutus is proficient at burrowing and will frequently do so during dry summer and fall months. Most burrows are simple, vertical, or steeply angling shafts. Chimneys of earthen pellets may or may not be present.

CRAYFISH ASSOCIATES – The following species have been collected from habitats containing *Procambarus acutus*: *Cambarellus puer, C. shufeldtii, Cambarus diogenes, C. tenebrosus, Fallicambarus fodiens, Orconectes burri, O. durelli, O. immunis, O. palmeri palmeri, O. placidus, O. tricuspis, P. clarkii*, and *P. viaeviridis*.

TAXONOMIC NOTES – The taxonomic status of *Procambarus acutus* has received considerable attention, and at the same time has caused much confusion. Morphological diversity within this wide-ranging species has long been recognized, with the description of "varieties" and subspecies being published over the past 100 years. Using mostly male gonopod morphology, recent published and unpublished data suggest that populations formerly identified as *P. acutus* are assignable to at least three different taxa (Hobbs and Hobbs 1990, Huner 2002). Those populations occurring in the Gulf Coastal Plains of Texas, Louisiana, Mississippi, and Alabama are assignable to *P. zonangulus*, a species described by Hobbs and Hobbs (1990). Populations from the Atlantic Slope and upper Gulf Coastal Plain east of the Mississippi River belong to *P. acutus acutus* while those from the middle and upper Mississippi River drainage belong to *P. acutus cuevachicae*. As such, Kentucky populations would be assigned to *P. a. cuevachicae*. Future data will be needed to determine the exact ranges of the *P. acutus* subspecies and if full specific status is warranted. This revisionary work has also resulted in the name "White River Crawfish" being applied to two different species, *P. acutus* and *P. zonangulus*. Until more data become available we use the Latin name *P. acutus* and the common name White River Crawfish for Kentucky individuals.

Procambarus clarkii **(Girard)**
Red Swamp Crawfish

Fig. 175A: a large mature form I male.

Fig. 175B: a smaller form I adult male.

LIFE COLORS – (Figs. 175A, B) Base color of dorsal and lateral surfaces of chelae, carapace, and abdomen of large, mature adults dark red. Abdomen with long, dark black wedge running down center of dorsum. Dorsal and lateral surfaces of smaller individuals grayish or light brown and mottled with black and white spots. Venter of all individuals cream to whitish. Venter of abdomen with longitudinal blue stripe running length-wise down center.

CHARACTERS – (Fig. 176) Rostrum deeply excavated, margins converging anteriorly and terminating in small spines at base of short acumen. Acumen of equal length or shorter than width of rostrum at marginal spines. Carapace laterally compressed and moderately arched dorsoventrally, dorsal and lateral surfaces of larger individuals covered with tubercles. Cervical spines present. Areola usually obliterated; if open, only slightly and never with room for

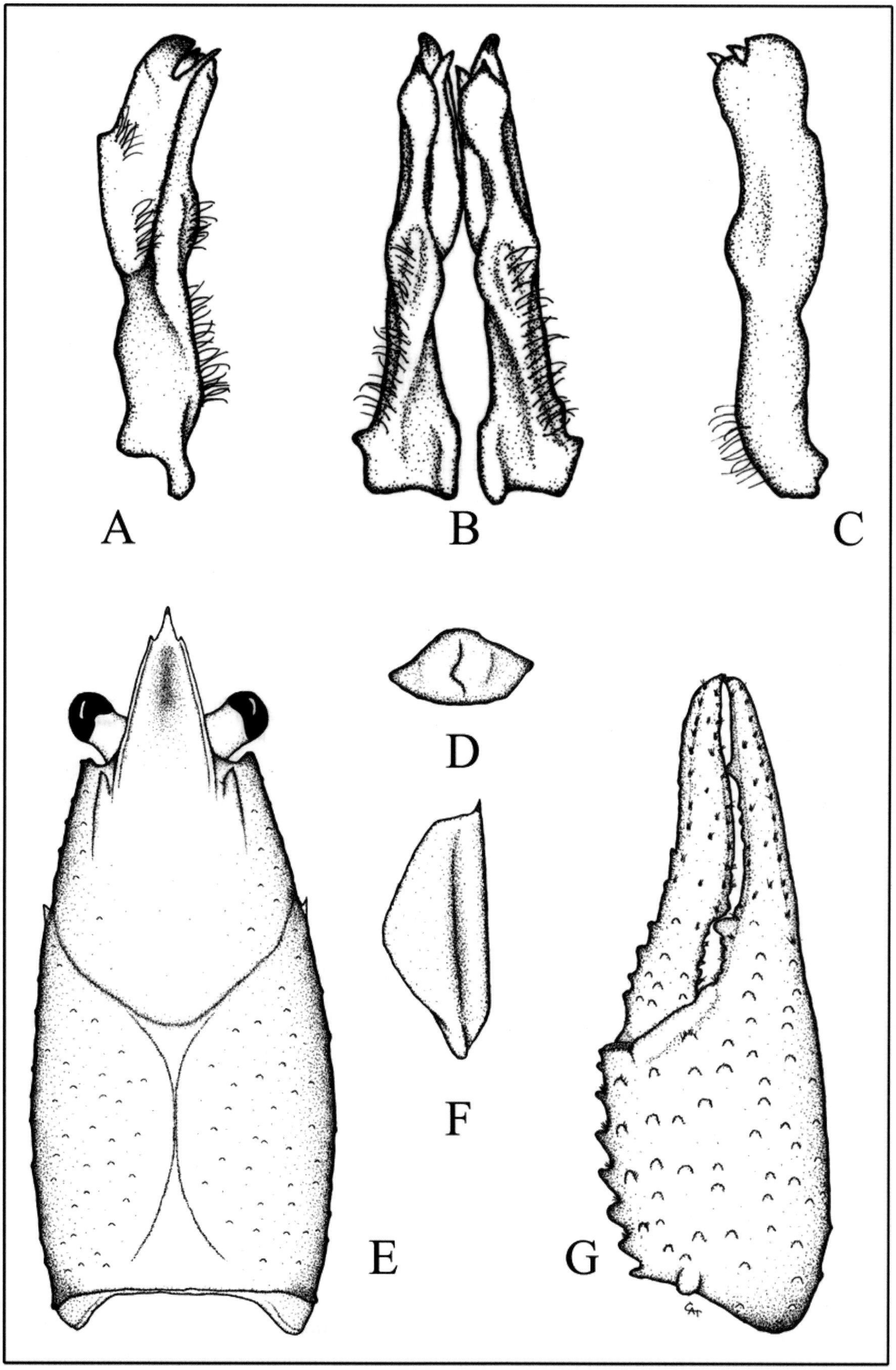

Fig. 176. *Procambarus clarkii*: A, mesial view of form I gonopod; B, ventral view of form I gonopods; C, lateral view of form II gonopod; D, ventral view of annulus ventralis; E, dorsal view of carapace; F, dorsal view of right antennal scale; G, dorsal view of right chela.

punctations. Antennal scale widest proximal to midpoint. Palm and fingers of chelae relatively narrow, dorsal surface of palm covered with tubercles. Dactyl with concave mesial margin and basal third with scattered tubercles on dorsal surface. Mesial margin of propodus with greatly enlarged third or fourth tubercle and with enlarged and pointed terminal tubercle. Mesial margin of palm of chela with single row of seven strong, anteriorly directed tubercles. Mesial margin of dactyl with five or six tubercles. Males with slightly longer chelae than females of same carapace length.

Gonopod of form I male terminating in four short elements: a corneous, short, and pointed central projection; a shorter corneous, bladelike caudal process; an elongated, arching, and pointed mesial process; and an arching, bladelike, corneous cephalic process. Caudal surface of gonopod with strong, angular shoulder. Annulus ventralis of female oval or subrhomboidal with sinus running through an elevated central section; lacking well-defined fossa.

MAXIMUM SIZE – The maximum known size for the species is 160.0 mm total length.

MOST LIKE – In possessing four terminal elements on the form I male gonopod and a dark wedge on the dorsal surface of its abdomen, *P. clarkii* is most similar to the other two members of the genus *Procambarus* found in Kentucky, *P. acutus* and *P. viaeviridis*. *Procambarus clarkii* differs from those two species in having an obliterated areola and a strong, angular shoulder on the caudal surface of the male gonopod.

DISTRIBUTION AND HABITAT – (Fig. 177) *Procambarus clarkii* occurs in the Gulf Coastal Plain from extreme southern Illinois south and west to northeastern Mexico and east to the Florida panhandle (Hobbs 1989). The species has been widely introduced in North America, Europe, Asia, and Africa. In the United States established non–native populations now occur in California, Nevada, Idaho, Illinois, New York, North Carolina, Ohio, Oregon, and Washington. In Kentucky the species has a distribution almost identical to that of the vernal crayfish, *P. viaeviridis*. It is known from the floodplains of the extreme lower Tennessee, lower Ohio, and Mississippi rivers and from tributaries of the Mississippi River in the extreme western portion of the state.

In Kentucky *Procambarus clarkii* is found in swamps, vegetated ponds, seasonally flooded ditches, and creeks. Substrate in lentic habitats is usually soft mud while in flowing creeks substrates are compacted sand and mud. The species is frequently collected from under woody debris and among rooted vegetation in creeks. As water levels recede in seasonally flooded habitats such as ponds and ditches during summer and fall months, individuals will be found in deep bur-

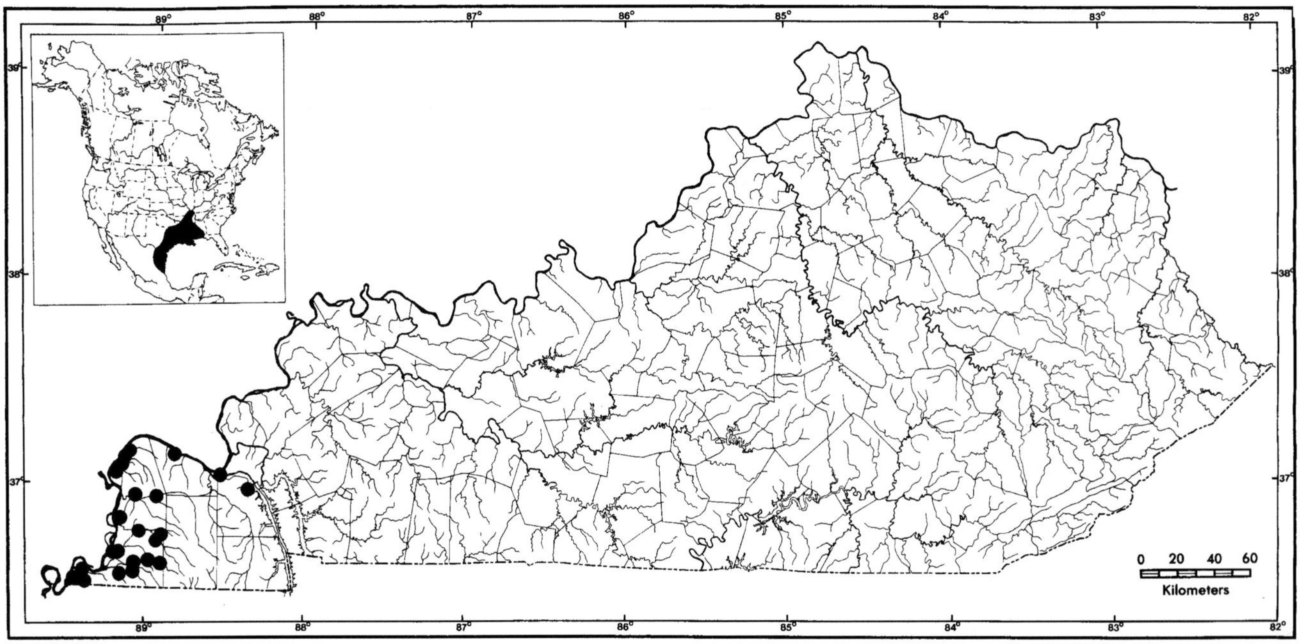

Fig. 177. Distribution of *Procambarus clarkii* in Kentucky.

rows. Burrows of up to 2 m in depth have been reported for *P. clarkii* (Huner 2002).

LIFE HISTORY – Economically *Procambarus clarkii* is the most important species of crayfish in North America, if not the world. It is commonly consumed by humans in the southern United States and is marketed as a delicacy in other regions of the country and world. Total annual harvest from native and pond-cultivated populations in the United States averages between 30,000 and 40,000 metric tons (Huner 2002). Given its economic importance, more is known about the life history of *P. clarkii* than of any other species. In Kentucky form I males have been collected in May, June, September, and October while juveniles less than 20 mm in total length have been collected in May and September. No ovigerous females or females with young have been collected or observed in Kentucky; however, in southern Illinois Page (1985) collected a female with young attached in February while Brown (1955) collected females with eggs and females with young from burrows in September. These data suggest that the species is capable of mating twice per year. Huner (2002) stated that at least two generations per year were possible for populations of the species at "lower latitudes." Huner (2002) summarized life history information for *P. clarkii* and reported the following attributes: size at sexual maturity ranged from 45 to 125 mm total length; fecundity is correlated with female size with individuals of 60 mm total length carrying 50 eggs and over 600 eggs on 120-mm individuals; under laboratory conditions the species can live to four years of age, while in nature it rarely exceeds 12 to 18 months.

Procambarus clarkii is proficient at burrowing and will frequently do so during dry summer and fall months. Huner (2002) described most burrows as simple, steeply angling shafts. Simple chimneys of earthen pellets may or may not be present. Detailed life history information for the species is also available from Penn (1943) and Huner and Barr (1991), while an excellent summary of culturing practices is provided by Huner (2002).

CRAYFISH ASSOCIATES – The following species have been collected from habitats containing *Procambarus clarkii*: *Cambarellus shufeldtii, Cambarus diogenes, Fallicambarus fodiens, Orconectes palmeri palmeri, P. acutus,* and *P. viaeviridis*.

Procambarus viaeviridis (Faxon)
Vernal Crayfish

Fig. 178

LIFE COLORS – (Fig. 178) Base color of dorsal surfaces of chelae, carapace, legs, and abdomen light pink, reddish brown or light brown. Tubercles on dorsal surfaces of chelae may be dark brown or black, giving the appearance of dark flecks. Abdomen with long, dark black wedge extending down center of dorsum. Venter cream to whitish.

CHARACTERS – (Fig. 179) Rostrum flat, margins gradually converging into short acumen. Carapace moderately arched dorsoventrally; with lateral surfaces covered with small tubercles. Small cervical spines or tubercles present. Areola narrow, 10.5 to 18.0 times longer than wide at its narrowest point, with room for one to three punctations. Antennal scale widest at midpoint. Palm and fingers of chelae relatively narrow, dorsal surface of palm covered with small tubercles. Dorsal surface of propodus of chelae with weakly developed longitudinal ridge. Mesial margin of palm of chela with single row of eight or nine anteriorly directed tubercles, mesial margin of dactyl usually with two or three anteriorly directed tubercles near base. Males with significantly longer chelae than females of same carapace length.

Gonopod of form I male terminating in four short elements: a corneous, caudodistally arched and pointed central projection; a shorter corneous, bladelike caudal process; a bifurcating mesial process capped with setae; and a knoblike cephalic process capped with long setae. Annulus ventralis of female subcircular; with central,

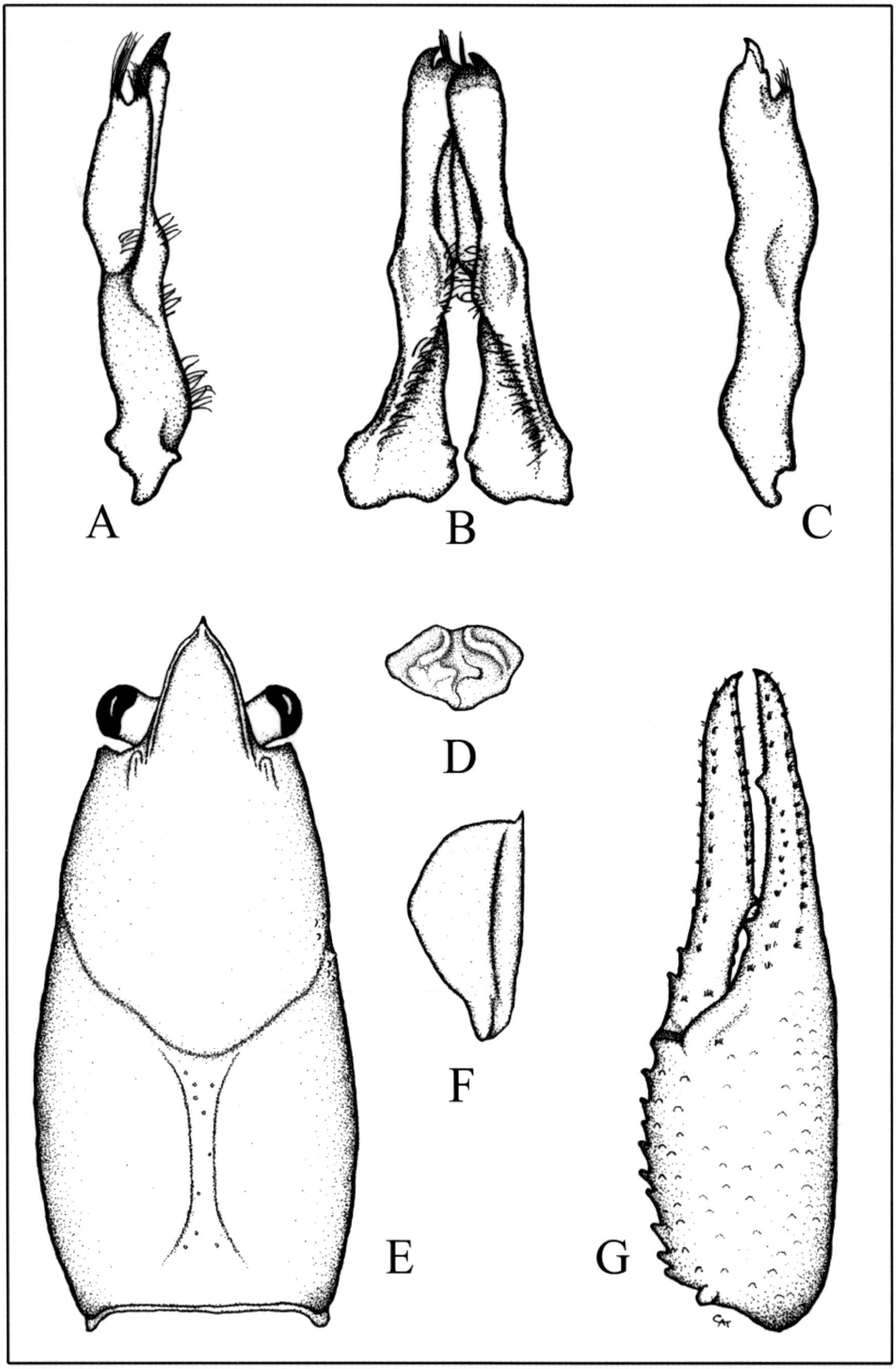

Fig. 179. *Procambarus viaeviridis*: A, mesial view of form I gonopod; B, ventral view of form I gonopods; C, lateral view of form II gonopod; D, ventral view of annulus ventralis; E, dorsal view of carapace; F, dorsal view of right antennal scale; G, dorsal view of right chela.

deeply recessed and wide trough running through cephalic margin; fossa poorly developed; sinus running from fossa to caudal margin.

MAXIMUM SIZE – The maximum known size for the species is 73.6 mm total length.

MOST LIKE – In possessing four terminal elements on the form I male gonopod and a dark wedge on the dorsal surface of its abdomen, *Procambarus viaeviridis* is most similar to the other two members of the genus *Procambarus* found in Kentucky, *P. acutus* and *P. clarkii*. *Procambarus viaeviridis* differs from those two species in having a rostrum with gradually converging margins that lacks marginal spines.

DISTRIBUTION AND HABITAT – (Fig. 180) *Procambarus viaeviridis* occurs in the Gulf Coastal Plain from extreme southern Illinois to northeastern Louisiana and eastward to Alabama (Hobbs 1989). In Kentucky a handful of records for the species are known from the lower Ohio, lower Cumberland, lower Tennessee, and Mississippi River drainages west of the Cumberland River.

In Kentucky *Procambarus viaeviridis* is usually found in seasonally flooded ditches and vernal ponds in the floodplains of small creeks and in lowland lakes. Substrates in these ditches, lakes, and ponds are usually soft mud or soils with abundant grass and sedge growth. As water levels recede in seasonally flooded habitats during summer and fall months, individuals will be found in shallow burrows. Page (1985) reported collecting the species from burrows ranging in depth from 15 to 30 cm. In early spring adults of the species can occasionally be found among woody debris in flowing lowland creeks.

LIFE HISTORY – Due to its burrowing behavior and preference for seasonally flooded habitats, very little is known about the life history of *P. viaeviridis*. In Kentucky the species has only been collected in the months of April and May. All of those collections contain form I males, form II males, and females. No ovigerous females, females with young, or juveniles have been collected or observed in Kentucky. Neither Page (1985) nor Pflieger (1996) reported ovigerous females from Illinois and Missouri, respectively. Page (1985) did report the presence of small juveniles in January and February in extreme southern Illinois. Given the proximity of those southern Illinois populations to those in western Kentucky, particularly those near the Ohio River in Livingston County, it seems reasonable to assume that juveniles of the species could be found during January and February in Kentucky. This would suggest egg–laying occurs either in the late spring or fall. Page (1985) also presented length frequency data for the species in southern Illinois that suggest a two-year life span.

CRAYFISH ASSOCIATES – The following species have been collected from habitats containing *Procambarus viaeviridis*: *Cambarellus shufeldtii, Fallicambarus fodiens, Cambarus diogenes, Orconectes palmeri palmeri, P. acutus,* and *P. clarkii.*

CONSERVATION STATUS – KSNPC Threatened.

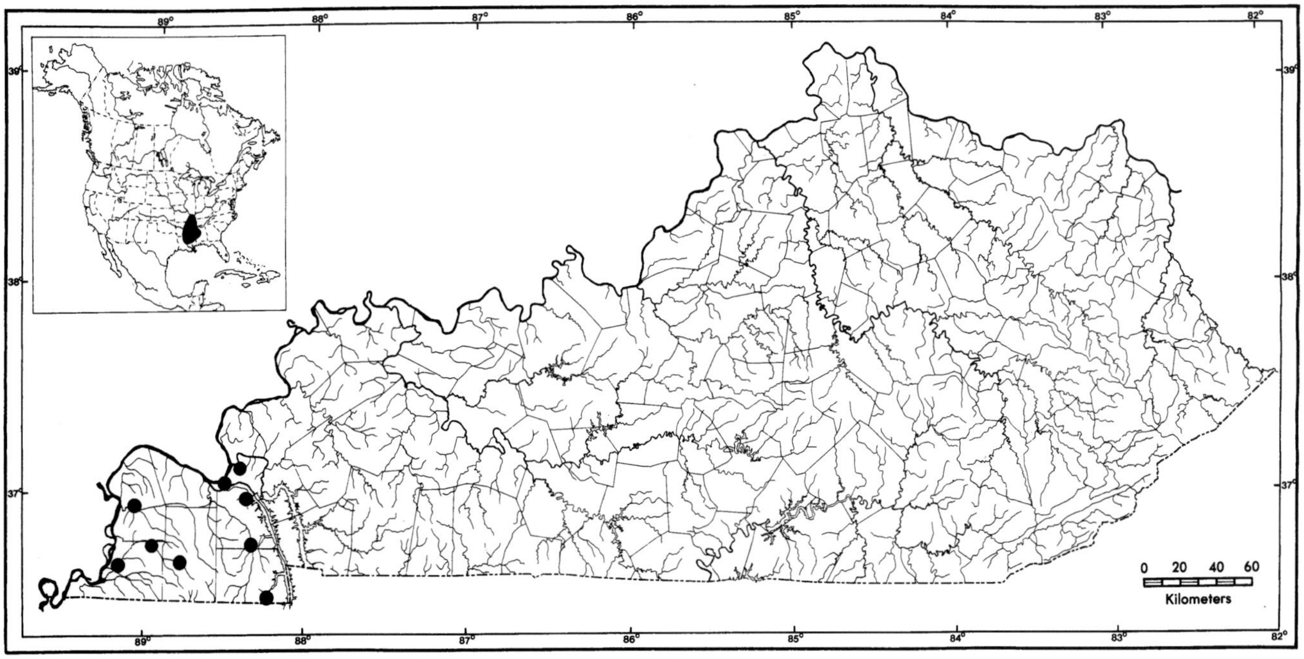

Fig. 180. Distribution of *Procambarus viaeviridis* in Kentucky.

Glossary

Acumen – the pointed terminal extension of the rostrum.

Annulus ventralis – the pocketlike structure with an elevated opening found on the underside of the female carapace; used for the storage of sperm.

Antennal scale – the bladelike structure terminating in a spine found at the base of the antenna.

Areola – an area on the dorsal surface of the carapace located behind the cervical groove and forward of the abdomen and defined by a pair of shallow, gradually arching ridges.

Carapace – the hardened, unsegmented covering of the head and thoracic sections of the crayfish body.

Caudal process – one of the terminal elements of the gonopod, usually found on members of genera *Procambarus* and *Cambarellus*.

Caudodistally – the curving or arching of the terminal elements of an appendage towards the tail end of the crayfish when the appendage is hanging perpendicular to the main axis of the body. When the gonopod of the male is lying in its normal position, caudodistally arched elements are directed ventrally.

Central projection – one of the terminal elements of the gonopod, usually the largest. The central projection is corneous in the form I state.

Cephalic process – one of the terminal elements of the gonopod, found only on members of the genus *Procambarus*.

Cephalodistally – the curving or arching of the terminal elements of an appendage towards the head end of the crayfish when the appendage is hanging perpendicular to the main axis of the body. When the gonopod of the male is lying in its normal position cephalodistally arched elements are directed dorsally.

Cervical groove – the transverse, shallow groove on the carapace that roughly separates the head region from the thoracic region.

Cervical Spine – the sharp spine occurring on the cervical groove on the lateral surface of the carapace.

Chela – the enlarged clawlike terminal end of the first walking leg.

Corneous – sclerotized, amber–colored. In reference to the appearance of certain terminal elements of the form I gonopod.

Dactyl – the movable, inner "finger" of the chela.

Epistome – the flat, tonguelike structure found on the ventral surface of the head between the bases of the antennae.

Epistomal zygoma – the posterior region of the epistome.

Fossa – the small, usually centrally located, depression or opening found on the ventral surface of the annulus ventralis.

Gonopod – the first pleopod of the male crayfish modified for sperm transfer.

Ischium – the third most basal segment of the walking leg.

Lentic – pertaining to still, nonflowing waters such as lakes, reservoirs, and swamps.

Lotic – pertaining to flowing waters such as creeks and rivers.

Median carina – a longitudinal, elevated ridge found in the center of the floor of the rostrum.

Mesial process – one of the terminal elements of the gonopod that is usually smaller than the central projection and noncorneous in the form I state.

Palm – the region of the chela running from the base of the dactyl to the base of the chela.

Pereiopod – the jointed walking leg of a crayfish. Crayfish have five pairs of pereiopods, the first of which is greatly enlarged and is known as the chela.

Pleopod – the jointed appendage found on the ventral surface of each abdominal segment. Crayfish have five pairs of pleopods. See also **Gonopod**.

Pleuron – the nearly verticle side, or lateral, plate found on each abdominal segment.

Propodus – the immovable outer "finger" of the chela.

Punctation – a small, shallow, usually circular depression found on the external surface of the exoskeleton.
Rostrum – the triangular shaped dorsal extension of the carapace located between the eyes.
Shoulder – a protruding, angular or rounded bulge found on the caudal surface of the gonopod, usually at the base of the central projection.
Sinus – the thin, usually curving groove found on the caudal half of the ventral surface of the annulus ventralis.
Spine – a sharply pointed and elevated protrusion found on the surface of the exoskeleton.
Suborbital angle – the angle formed by a subsection of the anterior edge of the carapace just below the eye.
Subpalmar – the region at the base of the dactyl on the ventral, or underside, of the chela.
Tubercle – a hard, rounded and slightly elevated bump found on the surface of the exoskeleton.

Literature Cited

Black, J.B. 1966. Cyclic male reproductive activities in the dwarf crawfishes *Cambarellus shufeldtii* (Faxon) and *Cambarellus puer* Hobbs. American Microscopial Society Transactions 85:214–232.

Black, J.B. 1972. Life history notes on the crawfish *Orconectes lancifer*. Proceedings of the Louisiana Academy of Sciences 35:7–9.

Bouchard, R.W. 1972. A contribution to the knowledge of Tennessee crayfish. Doctoral Dissertation, University of Tennessee, Knoxville.

Bouchard, R.W. 1973. *Cambarus buntingi*, a new species of *Puncticambarus* (Decapoda, Astacidae) from Kentucky and Tennessee. American Midland Naturalist 90:406–412.

Bouchard, R.W. 1976a. Investigations on the status of fourteen species of freshwater decapod crustaceans in the United States. U.S. Department of Interior, Fish and Wildlife Service, Office of Endangered Species, Washington, D.C.

Bouchard, R.W. 1976b. Geography and ecology of crayfishes of the Cumberland Plateau and Cumberland Mountains, Kentucky, Virginia, Tennessee, Georgia, and Alabama, Part II: The genera *Fallicambarus* and *Cambarus*. Pages 585–605 *in* J.W. Avault, Jr., ed. *Freshwater Crayfish*. Louisiana State University Division of Continuing Education, Baton Rouge.

Bouchard, R.W. 1978. Taxonomy, ecology and phylogeny of the subgenus *Depressicambarus*, with the description of a new species from Florida and redescriptions of *Cambarus graysoni*, *Cambarus latimanus* and *Cambarus striatus* (Decapoda: Cambaridae). Alabama Museum of Natural History Bulletin 3:27–60.

Bouchard, R.W., and J.W. Bouchard. 1995. Two new species and subgenera (*Cambarus* and *Orconectes*) of crayfishes (Decapoda: Cambaridae) from the eastern United States. Notulae Naturae 471:1–21.

Bouchard, R.W., and D.A. Etnier. 1979. A new species of primary burrowing crayfish of the genus *Cambarus* from the Ridge and Valley Province in Tennessee. Proceedings of the Biological Society of Washington 92:589–600.

Bovbjerg, R.V. 1952. Comparative ecology and physiology of the crayfish *Orconectes propinquus* and *Cambarus fodiens*. Physiological Zoology 25:34–56.

Boyd, J.A., and L.M. Page. 1978. The life history of the crayfish *Orconectes kentuckiensis* in Big Creek, Illinois. American Midland Naturalist 99:398–414.

Brown, P.L. 1955. The biology of crayfishes of central and southeastern Illinois. Doctoral Dissertation, University of Illinois, Urbana-Champaign.

Burr, B.M., and M.L. Warren, Jr. 1986. A distributional atlas of Kentucky fishes. Kentucky Nature Preserves Commission Scientific and Technical Series No. 4.

Butler, R.S., R.J. DiStefano, and G.A. Schuster. 2003. Crayfish: an overlooked fauna. Endangered Species Bulletin 28:10–11.

Caldwell, M.J., and R.V. Bovbjerg. 1969. Natural history of the two crayfish of northwestern Iowa, *Orconectes virilis* and *Orconectes immunis*. Iowa Academy of Science Proceedings 76:463–472.

Cooper, J.E. 1975. Ecological and behavioral studies in Shelta Cave, Alabama, with emphasis on decapod crustaceans. Doctoral Dissertation, University of Kentucky, Lexington.

Corey, S. 1990. Life history of *Cambarus robustus* Girard in the Eramosa-Speed River system of southwestern Ontario, Canada (Decapoda: Astacidae). Crustaceana 59:225–230.

Creaser, E.P. 1931. The Michigan decapod crustaceans. Papers of the Michigan Academy of Sciences Arts and Letters 13:257–276.

Crocker, D.W. 1957. The crayfishes of New York State (Decapoda, Astacidae). New York State Museum and Science Service Bulletin 355.

Crocker, D.W., and D.W. Barr. 1968. Handbook of the crayfishes of Ontario. Royal Ontario Museum, University of Toronto Press, Toronto.

Cummins, H. 1921. Spring migrations in the crayfish, *Cambarus argillicola* Faxon. American Microscopical Society Transactions 40:28–30.

Dewees, J.P. 1972. Geographic variation in the primary burrowing crayfish, *Cambarus dubius* Faxon and *Cambarus carolinus* (Erichson) (Decapoda: Astacidae) in Tennessee with notes on ecology and life history. Master's Thesis, University of Tennessee, Knoxville.

Eigenmann, C.H. 1899. Explorations in the caves of Missouri and Kentucky. Proceedings of the Indiana Academy of Science 1898:58–61.

Fielder, D.D. 1972. Some aspects of the life histories of three closely related crayfish species, *Orconectes obscurus*, *O. s. sanbornii*, and *O. propinquus*. Ohio Journal of Science 72:129–145.

Hamr, P., and M. Berrill. 1985. The life histories of north–temperate populations of the crayfish *Cambarus robustus* and *Cambarus bartoni*. Canadian Journal of Zoology 63:2313–2322.

Hay, W.P. 1896. The crawfishes of the state of Indiana. Pages 475–507 *in* 20th Annual Report of the Department of Geology and Natural Resources of Indiana.

Hobbs, H. H., Jr. 1942. The crayfishes of Florida. University of Florida Publications, Biological Science Series 3(2).

Hobbs, H.H., Jr. 1970. New crayfishes of the genus *Cambarus* from Tennessee and Georgia (Decapoda, Astacidae). Proceedings of the Biological Society of Washington 83:241–260.

Hobbs, H.H., Jr. 1981. The crayfishes of Georgia. Smithsonian Contributions to Zoology No. 318.

Hobbs, H.H., Jr. 1989. An illustrated checklist of the American crayfishes (Decapoda: Astacidae, Cambaridae, and Parastacidae). Smithsonian Contributions to Zoology No. 480.

Hobbs, H.H., Jr., and T.C. Barr, Jr. 1972. Origins and affinities of the troglobitic crayfishes of North America (Decapoda: Astacidae) II. Genus *Orconectes*. Smithsonian Contributions to Zoology, No. 105.

Hobbs, H.H., Jr., and R.W. Bouchard. 1973. A new crayfish from the Cumberland River system with notes on *Cambarus carolinus* (Erichson). Proceedings of the Biological Society of Washington 86:41–68.

Hobbs, H.H., Jr., and H.H. Hobbs, III. 1990. A new crayfish (Decapoda: Cambaridae) from southeastern Texas. Proceedings of the Biological Society of Washington 103:608–613.

Hobbs, H.H., Jr., and L.J. Marchand. 1943. A contribution toward a knowledge of the crayfishes of the Reelfoot Lake area. Journal of the Tennessee Academy of Science 18:6–35.

Hobbs, H.H., Jr., and H.W. Robison. 1989. On the crayfish genus *Fallicambarus* (Decapoda: Cambaridae) in Arkansas, with notes on the *fodiens* complex and descriptions of two new species. Proceedings of the Biological Society of Washington 102:651–697.

Hobbs, H.H., Jr., and M. Whiteman. 1991. Notes on the burrow, behavior, and color of the crayfish *Fallicambarus* (*F.*) *devastator* (Decapoda: Cambaridae). Southwestern Naturalist 36:127–135.

Hobbs III, H.H. 1976. Observations on the cave–dwelling crayfishes of Indiana. Pages 405–414 *in* Freshwater Crayfish, J. W. Avault, Jr., ed. Louisiana State University Division of Continuing Education, Baton Rouge.

Hobb III, H.H. 1978. Studies of the cave crayfish, *Orconectes inermis inermis* Cope (Decapoda, Cambaridae). Part IV: Mark–recapture procedures for estimating population size and movements of individuals. International Journal of Speleology 10:303–322.

Hobbs, H.H., III, and J.P. Jass. 1988. The crayfishes of Wisconsin. Milwaukee Public Museum, Special Publications in Biology and Geology No. 5.

Huner, J.V. 2002. *Procambarus*. Pages 541–584, *in* Biology of freshwater crayfish, D. M. Holdich, ed. Blackwell Science, Oxford, UK.

Huner, J.V., and J.E. Barr. 1991. Red swamp crawfish, biology and exploitation, 3rd Ed. Louisiana Seagrant College Program, Louisiana State University, Baton Rouge.

Huxley, T.H. 1880. The Crayfish: An Introduction to the Study of Zoology. International Science Series, Vol. XXVIII. C. Kegan Paul and Co., London, UK.

Jegla, T.C. 1964. Studies of the eyestalk, metabolism and molting and reproductive cycles in a cave crayfish. Doctoral Dissertation, University of Illinois, Urbana–Champaign.

Jegla, T.C. 1969. Cave crayfish: annual periods of molting and reproduction. Pages 135–137 *in* numbers 4–5 Actes du IV Congress International de Sp l ologie en Yugoslavie (12–26 IX 1965).
Jezerinac, R.F. 1993. A new subgenus and new species of crayfish (Decapoda: Cambaridae) of the genus *Cambarus*, with an amended description of the subgenus *Lacunicambarus*. Proceedings of the Biological Society of Washington 106:532–544.
Jezerinac, R.F., and G.W. Stocker. 1987. *Fallicambarus* (*Creaserinus*) *fodiens* (Cottle, 1863) (Decapoda: Cambaridae) in West Virginia: a new state record. Ohio Journal of Science 87:46–47.
Jezerinac, R.F., G.W. Stocker, and D.C. Tarter. 1995. The crayfishes (Decapoda: Cambaridae) of West Virginia. Bulletin of the Ohio Biological Survey 10(1).
Kentucky State Nature Preserves Commission. 2000. Rare and extirpated biota of Kentucky. Journal of the Kentucky Academy of Science 61:115–132.
Lawson, L.H. 2003. The crayfishes (Decapoda: Cambaridae) of the Green River Basin, Kentucky. Master's Thesis, Eastern Kentucky University, Richmond.
Lodge, D.M., C.A. Taylor, D.M. Holdich, and J. Skudal. 2000. Nonindigenous crayfishes threaten North American freshwater biodiversity: lessons from Europe. Fisheries 25(8):7–20.
Lowe, M.E. 1961. The female reproductive cycle of the crayfish *Cambarellus shufeldtii*: the influence of environmental factors. Tulane Studies in Zoology 8:157–176.
Martin, J.W., and G.E. Davis. 2001. An updated classification of the recent Crustacea. Natural History Museum of Los Angeles County, Science Series No. 39.
Mengel, R.M. 1965. The birds of Kentucky. American Ornithologists Union, Ornithological Monographs No. 3.
Mohr, C.E., and T.L. Poulson. 1966. The life of the cave. Our Living World of Nature Series. McGraw Hill, New York, NY.
Momot, W.T. 1995. Redefining the role of crayfish in aquatic ecosystems. Reviews in Fisheries Science 3:33–63.
Newcombe, C.L. 1929. The crayfishes of West Virginia. Ohio Journal of Science 29:267–288.
Norrocky, M.J. 1989. New distribution records for the crayfish *Cambarus* (*Cambarus*) *ortmanni* Williamson (Decapoda: Cambaridae) with life history notes. Ohio Journal of Science 89:67–69.
O'Bara, C.J. 1988. The distribution of the Big South Fork crayfish, *Cambarus bouchardi*, with general notes on its habitat. Transactions of the Kentucky Academy of Science 49:117–119.
Omernik, J.M. 1987. Ecoregions of the conterminous United States. Annals of the Association of American Geography 77:118–125.
Ortmann, A.E. 1931. Crawfishes of the southern Appalachians and the Cumberland Plateau. Annuls of the Carnegie Museum 20:61–160.
Page, L.M. 1985. The crayfishes of Illinois. Illinois Natural History Survey Bulletin 33:335–448.
Payne, J.F., and J.O. Price. 1983. Studies on the life history and ecology of *Orconectes palmeri palmeri* (Faxon). Pages 183–191, *in* Freshwater Crayfish V, C. R. Goldman, ed. AVI Publishing, Westport, CT.
Pearse, A.S. 1910. The crawfishes of Michigan. Michigan State Biological Survey Publication 1:9–22.
Penn, G.H. 1942. Observations on the biology of the dwarf crawfish *Cambarellus shufeldtii* (Faxon). American Midland Naturalist 28:644–647.
Penn, G.H. 1943. A study of the life history of the Louisiana red–crawfish, *Cambarus clarkii* Girard. Ecology 24:1–18.
Penn, G.H. 1950. The genus *Cambarellus* in Louisiana (Decapoda, Astacidae). American Midland Naturalist 44:421–426.
Penn, G.H., and J.F. Fitzpatrick, Jr. 1962. Interspecific competition between crawfishes. American Zoologist 2:436.
Penn, G.H., and J.F. Fitzpatrick, Jr. 1963. Interspecific competition between two sympatric species of dwarf crawfishes. Ecology 44:793–797.
Penn, G.H., and G. Marlow. 1959. The genus *Cambarus* in Louisiana. American Midland Naturalist 61:191–203.

Perry, W. L., J. L. Feder, and D. M. Lodge. 2001. Implications of hybridization between introduced and resident *Orconectes* crayfishes. Conservation Biology 15:1656–1666.

Pflieger, W.L. 1996. The Crayfishes of Missouri. Missouri Department of Conservation, Jefferson City.

Prins, R. 1968. Comparative ecology of the crayfishes *Orconectes rusticus* and *Cambarus tenebrosus* in Doe Run, Meade County, Kentucky. International Revue Gesamten Hydrobiologie 53:667–714.

Rhoades, R. 1944. The crayfishes of Kentucky, with notes on variation, distribution and descriptions of new species and subspecies. American Midland Naturalist 31:111–149.

Roell, J.M., and D.J. Orth. 1993. Trophic basis of production of stream–dwelling smallmouth bass, rock bass, and flathead catfish in relation to invertebrate bait harvest. Transactions of the American Fisheries Society 122:46–62.

Schuster, G.A. 1976. A new primary burrowing crayfish of the subgenus *Jugicambarus* (Decapoda, Cambaridae) from Kentucky, with notes on its life history. American Midland Naturalist 95:225–230.

St. John, F.L. 1976. Growth rate, life span and molt cycle of the crayfish *Orconectes sanbornii.* Ohio Journal of Science 76:73–77.

Tack, P.I. 1941. The life history and ecology of the crayfish *Cambarus immunis* Hagen. American Midland Naturalist 25:420–446.

Taylor, C.A. 1997. Taxonomic status of members of the subgenus *Erebicambarus*, genus *Cambarus* (Decapoda: Cambaridae), east of the Mississippi River. Journal of Crustacean Biology 17:352–360.

Taylor, C.A. 2000. Systematic studies of the *Orconectes juvenilis* complex (Decapoda: Cambaridae), with descriptions of two new species. Journal of Crustacean Biology 20:132–152.

Taylor, C.A., and M.H. Sabaj. 1998. A new crayfish of the genus *Orconectes* from the Blood River drainage of western Kentucky and Tennessee (Decapoda:Cambaridae). Proceedings of the Biological Society of Washington 111:645–652.

Taylor, C.A., M.L. Warren, Jr., J.F. Fitzpatrick, Jr., H.H. Hobbs, III, R.F. Jezerinac, W.L. Pflieger, and H.W. Robison. 1996. Conservation status of crayfishes of the United States and Canada. Fisheries 21 (4):25–38.

Tellkampf, T.G. 1844. Ueber den blinden Fisch der Mammuth–h hle in Kentucky, mit Bemerkungen Ÿber einige andere in dieser H hle lebende Thiere. (Taf. IX). Archiv fŸr Anatomie, Physiologie und Wissenschaftliche Medicin 4:381–394.

Thoma, R.F. 2000. *Cambarus (Jugicambarus) jezerinaci* (Crustacea: Decapoda: Cambaridae), a new species of crayfish from the Powell River drainage of Tennessee and Virginia. Proceedings of the Biological Society of Washington 113:731–738.

Thoma, R.F., and R.F. Jezerinac. 2000. Ohio crayfish and shrimp atlas. Ohio Biological Survey Miscellaneous Contributions No. 7.

Turner, C.L. 1926. The crayfishes of Ohio. Ohio State University Bulletin 30:145–195.

Volpe, E.P., and G.H. Penn. 1957. Dimorphism of chromatophore pattern in the dwarf crayfish. Journal of Heredity 48:90–96.

Walker, D., B.A. Porter, and J.C. Avise. 2002. Genetic parentage assessment in the crayfish *Orconectes placidus*, a high fecundity invertebrate with extended maternal brood care. Molecular Ecology 11:2115–2122.

Wetzel, J.E. 2002. Form alteration of adult female crayfishes of the genus *Orconectes* (Decapoda: Cambaridae). American Midland Naturalist 147:326–337.

Williams, A.B., L.G. Abele, D.L. Felder, H.H. Hobbs Jr., R.B. Manning, P.A. McLaughlin, and I.P. Farante. 1989. Common and scientific names of aquatic invertebrates from the United States and Canada: decapod crustaceans. American Fisheries Society Special Publication 17.

Williamson, E.B. 1907. Notes on the crayfish of Wells County, Indiana, with the description of a new species. Pages 749–763 *in* 31st Annual Report of the Department of Geology and Natural Resources, Indiana (1906).

Index